NestJS
全栈开发解析
快速上手与实践

温健民 编著

清华大学出版社
北京

内 容 简 介

本书旨在帮助读者快速掌握NestJS（简称Nest）开发，并应用于实战项目。本书共10章，首先介绍基本概念，为读者打下坚实的知识基础。接着，通过简洁的代码示例进行知识点的串联讲解，帮助读者快速克服学习瓶颈。最终，通过实践能力和工程思维的培养，帮助读者将知识从线性结构转变为网状结构，形成以Nest为基础的全栈知识体系。

本书采用通俗易懂的点线面知识构建方式进行讲解，适合从事前端开发和 Node.js 开发的工程师学习，同时也适合有意向学习 Nest 全栈知识的开发者。

本书封面贴有清华大学出版社防伪标签，无标签者不得销售。
版权所有，侵权必究。举报：010-62782989，beiqinquan@tup.tsinghua.edu.cn。

图书在版编目（CIP）数据

NestJS 全栈开发解析：快速上手与实践 / 温健民编著.
北京：清华大学出版社，2024.8. -- ISBN 978-7-302-67100-8
Ⅰ. TP393.092.2
中国国家版本馆 CIP 数据核字第 20244LB587 号

责任编辑：赵　军
封面设计：王　翔
责任校对：闫秀华
责任印制：丛怀宇

出版发行：清华大学出版社
　　　网　　址：https://www.tup.com.cn，https://www.wqxuetang.com
　　　地　　址：北京清华大学学研大厦A座　　　邮　　编：100084
　　　社 总 机：010-83470000　　　邮　　购：010-62786544
　　　投稿与读者服务：010-62776969，c-service@tup.tsinghua.edu.cn
　　　质 量 反 馈：010-62772015，zhiliang@tup.tsinghua.edu.cn
印 装 者：北京嘉实印刷有限公司
经　　销：全国新华书店
开　　本：190mm×260mm　　　印　张：21.5　　　字　数：580千字
版　　次：2024年9月第1版　　　印　次：2024年9月第1次印刷
定　　价：99.00元

产品编号：106315-01

前　　言

互联网技术迅猛发展，前端技术更新迅速，业务场景变得越来越复杂。JavaScript 语言从早期的网页脚本语言逐渐发展，基于 Node.js 的服务端框架和库为前端开发打开了新的视野。不同框架和库的设计思想和理念各有特色，它们针对不同场景设计，推动了 Node.js 领域的技术发展。

Nest 是基于 TypeScript 的开源企业级框架，建立在 Express 之上，提供了一层抽象。它凭借模块化、强大的依赖注入系统和优秀的设计思想，成为许多 Node.js 开发者的首选。

本书致力于普及 Nest 从基础到实战的核心知识，对每个核心知识点通过理论结合代码示例的方式进行讲解。本书通过完整的项目实践，将带你快速上手 Nest 并将其应用于实际项目中，帮助开发者构建坚实的独立开发基础。

本书内容

本书采用循序渐进的行文模式，从前端基础知识到后端中间件的使用和开发，全面覆盖。内容分为基础篇、进阶篇、扩展篇和项目实战篇，每篇包含若干章节。

基础篇介绍了学习 Nest 所需的前置知识，包括 Node.js 的请求响应对象和 TypeScript 基础知识，然后介绍 Nest 的核心设计理念、创建和调试 Nest 应用，并结合实际代码示例深入解读 Nest 的核心概念。

进阶篇详细讲解了后端中间件服务，如 MySQL 数据库、Redis 缓存等，并指导如何在 Nest 中集成和使用这些服务。此外，还介绍了企业级应用中实现身份验证与授权的方法，以及如何通过 Docker 高效部署和管理中间件服务。

扩展篇讨论了对系统稳定性至关重要的系统测试与日志管理，包括开发阶段的单元测试、集成测试、端对端测试等，以及生产应用阶段的日志统计实践。

项目实战篇提供了一个完整的数字门店管理平台开发实战案例。

配书资源

为了让读者能够更好地理解和实践所学的知识，本书提供了丰富的配书资源，包括源代码（本书中的代码示例都可以直接用于实际工作中）、PPT 教学课件。扫码下述二维码可免费下载：

源代码

PPT

如果在学习和资源下载的过程中遇到问题，可以发送邮件至 booksaga@126.com，邮件主题为"NestJS 全栈开发解析：快速上手与实践"。

面向的阅读群体

本书主要面向所有前端和 Node.js 开发工程师，以及有意向学习全栈知识的 IT 专业人员。要求读者具备 Node.js 和 TypeScript 语言的基础。

如何使用这本书

希望本书能为你带来一段愉快的学习之旅。

如果你是 Nest 初学者或者想全面了解本书内容，建议按照章节顺序进行阅读和实践。对于有实际后端开发经验并且了解 Nest 开发的读者，可以选择感兴趣的章节阅读，因为每个章节的设计相对独立。

在学习本书的过程中，建议在阅读完每个知识点后，按照书中的代码案例进行实践，将其转化为自己的储备知识。如果遇到问题，建议首先调试分析问题所在，总结经验，然后可以与作者沟通解答疑惑。

希望本书能给你带来一段愉快的学习之旅。

致　　谢

感谢清华大学出版社提供这次创作机会，编辑们花了大量时间与我沟通和修改。感谢这次相遇，没有这次相遇，我也无法以图文形式与大家交流，这本书也不会诞生。

感谢我的妻子 Minnie 在背后一如既往地支持我，她把家庭打理得井然有序，给了我足够的时间来完成创作。同时感谢我的儿子甜筒，他四个月大时我开始创作，他的懂事和健康成长让我深感欣慰。

感谢自己，成为时间的主人，在保证工作和休息的同时，利用每天 3 小时的通勤时间、1 小时的午休时间，以及陪伴家人的间隙完成了本次创作；不断改变，拥抱不确定性，是为了能够有更多时间陪伴身边亲近的人。

感谢每一位阅读本书的读者朋友，你们的支持与反馈是我不断进步、持续创作的最大动力。

最后，由于笔者能力有限，书中知识点可能存在疏忽和遗漏，恳请读者指正。

编　者

2024 年 4 月

目　　录

第 1 部分　基础篇

第 1 章　需要提前掌握的知识 ··· 2
- 1.1　Node 中的请求与响应对象 ··· 2
  - 1.1.1　原生 Node 处理 HTTP 请求 ·· 2
  - 1.1.2　Express 处理 HTTP 请求 ··· 4
  - 1.1.3　Nest 处理 HTTP 请求 ·· 7
- 1.2　TypeScript 基础与应用 ·· 7
  - 1.2.1　TypeScript 编译 ··· 8
  - 1.2.2　TypeScript 类型系统 ·· 9

第 2 章　Nest 初识 ·· 15
- 2.1　什么是 Nest ·· 15
  - 2.1.1　Nest 概述 ··· 15
  - 2.1.2　Nest 的主要特点 ··· 15
  - 2.1.3　Nest 的应用场景 ··· 16
- 2.2　快速上手 Nest CLI ··· 16
  - 2.2.1　Nest CLI 的安装 ·· 16
  - 2.2.2　创建项目 ··· 17
  - 2.2.3　生成指定的代码片段 ·· 19
  - 2.2.4　构建应用 ··· 22
  - 2.2.5　启动开发调试 ·· 24
  - 2.2.6　查看项目信息 ·· 25
- 2.3　创建第一个 Nest 应用 ··· 25
  - 2.3.1　生成后端项目 ·· 26
  - 2.3.2　生成前端项目 ·· 26
  - 2.3.3　准备工作 ··· 27
  - 2.3.4　运行结果 ··· 29
  - 2.3.5　模块化开发 ·· 30
- 2.4　Nest 的 AOP 架构理念 ·· 32
  - 2.4.1　MVC 架构概述 ·· 32
  - 2.4.2　AOP 解决的问题 ··· 33
  - 2.4.3　AOP 在 Nest 中的应用 ·· 33
- 2.5　IoC 思想解决了什么问题 ·· 40

	2.5.1	IoC 核心思想概述	40
	2.5.2	IoC 在 Nest 中的应用	41
2.6	学会调试 Nest 应用		44
	2.6.1	Chrome DevTools 调试	44
	2.6.2	VS Code 调试	46
	2.6.3	扩展调试技巧	48

第 3 章 Nest 核心概念介绍 51

3.1	贯穿全书的装饰器		51
	3.1.1	基本概念	51
	3.1.2	装饰器的种类	51
	3.1.3	Nest 中的装饰器	59
3.2	井然有序的模块化		60
	3.2.1	基本概念	60
	3.2.2	创建模块	61
	3.2.3	共享模块	62
	3.2.4	全局模块	64
	3.2.5	动态模块	65
3.3	控制器与服务的默契配合		66
	3.3.1	基本概念	66
	3.3.2	Controller 管理请求路由	67
	3.3.3	Controller 处理请求参数与请求体	68
	3.3.4	Service 处理数据层	70
	3.3.5	服务与服务提供者	71
3.4	耳熟能详的中间件		71
	3.4.1	类中间件	71
	3.4.2	函数式中间件	74
	3.4.3	局部中间件	74
	3.4.4	全局中间件	74
3.5	拦截器与 RxJS 知多少		75
	3.5.1	基本概念	75
	3.5.2	创建项目	76
	3.5.3	拦截器的基本使用方法	76
3.6	数据之源守护者：管道		79
	3.6.1	基本概念	79
	3.6.2	内置管道	80
	3.6.3	自定义管道	88
3.7	Nest 实现文件上传		89
	3.7.1	初识 Multer	89
	3.7.2	单文件上传	91

3.7.3　多文件上传 …………………………………………………………… 94
　　　3.7.4　上传任意文件 ………………………………………………………… 98
　　　3.7.5　文件验证 ……………………………………………………………… 99

第 2 部分　进阶篇

第 4 章　Nest 与数据库 ……………………………………………………………… 102
4.1　快速上手 MySQL ……………………………………………………………… 102
　　　4.1.1　安装和运行 …………………………………………………………… 102
　　　4.1.2　MySQL 的常用命令 …………………………………………………… 105
　　　4.1.3　可视化操作 MySQL …………………………………………………… 107
4.2　MySQL 表之间的关系 ………………………………………………………… 112
　　　4.2.1　一对一关系 …………………………………………………………… 112
　　　4.2.2　一对多/多对一关系 …………………………………………………… 119
　　　4.2.3　多对多关系 …………………………………………………………… 121
4.3　快速上手 TypeORM …………………………………………………………… 126
　　　4.3.1　基本概念 ……………………………………………………………… 126
　　　4.3.2　项目准备 ……………………………………………………………… 126
　　　4.3.3　创建模型及实体 ……………………………………………………… 126
　　　4.3.4　定义数据列及类型 …………………………………………………… 127
　　　4.3.5　连接数据库 …………………………………………………………… 128
　　　4.3.6　使用 Repository 操作 CRUD ………………………………………… 129
　　　4.3.7　使用 QueryBuilder 操作 CRUD ……………………………………… 131
4.4　使用 TypeORM 处理多表关系 ………………………………………………… 134
　　　4.4.1　一对一关系 …………………………………………………………… 134
　　　4.4.2　一对多/多对一关系 …………………………………………………… 140
　　　4.4.3　多对多关系 …………………………………………………………… 142
4.5　在 Nest 中使用 TypeORM 操作 MySQL ……………………………………… 144
　　　4.5.1　项目准备 ……………………………………………………………… 144
　　　4.5.2　使用 EntityManager 操作实体 ……………………………………… 147
　　　4.5.3　使用 Repository 操作实体 …………………………………………… 148
　　　4.5.4　使用 QueryBuilder 操作实体 ………………………………………… 151

第 5 章　性能优化之数据缓存 ……………………………………………………… 154
5.1　快速上手 Redis ………………………………………………………………… 154
　　　5.1.1　安装和运行 …………………………………………………………… 154
　　　5.1.2　Redis 的常用命令 ……………………………………………………… 155
5.2　在 Nest 中使用 Redis 缓存 …………………………………………………… 162
　　　5.2.1　项目准备 ……………………………………………………………… 162
　　　5.2.2　Redis 初始化 …………………………………………………………… 164
　　　5.2.3　建表并构建缓存 ……………………………………………………… 165

第 6 章　身份验证与授权 171

- 5.2.4　运行代码 167
- 5.2.5　设置缓存有效期 169
- 5.2.6　选择合理的有效期 170

6.1　Cookie、Session、Token、JWT、SSO 详解 171
- 6.1.1　什么是身份验证 171
- 6.1.2　什么是授权 172
- 6.1.3　什么是凭证 172
- 6.1.4　什么是 Cookie 172
- 6.1.5　什么是 Session 173
- 6.1.6　Session 与 Cookie 的区别 173
- 6.1.7　什么是 Token 174
- 6.1.8　什么是 JWT 176
- 6.1.9　JWT 与 Token 的区别 177
- 6.1.10　什么是 SSO 177

6.2　基于 Passport 和 JWT 实现身份验证 180
- 6.2.1　基本概念 181
- 6.2.2　项目准备 181
- 6.2.3　用本地策略实现用户登录 182
- 6.2.4　用 JWT 策略实现接口校验 184
- 6.2.5　代码优化 188

6.3　基于 RBAC 实现权限控制 191
- 6.3.1　基本概念 191
- 6.3.2　数据表设计 192
- 6.3.3　项目准备 193
- 6.3.4　创建实体 194
- 6.3.5　启动服务 196
- 6.3.6　实现角色守卫控制 196
- 6.3.7　生成测试数据 200
- 6.3.8　测试效果 202

第 7 章　系统部署与扩展 203

7.1　快速上手 Docker 203
- 7.1.1　初识 Docker 203
- 7.1.2　安装 Docker 204
- 7.1.3　Docker 的使用 205

7.2　快速上手 Dockerfile 209
- 7.2.1　Docker 的基本概念 209
- 7.2.2　Dockerfile 的基本语法 210
- 7.2.3　Dockerfile 实践 210

第 3 部分　扩展篇

第 8 章　单元测试与端到端测试 216
8.1　重新认识单元测试 216
8.1.1　什么是单元测试 216
8.1.2　为什么大部分公司没有进行单元测试 217
8.1.3　为什么要编写单元测试 217
8.1.4　先编写单元测试还是先编写代码 218
8.1.5　测试驱动开发 219
8.2　在 Nest 中使用 Jest 编写单元测试 220
8.2.1　初识 Jest 220
8.2.2　项目准备 223
8.2.3　编写测试用例 224
8.2.4　实现业务代码 225
8.2.5　重构代码 229
8.3　集成测试 230
8.3.1　编写测试用例 230
8.3.2　测试效果 232
8.4　端到端测试 232
8.4.1　编写测试用例 233
8.4.2　实现业务代码 235

第 9 章　日志与错误处理 237
9.1　如何在 Nest 中记录日志 237
9.1.1　为什么要记录日志 238
9.1.2　内置日志器 Logger 238
9.1.3　定制日志器 240
9.1.4　记录日志的正确姿势 241
9.1.5　第三方日志器 Winston 241
9.2　Winston 日志管理实践 241
9.2.1　Winston 的基础使用 242
9.2.2　本地持久化日志 244
9.3　面向切面日志统计实践 248
9.3.1　中间件日志统计 248
9.3.2　拦截器日志统计 249
9.3.3　过滤器日志统计 251

第 4 部分　Nest 项目实战篇

第 10 章　数字门店管理平台开发 254
10.1　产品需求分析与设计 254

- 10.1.1 产品需求说明 ·············· 254
- 10.1.2 功能原型图 ·············· 255
- 10.2 技术选型与项目准备 ·············· 262
  - 10.2.1 前端技术选型 ·············· 262
  - 10.2.2 初始化前端项目 ·············· 262
  - 10.2.3 前端架构设计 ·············· 263
  - 10.2.4 后端技术选型 ·············· 264
  - 10.2.5 初始化后端项目 ·············· 265
  - 10.2.6 后端架构设计 ·············· 266
- 10.3 API 接口及数据库表设计 ·············· 268
  - 10.3.1 API 接口功能划分 ·············· 268
  - 10.3.2 数据库设计 ·············· 269
- 10.4 实现注册登录 ·············· 273
  - 10.4.1 页面效果展示 ·············· 273
  - 10.4.2 接口实现 ·············· 281
- 10.5 实现用户与角色模块 ·············· 290
  - 10.5.1 页面效果展示 ·············· 290
  - 10.5.2 表关系设计 ·············· 295
  - 10.5.3 接口实现 ·············· 299
- 10.6 实现商品与订单模块 ·············· 303
  - 10.6.1 页面效果展示 ·············· 304
  - 10.6.2 表关系设计 ·············· 306
  - 10.6.3 接口实现 ·············· 307
- 10.7 基于 Redis 实现商品热销榜 ·············· 315
  - 10.7.1 页面效果展示 ·············· 316
  - 10.7.2 接口实现 ·············· 317
- 10.8 实现活动模块与定时任务 ·············· 320
  - 10.8.1 页面效果展示 ·············· 320
  - 10.8.2 表关系设计 ·············· 322
  - 10.8.3 接口实现 ·············· 323
- 10.9 使用 Docker Compose 部署项目 ·············· 325
  - 10.9.1 编写后端 Docker Compose 文件 ·············· 325
  - 10.9.2 编写 Dockerfile 文件 ·············· 330

完结语：是终点，更是新的起点 ·············· 334
- 一个小小的决定 ·············· 334
- 时间的杠杆 ·············· 334

结语 ·············· 334

第 1 部分

基 础 篇

第 1 部分将深入探讨 Nest 框架的基石，涵盖以下关键领域：Nest 的先决条件知识、初步了解 Nest 及其核心概念。前 3 章介绍的核心概念对于理解 Nest 框架与各种中间件进行交互至关重要，也是深入学习本书后续章节和进行项目实战开发的前提条件。

第 1 章

需要提前掌握的知识

本章将引导你学习 Nest 的前置知识：Node 的请求响应对象与 TypeScript 基础。不必过分担心，本章不涉及复杂概念，即使你对这两方面的内容了解不多，也能理解它们的重要性。

首先，Nest 底层默认采用 Express 作为 HTTP 服务器框架。这可能会引发疑问：原生 HTTP 模块、Express 与 Nest 之间有何联系？

其次，TypeScript 在普及过程中受到众多开发者的青睐，但也面临一些质疑。例如，学习曲线较陡、额外的开发成本、类型系统并非完美等问题让一些初学者犹豫不决。

因此，在正式学习 Nest 之前，有必要先回顾一些必要的前置知识，以便在后续章节中更好地理解和实践。

1.1 Node 中的请求与响应对象

在 Node 应用中，请求和响应对象是处理 HTTP 请求和发送 HTTP 响应的核心工具，主流框架如 Express、Koa、Nest、Egg 都实现了不同抽象级别的中间件。本节将开始学习这些内容。

1.1.1 原生 Node 处理 HTTP 请求

在 Node 的内置 HTTP 模块中有两个核心对象：请求对象（Request）和响应对象（Response）。请求对象表示客户端向服务端发送的请求信息，而响应对象则表示服务端向客户端返回的响应数据。

请求对象通常在客户端发送请求到服务端的过程中使用，携带的信息包含 URL、请求方式、请求头、请求体等。常见的属性和方法说明如下。

- req.url：包含客户端请求的 URL，不包括协议、主机名和端口号。
- req.method：包含客户端使用的 HTTP 请求方法，例如 GET、POST 等。
- req.headers：一个包含所有请求头的对象，可以通过属性名访问具体的请求头信息。

- req.params：包含路由中匹配的参数，对于包含参数的路由非常有用。
- req.query：包含 URL 查询参数的对象，用于解析 URL 中的键-值对（Key-Value Pair）。
- req.body：对于 POST 请求，包含请求体的数据。在 Express 中需要使用中间件（例如 body-parser）来解析请求体。

相反，响应对象是向客户端发送的 HTTP 响应，设置响应状态、响应头和响应体等信息。常见的属性和方法说明如下。

- res.status(code)：设置 HTTP 响应状态码。
- res.setHeader(name, value)：设置响应头的值。
- res.send(body)：发送响应体内容给客户端，可以是字符串、JSON 对象、缓冲区等。
- res.json(obj)：发送 JSON 格式的响应体给客户端。
- res.sendFile(path, options, callback)：发送文件作为响应。
- res.cookie(name, value, option])：设置 Cookie 信息。

由此可见，Node 提供了非常丰富的基础 API 操作对象，但也带来了一些问题，例如开发者需要关注如何正确处理客户端发送过来的请求数据。请看下面的示例。

原生 Node 处理请求 URL 代码示例：

```javascript
const http = require("http");
const url = require("url");

const server = http.createServer((req, res) => {
  // 解析请求的 URL
  const parsedUrl = url.parse(req.url, true);

  // 获取路径和查询参数
  const path = parsedUrl.pathname;
  const queryParams = parsedUrl.query;

  // 设置响应头
  res.writeHead(200, { "Content-Type": "text/plain" });
  res.end(`Path: ${path}, Query Parameters: ${JSON.stringify(queryParams)}`);
});

const PORT = 3000;
server.listen(PORT, () => {
  console.log(`Server is listening on port ${PORT}`);
});
```

原生 Node 处理请求体代码示例：

```javascript
const http = require("http");

const server = http.createServer((req, res) => {
  let body = "";

  // 监听请求的数据
  req.on("data", (chunk) => {
    body += chunk;
  });
```

```
  // 请求结束时处理请求体
  req.on("end", () => {
    // 解析 JSON 请求体
    const jsonData = JSON.parse(body);

    res.writeHead(200, { "Content-Type": "application/json" });
    res.end(
      JSON.stringify({ message: "Request Body Processed", data: jsonData })
    );
  });
});

const PORT = 3000;
server.listen(PORT, () => {
  console.log(`Server is listening on port ${PORT}`);
});
```

从上述代码可以看出，Node 的原生 HTTP 模块在处理请求 URL 时，需要使用 Node 内置的 url 模块中的 url.parse 方法进行解析。如果是 POST 请求，则需要通过 JSON.parse 方法来解析请求体。在构造响应时，每个请求都需要设置响应头，最终通过 send 方法完成发送操作。显然，这种方法在实际的生产开发中显得相当烦琐。

1.1.2　Express 处理 HTTP 请求

然而，Express 的诞生简化并抽象了 HTTP 模块，提供了更加灵活的路由系统。结合中间件，它使得路由的定义和处理变得更加简单。请看下面的例子。

Express 处理请求 URL 代码示例：

```
const express = require("express");
const app = express();

// 处理带参数的路由
app.get("/users/:id", (req, res) => {
  const userId = req.params.id;
  res.send(`User ID: ${userId}`);
});

// 处理查询字符串参数
app.get("/search", (req, res) => {
  const query = req.query.q;
  res.send(`Search Query: ${query}`);
});

const PORT = 3000;
app.listen(PORT, () => {
  console.log('Server is listening on port ${PORT}');
});
```

Express 处理请求体代码示例：

```
const express = require("express");
const app = express();
```

```
// 使用 express.json() 中间件
app.use(express.json());
// 使用 express.urlencoded() 中间件
app.use(express.urlencoded({ extended: true }));

// 处理 Ajax 发送的 JSON 数据
app.post("/api/data", (req, res) => {
  const data = req.body; // 获取 JSON 请求体 res.json(data);
});

// 处理 HTML 表单提交的数据
app.post("/api/data2", (req, res) => {
  const data = req.body; // 获取 URL 编码请求体
  res.json(data);
});

const PORT = 3000;
app.listen(PORT, () => {
  console.log('Server is listening on port ${PORT}');
});
```

从上述代码可以看出，在 Express 中，GET 请求方式可以直接通过 req 对象获取对应参数。对于 POST 请求，通过设置路由中间件，express.json()用于解析请求体中的 JSON 数据，而 express.urlencoded()则用于处理请求体中的 URL 编码参数，从而简化了烦琐的数据转换操作。

此时，Express 的优势已经显而易见。然而，这还不是全部。除了提供强大的路由系统和中间件之外，它还集成了模板引擎、静态文件服务、错误处理机制等功能。但这并不是本书的重点，这里不作过多扩展，感兴趣的读者可以自行了解。

既然 Express 拥有如此多的优势，Nest 又是凭借什么在市场中占据一席之地呢？答案是它解决了架构问题。

再来看 Express 的路由管理，它可能是这样的：

```
const express = require("express");
const app = express();

// 定义/users 路由
app.get("/users", (req, res) => {
  res.send("Welcome to users page");
});

// 定义带参数的路由
app.get("/users/:id", (req, res) => {
  const userId = req.params.id;
  res.send(`User ID: ${userId}`);
});

// 使用 Router 对象定义更复杂的路由
const adminRouter = express.Router();

adminRouter.get("/", (req, res) => {
  res.send("Welcome to admin home");
});
```

```javascript
adminRouter.get("/dashboard", (req, res) => {
  res.send("Welcome to admin dashboard");
});

// 将/admin 路由挂载到/app 路径下
app.use("/app", adminRouter);

const PORT = 3000;
app.listen(PORT, () => {
  console.log(`Server is listening on port ${PORT}`);
});
```

有经验的读者可能会发现,这种方式的可维护性太差了,于是有了下面的路由管理方式。在主应用中使用模块化管理路由,代码如下:

```javascript
// app.js
const express = require('express');
const app = express();

// users 模块
const usersRouter = require('./routes/users');
app.use('/users', usersRouter);

const PORT = 3000;
app.listen(PORT, () => {
  console.log(`Server is listening on port ${PORT}`);
});
```

对 users 路由进行模块化管理,代码如下:

```javascript
// users.js
const express = require('express');
const router = express.Router();

// RESTful 风格的路由
app.get('/', getAllUsers);
app.get('/users/:id', getUserById);
app.post('/users', createUser);
app.put('/users/:id', updateUser);
app.delete('/users/:id', deleteUser);

function getAllUsers(req, res) {}
function getUserById(req, res) {}
function createUser(req, res) {}
function updateUser(req, res) {}
function deleteUser(req, res) {}

module.exports = router;
```

显然,在对比之下,第二种方式的可维护性和扩展性更高。存在这种差异的原因在于 Express 没有规定开发者必须遵循哪种方式编写代码。开发者之间的水平不一致导致了这种差异化。从设计者的角度来看,这显然是不可接受的。因此,更高层次的框架——Nest 应运而生。

1.1.3　Nest 处理 HTTP 请求

在 Nest 中是如何处理请求的呢？下面一起来看看。

Nest 处理请求 URL 代码示例：

```
// 通过路由参数
import { Controller, Get, Param } from '@nestjs/common';

@Controller('cats')
export class CatsController {
  // 使用路由参数装饰器（如@Param）来获取请求 URL 中的参数
  @Get(':id')
  findOne(@Param('id') id: string): string {
    return 'This action returns a cat with the ID: ${id}';
  }

  // 使用查询参数装饰器（如@Query）来获取请求 URL 中的查询参数
  @Get()
  findAll(@Query('page') page: number): string {
    return 'This action returns all cats on page ${page}';
  }
}
```

Nest 处理请求体代码示例：

```
import { Controller, Post, Body } from '@nestjs/common';

@Controller('cats')
export class CatsController {
  // 使用@Body 装饰器来获取 POST 请求中的请求体数据
  @Post()
  create(@Body() data): string {
    return 'This action creates a new cat with the name: ${data}';
  }
}
```

是不是简单多了？没错，这就是在 Nest 中处理请求的常用方式。它提供了强大的装饰器来操作请求头数据，并且自动判断返回的数据类型来设置合理的 Content-type，使得请求处理变得非常灵活。

至此，回顾一下开篇的问题：原生 HTTP 模块、Express 与 Nest 之间有何联系？相信到这里读者已经有了答案，至于 Nest 是如何解决架构问题的，会在后续的章节中详细讲解。

1.2　TypeScript 基础与应用

Nest 是基于 TypeScript 进行构建的，它提供了类型安全系统，并引入了面向对象编程、装饰器和元数据等概念。这些概念贯穿于 Nest 的源码设计和项目实践开发中。本节将从 TypeScript 的编译配置和类型系统两大板块着手，回顾 TypeScript 的基础知识。

1.2.1　TypeScript 编译

1. 编译上下文

TypeScript 被设计为 JavaScript 的超集，在使用时需要通过特定的编译器将其编译为普通的 JavaScript 才能被浏览器识别。编译器运行的环境被称为 TypeScript 的编译上下文，它包含编译器需要用到的各种配置选项、输入输出文件等信息。而 tsconfig.json 文件就是用来指定编译器运行时的行为，即编译选项。

2. 编译选项

通常用 compilerOptions 来定制编译选项，它包含丰富的选项，示例如下：

```
{
  "compilerOptions": {

    /* 基本选项 */
    "target": "es5",                       // 指定 ECMAScript 目标版本: 'ES3' (default), 'ES5', 'ES6'/'ES2015', 'ES2016', 'ES2017','ESNEXT'
    "module": "commonjs",                  // 指定使用的模块: 'commonjs', 'amd', 'system', 'umd' or 'es2015'
    "lib": [],                             // 指定要包含在编译中的库文件
    "allowJs": true,                       // 允许编译 JavaScript 文件
    "checkJs": true,                       // 报告 JavaScript 文件中的错误
    "jsx": "preserve",                     // 指定 JSX 代码的生成: 'preserve', 'react-native', or 'react'
    "declaration": true,                   // 生成相应的'.d.ts'文件
    "sourceMap": true,                     // 生成相应的'.map'文件
    "outFile": "./",                       // 将输出文件合并为一个文件
    "outDir": "./",                        // 指定输出目录
    "rootDir": "./",                       // 用来控制输出目录结构 --outDir.
    "removeComments": true,                // 删除编译后的所有注释
    "noEmit": true,                        // 不生成输出文件
    "importHelpers": true,                 // 从 tslib 导入辅助工具函数
    "isolatedModules": true,               // 将每个文件作为单独的模块 （与'ts.transpileModule'类似）

    /* 严格的类型检查选项 */
    "strict": true,                        // 启用所有严格类型的检查选项
    "noImplicitAny": true,                 // 在表达式和声明上有隐含的 any 类型时报错
    "strictNullChecks": true,              // 启用严格的 null 检查
    "noImplicitThis": true,                // 当 this 表达式值为 any 类型时，生成错误
    "alwaysStrict": true,                  // 以严格模式检查每个模块，并在每个文件中加入 'use strict'

    /* 额外的检查 */
    "noUnusedLocals": true,                // 有未使用的局部变量时抛出错误
    "noUnusedParameters": true,            // 有未使用的参数时抛出错误
    "noImplicitReturns": true,             // 并不是所有函数都有返回值时抛出错误
    "noFallthroughCasesInSwitch": true,    // 报告 switch 语句的 fallthrough 错误（即不允许 switch 的 case 语句贯穿）

    /* 模块解析选项 */
```

```
        "moduleResolution": "node",              // 选择模块解析策略：'node' (Node.js) or
'classic' (TypeScript pre-1.6)
        "baseUrl": "./",                         // 用于解析非相对路径的模块名称的基目录
        "paths": {},                             // 模块名到基于 baseUrl 的路径映射的列表
        "rootDirs": [],                          // 根文件夹列表，其组合内容表示项目运行时的结构内容
        "typeRoots": [],                         // 包含类型声明的文件列表
        "types": [],                             // 需要包含的类型声明文件名列表
        "allowSyntheticDefaultImports": true,    // 允许从没有设置默认导出的模块中默认导入

        /* Source Map Options */
        "sourceRoot": "./",       // 指定调试器应该找到 TypeScript 文件而不是源文件的位置
        "mapRoot": "./",          // 指定调试器应该找到映射文件而不是生成文件的位置
        "inlineSourceMap": true,  // 生成单个 sourcemaps 文件，而不是将 sourcemaps 生成不同的文件
        "inlineSources": true,    // 将源代码与 sourcemaps 生成到一个文件中，要求同时设置了
  --inlineSourceMap 或--sourceMap 属性

        /* 其他选项 */
        "experimentalDecorators": true,          // 启用实验性的装饰器
        "emitDecoratorMetadata": true            // 为装饰器提供元数据支持
    }
}
```

3. 编译运行

TypeScript 内置了 TSC 编译器，可以通过以下方式运行它：

- 运行 tsc，它会在当前目录或父级目录中寻找 tsconfig.json 文件。
- 运行 tsc -p ./path-to-project-directory。这里的路径可以是绝对路径，也可以是相对于当前目录的相对路径。

4. 指定文件

可以用 include 和 exclude 选项来指定需要包含的文件和排除的文件：

```
{
  "include": [
    "./folder"
  ],
  "exclude": [
    "./folder/**/*.spec.ts",
    "./folder/someSubFolder"
  ]
}
```

1.2.2　TypeScript 类型系统

1. 基本注解

类型注解使用":TypeAnnotation"语法。在类型声明空间中，可用的任何内容都可以用作类型注解。

在下面这个例子中，使用了变量、函数参数以及函数返回值的类型注解：

```
const num: number = 123;
```

```
function identity(num: number): number {
  return num;
}
```

2. 原始类型

JavaScript 的原始类型同样适用于 TypeScript 的类型系统，因此 string、number、boolean 也可以用作类型注解：

```
let num: number;
let str: string;
let bool: boolean;

num = 123;
num = 123.456;
num = '123'; // Error
```

3. 数组

TypeScript 为数组提供了专用的类型语法，因此我们可以很容易地为数组添加类型注解，代码如下：

```
let boolArray: boolean[];

boolArray = [true, false];
console.log(boolArray[0]); // true
```

4. 接口

接口可以将多个类型声明合并为一个类型声明：

```
interface Name {
  first: string;
  second: string;
}

let name: Name;
name = {
  first: 'John',
  second: 'Doe'
};

name = {
  // Error: 'Second is missing'
  first: 'John'
};
```

在这里，我们把类型注解 first: string + last: string 合并到了一个新的类型注解 Name 中，这样能够强制对每个成员进行类型检查。

5. 特殊类型

除了前面提到的一些原始类型，TypeScript 中还存在一些特殊的类型，它们是 any、null、undefined 以及 void。

1) any

any 类型为我们提供了一个类型系统的"后门",使用它时,TypeScript 会关闭类型检查。在类型系统中,any 能够兼容所有的类型(包括它自己),代码如下:

```
let power: any;

// 赋值任意类型
power = '123';
power = 123;
```

2) null 和 undefined

在类型系统中,JavaScript 中的 null 和 undefined 字面量与其他被标注了 any 类型的变量一样,都可以赋值给任意类型的变量,但不会关闭类型检查,代码如下:

```
// strictNullChecks: false

let num: number;
let str: string;

// 这些类型可用于赋值
num = null;
str = undefined;
```

3) void

void 表示没有任何类型,通常用于那些不返回任何值的函数的返回类型。

```
function myFunction(): void {
    // 函数体
}
```

6. 泛型

泛型可以应用于 Typescript 中的函数(函数参数、函数返回值)、接口和类(类的实例成员、类的方法)。

下面我们来创建第一个使用泛型的示例函数:reverse()。这个函数会返回传入的任意值。在不使用泛型的情况下,这个函数可能是这样的:

```
function reverse(arg: any): any {
  return arg;
};
```

使用 any 类型会导致这个函数可以接收任何类型的 arg 参数,但我们希望传入的类型与返回的类型相同。

因此,需要一种方法来确保返回值的类型与传入参数的类型相同。这里,我们使用了类型变量,它是一种特殊的变量,只用于表示类型而不是值,代码如下:

```
function reverse<T>(arg: T): T {
  return arg;
}
```

我们给 reverse 添加了类型变量 T。T 帮助我们捕获用户传入的类型（比如 string），之后把 T 用作返回值类型，这样就可以跟踪函数中使用的类型信息了。

我们把这个版本的 reverse() 函数称为泛型，因为它适用于多个类型。

7. 联合类型

我们希望 JavaScript 属性能够支持多种类型，例如字符串或者字符串数组。在 TypeScript 中，可以使用联合类型来满足这一需求（联合类型使用"|"作为分隔符，例如 TypeA | TypeB）。以下是关于联合类型的例子：

```
function formatCommandline(command: string[] | string) {
  let line = '';
  if (typeof command === 'string') {
    line = command.trim();
  } else {
    line = command.join(' ').trim();
  }
}
```

8. 交叉类型

在 JavaScript 中，extend 是一种非常常见的模式。在这种模式下，我们可以通过两个对象创建一个新对象，使新对象继承这两个对象的所有功能。交叉类型可以让我们安全地使用这种模式：

```
function extend<T extends object, U extends object>(first: T, second: U): T & U {
  const result = {} as T & U;

  for (let id in first) {
    (result as T)[id] = first[id];
  }

  for (let id in second) {
    (result as U)[id] = second[id];
  }

  return result;
}
```

9. 类型别名

TypeScript 提供了一种便捷的语法来为类型注解设置别名。我们可以使用 type SomeName = someValidTypeAnnotation 来创建别名：

```
type StrOrNum = string | number;

// 使用
let sample: StrOrNum;
sample = 123;
sample = '123';

// 检查类型
sample = true; // Error
```

10. 类型断言

TypeScript 允许覆盖其类型推断,并可以按照我们期望的方式重新分析类型。这种机制被称为"类型断言"。类型断言用于告知编译器,我们对该类型的了解比编译器的推断更为准确,因此编译器不应再发出类型错误。以下是类型断言的常见用法示例:

```
function isArray(o: any): boolean {
  return Object.prototype.toString.call(o) === '[object Array]';
}

function checkAuth(auth: number | number[]) {
  if (isArray(auth)) {
    // 错误,Property 'some' does not exist on type 'number'
    return auth.some(auth => auth === 1);
  } else {
    return auth === 1;
  }
}
```

这里的代码发出了错误警告,因为 auth 的类型为 number|number[],其属性没有 some()方法。因此,我们不能使用 some()方法。虽然已经知道 auth 的实际类型是 number[],但 TypeScript 无法推导出这一点。在这种情况下,可以通过类型断言来避免此问题:

```
function isArray(o: any): boolean {
  return Object.prototype.toString.call(o) === '[object Array]';
}

function checkAuth(auth: number | number[]) {
  if (isArray(auth)) {
    return (auth as number[]).some(auth => auth === 1);
  } else {
    return auth === 1;
  }
}
```

11. 枚举

枚举是一种用于收集相关值的集合的方法,示例如下:

```
enum Status {
  Success,
  Error,
  Warning
}

enum Env {
  Null = '',
  Dev = 'development',
  Prod = 'production'
}

// 简单使用
let status = Status.Success;
```

```
let env = Env.Dev;
```

12. 重载

TypeScript 允许我们声明函数重载。这对于文档+类型安全非常实用。请看以下代码：

```
function padding(top: number, right?: number, bottom?: number, left?: number) {
  if (right === undefined && bottom === undefined && left === undefined) {
    right = bottom = left = top;
  } else if (bottom === undefined && left === undefined) {
    left = right;
    bottom = top;
  }

  return { top, right, bottom, left };
}
```

我们可以通过函数重载来强制执行并记录这些约束，只需多次声明函数头即可。最后一个函数头在函数体内实际处于活动状态，但这函数头不可在外部直接使用，如下所示：

```
// 重载
function padding(all: number);
function padding(topBottom: number, leftRight: number);
function padding(top: number, right: number, bottom: number, left: number);
// 函数主体的具体实现包含需要处理的所有情况
function padding(top: number, right?: number, bottom?: number, left?: number) {
  if (right === undefined && bottom === undefined && left === undefined) {
    right = bottom = left = top;
  } else if (bottom === undefined && left === undefined) {
    left = right;
    bottom = top;
  }

  return { top, right, bottom, left };
}
```

本节通过示例介绍了 TypeScript 的基本概念和用法，这些类型在 Nest 开发中都会频繁使用。掌握本节内容可以为我们以后编写 Nest 应用打下坚实的基础。

第 2 章

Nest 初识

本章主要介绍 Nest 框架的基础概念和主要特点，讲解如何通过 Nest 命令行界面（CLI）快速创建你的应用程序。此外，本章还将结合前端 React 框架，演示一个完整的前后端项目请求闭环流程。接下来，将介绍 Nest 中两个非常重要的软件设计思想：面向切面编程（Aspect-Oriented Programming，AOP）和控制反转（Inversion of Control，IoC），以及它们在 Nest 中的应用。最后，本章将介绍如何通过 Chrome 开发者工具（DevTools）和 VS Code 高效调试 Nest 项目。

2.1 什么是 Nest

在第 1 章中，我们了解到在 Nest 中可以更优雅地处理 HTTP 请求方式。它的优势远不止于此，还包括强大的依赖注入系统，以及对面向对象编程和函数式编程等多种范式的支持。虽然这听起来可能有些复杂，但不必担心。我们先来掌握 Nest 的基本概念，这有助于在后续的章节中更好地理解它。

2.1.1 Nest 概述

Nest 是一个用于构建高效、可扩展的 Node.js 服务器端应用的框架。它默认使用 Express 作为 HTTP 服务端框架，同时也支持 Fastify 作为替代选项。然而，Nest 并不依赖于这两种框架，而是采用了优雅的适配器模式来实现其可扩展性。简言之，即使未来出现更优秀的第三方模块，开发者仍然拥有自由选择的权利。

2.1.2 Nest 的主要特点

在架构设计方面，Nest 属于 MVC（Model-View-Controller，模型-视图-控制器）架构体系，具有模块化和松耦合的特性。它引入了依赖注入来管理对象之间的关系，使得 Model、View、

Controller 之间的交互变得简单且高效。

在项目管理方面，模块化结构将各个功能模块拆分为更细的粒度进行管理。每个模块中都包含独立的控制器、路由、服务、拦截器等，最后将这些模块聚合在一起，形成一个类似搭积木的过程。

在编码风格方面，Nest 支持 JavaScript 与 TypeScript 两种语言。使用 TypeScript 具备更多优势，例如静态类型检测、代码自动补全、智能提示以及函数重载等高级语法支持。

在扩展性方面，Nest 支持各种中间件和插件，包括身份验证、日志记录、缓存等。此外，Nest 还允许与各种数据库、通信协议以及其他第三方服务无缝集成，使开发者能够轻松扩展应用功能。

2.1.3　Nest 的应用场景

Nest 是一个多功能的 Node 后端框架。其灵活性使得它能够满足不同需求，从小型项目到大型应用等各种场景，都可以用 Nest 来构建高质量、可维护的后端服务。应用场景包括但不限于：

- Web 应用程序：Nest 提供强大的路由系统和 HTTP 模块，能够与各种模板引擎良好配合，用于构建博客系统、企业官网、电子商务网站等各种 Web 应用。
- 微服务架构：Nest 支持微服务架构，可创建多个微服务，并轻松实现不同微服务之间的交互和通信，适用于构建复杂的分布式系统。
- 服务端渲染（Server-Side Rendering，SSR）：Nest 支持与前端主流的服务端渲染框架（如 Next.js、Nuxt.js）配合使用，用于构建单页面应用（Single-Page Application，SPA）。
- 企业级应用：Nest 拥有强大的依赖注入和模块系统，与各种数据库、日志系统集成紧密，能够满足大型企业级应用的各种需求。

无论你需要构建小型应用还是大型复杂的企业级应用，Nest 都是一个值得选择的框架。

2.2　快速上手 Nest CLI

相信 CLI 对于读者来说并不陌生，它广泛应用于各个工程化领域，用于快速创建项目、运行编译、执行构建等操作，如前端的 Vue CLI、Vite、Create React App，后端的 Spring Boot CLI、express-generator 等。Nest 也不例外，提供了强大的 CLI 命令行工具，它被放在@nestjs/cli 包中。本节首先介绍 CLI 的安装并创建一个实际的项目，接着通过项目演示来介绍它生成各种代码片段的命令，帮助读者彻底掌握 CLI 的使用。

2.2.1　Nest CLI 的安装

执行以下命令可将 Nest CLI 安装为全局工具，方便日后使用：

```
npm install -g @Nestjs/cli
```

需要注意的是，这种方式需要不定时更新，以获取最新的模板代码：

```
npm update -g @Nestjs/cli
```

安装完成之后，执行"nest –h"命令，结果如图 2-1 所示。

```
$ nest -h
Usage: nest <command> [options]

Options:
  -v, --version                         Output the current version.
  -h, --help                            Output usage information.
Commands:
  new|n [options] [name]                Generate Nest application.
  build [options] [app]                 Build Nest application.
  start [options] [app]                 Run Nest application.
  info|i                                Display Nest project details.
  add [options] <library>               Adds support for an external library to your project.
  generate|g [options] <schematic> [name] [path]  Generate a Nest element.
    Schematics available on @nestjs/schematics collection:

    name             alias          description
    application      application    Generate a new application workspace
    class            cl             Generate a new class
    configuration    config         Generate a CLI configuration file
    controller       co             Generate a controller declaration
    decorator        d              Generate a custom decorator
    filter           f              Generate a filter declaration
    gateway          ga             Generate a gateway declaration
    guard            gu             Generate a guard declaration
    interceptor      itc            Generate an interceptor declaration
    interface        itf            Generate an interface
    library          lib            Generate a new library within a monorepo
    middleware       mi             Generate a middleware declaration
    module           mo             Generate a module declaration
    pipe             pi             Generate a pipe declaration
    provider         pr             Generate a provider declaration
    resolver         r              Generate a GraphQL resolver declaration
    resource         res            Generate a new CRUD resource
    service          s              Generate a service declaration
    sub-app          app            Generate a new application within a monorepo
```

图 2-1 "nest – h"命令的执行结果

由图 2-1 可见，Nest 提供了非常丰富的命令。

- nest new：用于创建项目。
- nest build：用于构建生产环境代码。
- nest start：用于启动本地开发服务。
- nest info：用于查看当前项目中的 Nest 包信息。
- nest add：用于添加官方插件或者第三方模块。
- nest generate：用于生成各种模块代码，如 Module、Controller、Service、Pipe、Middleware 等。

同时，部分命令也支持别名，如"nest n""nest i""nest g"，后续将分别进行测试。

2.2.2 创建项目

用户可以用"nest new"命令来创建项目，具体参数如图 2-2 所示。

```
$ nest new -h
Usage: nest new|n [options] [name]

Generate Nest application.

Options:
  --directory [directory]              Specify the destination directory
  -d, --dry-run                        Report actions that would be performed without writing out results. (default: false)
  -g, --skip-git                       Skip git repository initialization. (default: false)
  -s, --skip-install                   Skip package installation. (default: false)
  -p, --package-manager [packageManager]   Specify package manager.
  -l, --language [language]            Programming language to be used (TypeScript or JavaScript) (default: "TypeScript")
  -c, --collection [collectionName]    Schematics collection to use (default: "@nestjs/schematics")
  --strict                             Enables strict mode in TypeScript. (default: false)
  -h, --help                           Output usage information.
```

图 2-2　创建项目时可以使用的参数选项

下面挑选几个常用的参数进行说明。

- --skip-git 和--skip-install：这些参数用于跳过 Git 初始化和 npm 包安装步骤。
- --package-manager：此参数用于指定项目使用的包管理器（npm、yarn、pnpm）。推荐使用 pnpm，它作为继 npm、yarn 之后推出的包管理器，因其速度快、节省磁盘空间而受到青睐。
- --language：此参数决定使用 TypeScript 还是 JavaScript 进行编写。推荐使用默认的 TypeScript。
- --collection：用于指定工作流集合。默认是@nestjs/schematics，用于快速创建模块、控制器、服务等，与"nest generate"命令相关。通常使用默认值，后续将详细讲解。
- --strict：此参数用于指定 TypeScript 是否以严格模式运行。

运行"nest n cli-test"命令后，CLI 提供了交互式命令让你选择包管理器，如图 2-3 所示。

```
$ nest n cli-test
⚡  We will scaffold your app in a few seconds..

? Which package manager would you ❤ to use? (Use arrow keys)
❯ npm
  yarn
  pnpm
```

图 2-3　选择包管理器

当然，我们可以直接使用"nest n cli-test -p pnpm"命令来指定使用 pnpm，如图 2-4 所示。

```
$ nest n cli-test -p pnpm
⚡  We will scaffold your app in a few seconds..

CREATE cli-test/.eslintrc.js (663 bytes)
CREATE cli-test/.prettierrc (51 bytes)
CREATE cli-test/README.md (3347 bytes)
CREATE cli-test/nest-cli.json (171 bytes)
CREATE cli-test/package.json (1949 bytes)
CREATE cli-test/tsconfig.build.json (97 bytes)
CREATE cli-test/tsconfig.json (546 bytes)
CREATE cli-test/src/app.controller.spec.ts (617 bytes)
CREATE cli-test/src/app.controller.ts (274 bytes)
CREATE cli-test/src/app.module.ts (249 bytes)
CREATE cli-test/src/app.service.ts (142 bytes)
CREATE cli-test/src/main.ts (208 bytes)
CREATE cli-test/test/app.e2e-spec.ts (630 bytes)
CREATE cli-test/test/jest-e2e.json (183 bytes)

▶▶▶▶▶ Installation in progress... 🍗
```

图 2-4　指定 pnpm

由图 2-4 可见，CLI 直接创建默认的项目模板，然后执行安装依赖，完成后自动创建一个 pnpm-lock.yaml 文件，如图 2-5 所示。

图 2-5　自动创建 pnpm-lock.yaml 文件

"nest generate" 命令默认执行的是 @nestjs/schematics 包中的命令。而 Schematics 是一种用于创建、删除和更新 Angular 应用程序代码的工具。Nest 在此基础上进行了扩展和定制，以适应其框架的需求。

2.2.3　生成指定的代码片段

执行 "nest generate -h" 命令来生成指定的代码片段，可以查看其中包含的内容，如图 2-6 所示。

图 2-6　生成指定代码片段的命令参数

其中包含丰富的 Nest 元素（如 Controller、Decorator、Filter 等），同时提供了灵活的参数控制。

我们来试一下，进入刚刚创建的项目 cli-test，执行"nest g controller"命令，如图 2-7 所示。

```
$ nest g controller user
CREATE src/user/user.controller.spec.ts (478 bytes)
CREATE src/user/user.controller.ts (97 bytes)
UPDATE src/app.module.ts (446 bytes)
```

图 2-7　使用命令生成控制器（controller）

接着可以看见在 src 目录下创建了一个 user 控制器，如图 2-8 所示。

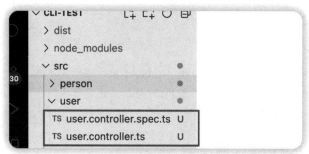

图 2-8　生成 user 控制器结果

其中，.spec.ts 是单元测试文件，可以通过设置 --no-spec 参数表明不生成测试文件。

- --flat 和 --not-flat 这两个参数表示是否使用扁平化结构。我们接着创建一个 user 过滤器来测试。
- --flat: 参数表示扁平化，会将生成的文件放到 src 目录下，而不生成对应的目录，如图 2-9 所示。

```
$ nest g filter user --flat
CREATE src/user.filter.spec.ts (164 bytes)
CREATE src/user.filter.ts (186 bytes)
```

图 2-9　使用扁平化参数对应的执行结果

- --not-flat: 参数表示非扁平化，会生成对应的目录，如图 2-10 所示。

```
$ nest g filter user --no-flat
CREATE src/user/user.filter.spec.ts (164 bytes)
CREATE src/user/user.filter.ts (186 bytes)
```

图 2-10　非扁平化参数执行结果

- --skip-import 表示是否跳过自动导入依赖，默认情况下会自动导入，以 user controller 为例，如图 2-11 所示。

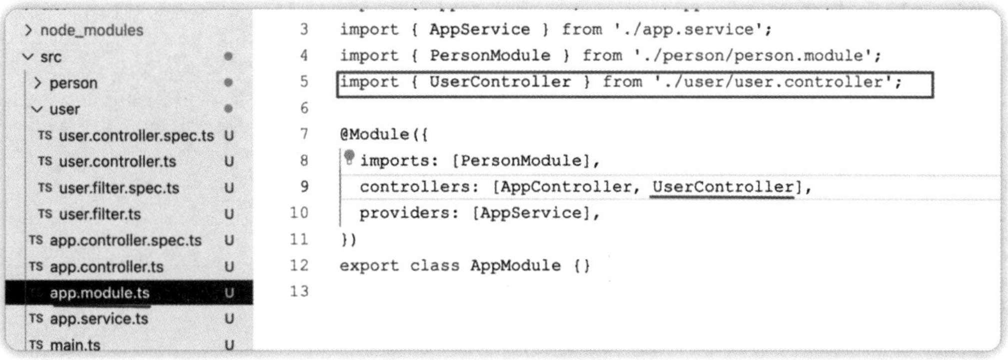

```
> node_modules            3   import { AppService } from './app.service';
∨ src                     4   import { PersonModule } from './person/person.module';
  > person                5   import { UserController } from './user/user.controller';
  ∨ user                  6
    TS user.controller.spec.ts U   7   @Module({
    TS user.controller.ts  U   8     imports: [PersonModule],
    TS user.filter.spec.ts U   9     controllers: [AppController, UserController],
    TS user.filter.ts      U  10     providers: [AppService],
  TS app.controller.spec.ts U 11   })
  TS app.controller.ts   U  12   export class AppModule {}
    app.module.ts        U  13
  TS app.service.ts      U
  TS main.ts             U
```

图 2-11　默认自动导入依赖的效果图

由图 2-11 可见，UserController 会自动导入应用的主模块中，并自动将其添加到 Controllers 依赖项列表中。同样的流程也适用于生成 Service 服务和 module 模块。

然而，逐个创建 Controller 或 Filters 可能不够方便。能否一次性生成所需的模板呢？当然可以。Nest 提供了 nest generate resource 命令，可以一键生成代码模板。执行该命令后，CLI 会询问我们选择使用哪种代码风格。我们选择 REST 风格的 API，如图 2-12 所示。

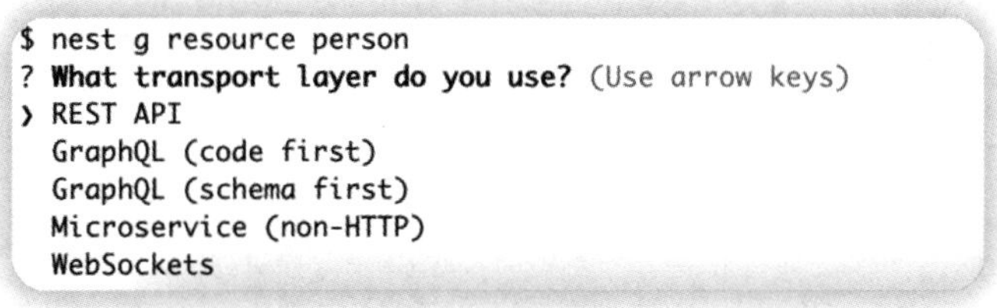

```
$ nest g resource person
? What transport layer do you use? (Use arrow keys)
> REST API
  GraphQL (code first)
  GraphQL (schema first)
  Microservice (non-HTTP)
  WebSockets
```

图 2-12　选择不同风格的 API

选择 yes 后，系统将自动生成与 CRUD（创建、读取、更新、删除）相关的代码，如图 2-13 所示。

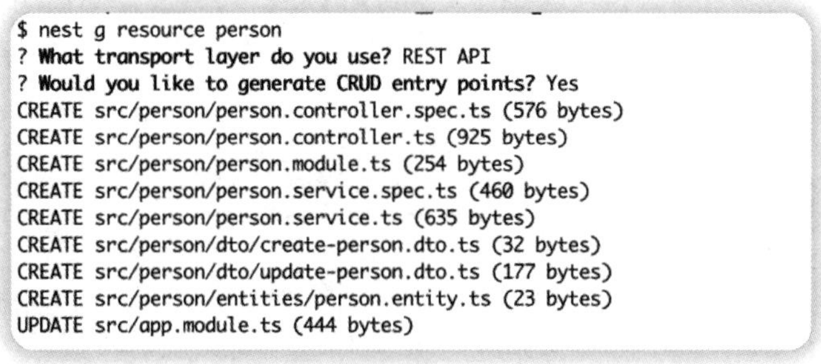

```
$ nest g resource person
? What transport layer do you use? REST API
? Would you like to generate CRUD entry points? Yes
CREATE src/person/person.controller.spec.ts (576 bytes)
CREATE src/person/person.controller.ts (925 bytes)
CREATE src/person/person.module.ts (254 bytes)
CREATE src/person/person.service.spec.ts (460 bytes)
CREATE src/person/person.service.ts (635 bytes)
CREATE src/person/dto/create-person.dto.ts (32 bytes)
CREATE src/person/dto/update-person.dto.ts (177 bytes)
CREATE src/person/entities/person.entity.ts (23 bytes)
UPDATE src/app.module.ts (444 bytes)
```

图 2-13　选择 CRUD 入口节点

创建完成后，生成 REST 风格的 API，内容如下：

```
import { Controller, Get, Post, Body, Patch, Param, Delete } from '@nestjs/common';
import { PersonService } from './person.service';
import { CreatePersonDto } from './dto/create-person.dto';
import { UpdatePersonDto } from './dto/update-person.dto';

@Controller('person')
```

```typescript
export class PersonController {
  constructor(private readonly personService: PersonService) {}

  @Post()
  create(@Body() createPersonDto: CreatePersonDto) {
    return this.personService.create(createPersonDto);
  }

  @Get()
  findAll() {
    return this.personService.findAll();
  }

  @Get(':id')
  findOne(@Param('id') id: string) {
    return this.personService.findOne(+id);
  }

  @Patch(':id')
  update(@Param('id') id: string, @Body() updatePersonDto: UpdatePersonDto) {
    return this.personService.update(+id, updatePersonDto);
  }

  @Delete(':id')
  remove(@Param('id') id: string) {
    return this.personService.remove(+id);
  }
}
```

dto 和 entities 是 CRUD 相关的代码，最终集成到 PersonModule 中，并被自动导入 AppModule，如图 2-14 所示。

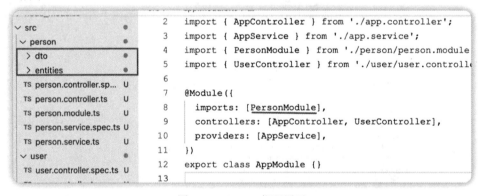

图 2-14 CRUD 代码展示

以上就是 Nest CLI 提供的用于快速创建项目代码的工具。

2.2.4 构建应用

前面介绍了 nest generate 命令，接下来使用 new build 命令来构建应用。

执行 nest build –h 命令，可以看到 build 命令提供了一些可选参数，如图 2-15 所示。

```
$ nest build -h
Usage: nest build [options] [app]

Build Nest application.

Options:
  -c, --config [path]      Path to nest-cli configuration file.
  -p, --path [path]        Path to tsconfig file.
  -w, --watch              Run in watch mode (live-reload).
  -b, --builder [name]     Builder to be used (tsc, webpack, swc).
  --watchAssets            Watch non-ts (e.g., .graphql) files mode.
  --webpack                Use webpack for compilation (deprecated option, use --builder instead).
  --type-check             Enable type checking (when SWC is used).
  --webpackPath [path]     Path to webpack configuration.
  --tsc                    Use typescript compiler for compilation.
  --preserveWatchOutput    Use "preserveWatchOutput" option when using tsc watch mode.
  -h, --help               Output usage information.
```

图 2-15　build 命令提供的参数

其中，各选项说明如下。

- --path：用于指定 tsconfig 文件的路径。
- --watch：开启实时监听模式，在文件发生变化时自动执行构建操作。
- --builder：选择使用指定的工具进行构建，可选的工具包括 tsc、webpack、swc 等。

默认情况下，Nest 使用 tsc 进行编译，运行 nest build 命令的效果如图 2-16 所示。

若要切换为使用 webpack 进行打包，可以运行 nest build-b webpack 命令，效果如图 2-17 所示。

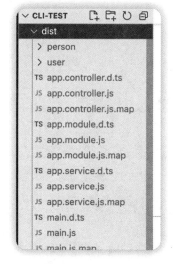

图 2-16　构建之后的目录文件

图 2-17　webpack 打包后的目录文件

--webpack 和--tsc 分别指定了不同的编译器，webpack 用于编译和打包，而 tsc 只用于编译，它们的运行效果与前文介绍的类似。

每次都需要在命令后面添加参数，这样有点麻烦，是否可以将这些参数写入配置文件中进行管理呢？答案是可以的。

- --config：指定 nest-cli 的配置文件路径，即 nest-cli.json 文件，可以用来配置打包参数，如图 2-18 所示。

```
CLI-TEST                  {} nest-cli.json > ...
∨ dist                    1  {
  JS main.js              2    "$schema": "https://json.schemastore.org/nest-cli",
> node_modules            3    "collection": "@nestjs/schematics",
> src                     4    "sourceRoot": "src",
> test                    5    "generateOptions": {
  .eslintrc.js      U     6      "flat": false,
  .gitignore        U     7      "spec": true
  .prettierrc       U     8    },
  nest-cli.json     U     9    "compilerOptions": {
  package.json      U     10     "webpack": false,
  pnpm-lock.yaml    U     11     "deleteOutDir": true,
  README.md         U     12     "builder":"tsc",
  tsconfig.build.json U   13     "watchAssets": false
  tsconfig.json     U     14   }
                          15 }
                          16
```

图 2-18 nest-cli 的配置文件

由图 2-18 可以看到，前面提到的 flat、spec 都可以在这里进行配置。通过编译选项也可以指定是否使用 webpack 进行构建，以及是否使用 builder 指定选择的编译器等。

2.2.5 启动开发调试

本小节将用 nest start 命令来启动开发调试，它的可选参数如图 2-19 所示。

```
$ nest start -h
Usage: nest start [options] [app]

Run Nest application.

Options:
  -c, --config [path]        Path to nest-cli configuration file.
  -p, --path [path]          Path to tsconfig file.
  -w, --watch                Run in watch mode (live-reload).
  -b, --builder [name]       Builder to be used (tsc, webpack, swc).
  --watchAssets              Watch non-ts (e.g., .graphql) files mode.
  -d, --debug [hostport]     Run in debug mode (with --inspect flag).
  --webpack                  Use webpack for compilation (deprecated option, use --builder instead).
  --webpackPath [path]       Path to webpack configuration.
  --type-check               Enable type checking (when SWC is used).
  --tsc                      Use typescript compiler for compilation.
  --sourceRoot [sourceRoot]  Points at the root of the source code for the single project in standard mode str
  --entryFile [entryFile]    Path to the entry file where this command will work with. Defaults to the one def
  -e, --exec [binary]        Binary to run (default: "node").
  --preserveWatchOutput      Use "preserveWatchOutput" option when using tsc watch mode.
  -h, --help                 Output usage information.
```

图 2-19 start 命令提供的参数

在开发阶段，nest start 命令用于开启本地服务。运行该命令的结果如图 2-20 所示。

```
○ $ nest start
  [Nest] 86023  - 2023/12/29 16:35:58   LOG [NestFactory] Starting Nest application...
  [Nest] 86023  - 2023/12/29 16:35:58   LOG [InstanceLoader] AppModule dependencies initialized +7ms
  [Nest] 86023  - 2023/12/29 16:35:58   LOG [InstanceLoader] PersonModule dependencies initialized +0ms
  [Nest] 86023  - 2023/12/29 16:35:58   LOG [RoutesResolver] AppController {/}: +5ms
  [Nest] 86023  - 2023/12/29 16:35:58   LOG [RouterExplorer] Mapped {/, GET} route +1ms
  [Nest] 86023  - 2023/12/29 16:35:58   LOG [RoutesResolver] UserController {/user}: +0ms
  [Nest] 86023  - 2023/12/29 16:35:58   LOG [RoutesResolver] PersonController {/person}: +0ms
  [Nest] 86023  - 2023/12/29 16:35:58   LOG [RouterExplorer] Mapped {/person, POST} route +0ms
  [Nest] 86023  - 2023/12/29 16:35:58   LOG [RouterExplorer] Mapped {/person, GET} route +1ms
  [Nest] 86023  - 2023/12/29 16:35:58   LOG [RouterExplorer] Mapped {/person/:id, GET} route +0ms
  [Nest] 86023  - 2023/12/29 16:35:58   LOG [RouterExplorer] Mapped {/person/:id, PATCH} route +0ms
  [Nest] 86023  - 2023/12/29 16:35:58   LOG [RouterExplorer] Mapped {/person/:id, DELETE} route +0ms
  [Nest] 86023  - 2023/12/29 16:35:58   LOG [NestApplication] Nest application successfully started +1ms
```

图 2-20 nest start 运行结果

--debug 参数用来调试。运行 nest start–d 命令后会启动一个 WebSocket 调试服务，通过调试工具链接到这个端口即可进行调试，如图 2-21 所示。

图 2-21　带--debug 参数运行 nest 命令的运行结果

详细的调试内容及技巧将在 2.6 节介绍。其他配置与 build 命令类似，这里不再赘述。

2.2.6　查看项目信息

nest info 命令用于查看 Node.js、npm 以及 Nest 依赖包的相关版本信息，如图 2-22 所示。

图 2-22　nest info 命令的执行结果

2.3　创建第一个 Nest 应用

了解了 Nest CLI 的使用之后，本节我们来小试牛刀，创建一个服务端应用，并使用 React 构建一个客户端应用。我们将实现客户端发送请求后，服务端接收请求、进行数据处理，并返回新的数据给客户端。如果你喜欢使用 Vue，也可以选择通过 Vue 来创建项目；但请注意，本案例不涉及复杂的前端交互。

2.3.1　生成后端项目

首先，运行 nest n web-app -p pnpm 命令生成服务端项目，如图 2-23 所示。

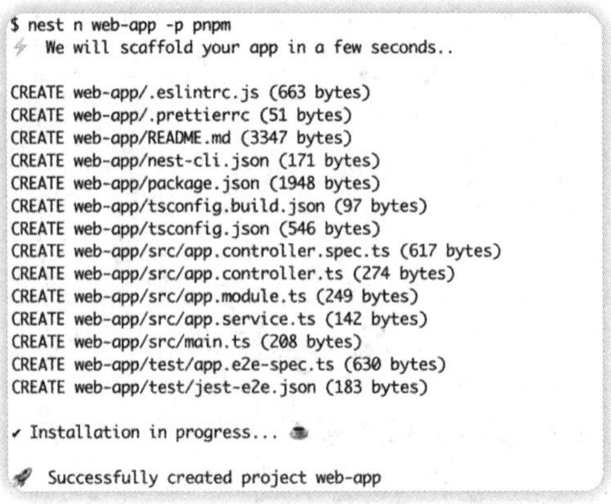

```
$ nest n web-app -p pnpm
⚡  We will scaffold your app in a few seconds..

CREATE web-app/.eslintrc.js (663 bytes)
CREATE web-app/.prettierrc (51 bytes)
CREATE web-app/README.md (3347 bytes)
CREATE web-app/nest-cli.json (171 bytes)
CREATE web-app/package.json (1948 bytes)
CREATE web-app/tsconfig.build.json (97 bytes)
CREATE web-app/tsconfig.json (546 bytes)
CREATE web-app/src/app.controller.spec.ts (617 bytes)
CREATE web-app/src/app.controller.ts (274 bytes)
CREATE web-app/src/app.module.ts (249 bytes)
CREATE web-app/src/app.service.ts (142 bytes)
CREATE web-app/src/main.ts (208 bytes)
CREATE web-app/test/app.e2e-spec.ts (630 bytes)
CREATE web-app/test/jest-e2e.json (183 bytes)

✔ Installation in progress... ☕

🚀  Successfully created project web-app
```

图 2-23　创建服务端项目

然后，启动该项目并保持热重载，如图 2-24 所示。

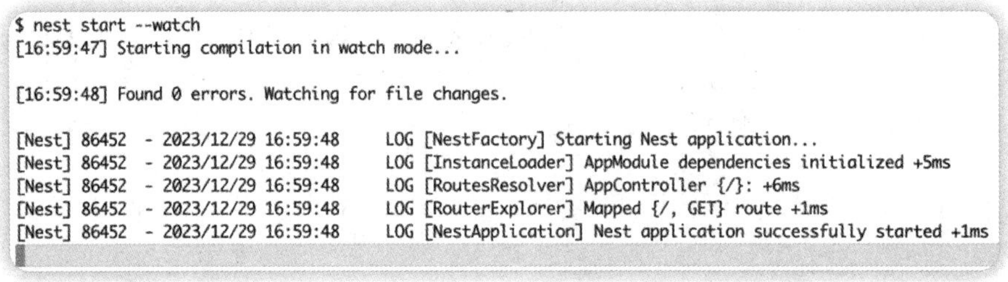

```
$ nest start --watch
[16:59:47] Starting compilation in watch mode...

[16:59:48] Found 0 errors. Watching for file changes.

[Nest] 86452  - 2023/12/29 16:59:48     LOG [NestFactory] Starting Nest application...
[Nest] 86452  - 2023/12/29 16:59:48     LOG [InstanceLoader] AppModule dependencies initialized +5ms
[Nest] 86452  - 2023/12/29 16:59:48     LOG [RoutesResolver] AppController {/}: +6ms
[Nest] 86452  - 2023/12/29 16:59:48     LOG [RouterExplorer] Mapped {/, GET} route +1ms
[Nest] 86452  - 2023/12/29 16:59:48     LOG [NestApplication] Nest application successfully started +1ms
```

图 2-24　运行服务端项目

2.3.2　生成前端项目

使用 React 提供的脚手架 create-react-app 生成项目，先全局安装一下：

```
npm install -g create-react-app
```

查看是否安装成功：

```
create-react-app --version
```

结果如图 2-25 所示。

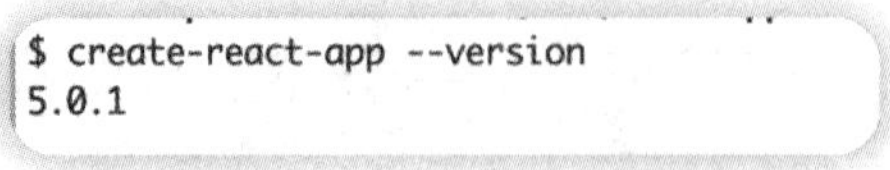

```
$ create-react-app --version
5.0.1
```

图 2-25　检查是否安装成功

执行 create-react-app web-app-front 命令，创建名为 web-app-front 的项目。创建成功后的效果如图 2-26 所示。

```
Inside that directory, you can run several commands:

  npm start
    Starts the development server.

  npm run build
    Bundles the app into static files for production.

  npm test
    Starts the test runner.

  npm run eject
    Removes this tool and copies build dependencies, configuration files
    and scripts into the app directory. If you do this, you can't go back!

We suggest that you begin by typing:

  cd web-app-front
  npm start

Happy hacking!
```

图 2-26　创建客户端项目

同样，启动该项目，执行 npm start 命令，结果如图 2-27 所示。

图 2-27　执行客户端项目

2.3.3　准备工作

在发送请求之前，我们需要进行一些准备工作。首先，在前端设置请求代理，以便能够访问 8088 端口下的服务。接着，通过 axios 发送一个 GET 请求，成功后将数据显示在页面上。

为了设置请求代理，在目录 web-app-front/src 下新建 setupProxy.js 文件，如图 2-28 所示。

图 2-28　新建 setupProxy 文件

设置代理逻辑,这里将端口自定义为 8088,代码如下:

```js
// 引入 http-proxy-middleware, react 脚手架已经安装
const proxy = require("http-proxy-middleware");

module.exports = function (app) {
  app.use(
    // 遇见/api 前缀的请求就会触发该代理配置
    proxy.createProxyMiddleware("/api", {
      // 请求转发给谁
      target: "http://localhost:8088",
      // 控制服务器收到的请求头中 Host 的值
      changeOrigin: true,
      // 重写请求路径
      pathRewrite: { "^/api": "" },
    })
  );
};
```

执行 npm install axios 命令来安装 axios 库。然后在 App.js 文件中引入 axios,并定义一个名为 getHello 的 API 接口,用于请求/123 路径的资源。以下是示例代码:

```js
import axios from 'axios';

function getHello() {
  return axios.get('/api/123');
}
```

通过单击按钮触发 AJAX 请求,并回显接收到的数据。以下是部分示例代码:

```js
// 头部导入 useState
import { useState } from "react";

// App 方法
function App() {
  const [nestData, setNestData] = useState({});

  const handleGetData = async () => {
    const { data } = await getHello();
    setNestData(data);
  };

  const handleCreateUser = async () => {
    const params = {
      name: "mouse",
      age: "22",
      address: "广州",
    };
    const { data } = await createUser(params);
    setNestData(data);
  };

  return (
    <div className="App">
      <header className="App-header">
        <img src={logo} className="App-logo" alt="logo" />
```

```
            <button onClick={handleGetData}>发送请求给 Nest</button>
            <br />
            <button onClick={handleCreateUser}>创建用户</button>
            <br />
            后端返回数据：{JSON.stringify(nestData)}
        </header>
     </div>
  );
}
```

切换到 web-app 服务端，在 main.ts 文件中将端口修改为 8088，以匹配前端代理端口，代码如下：

```
async function bootstrap() {
  const app = await NestFactory.create(AppModule);
  // 修改端口为 8088
  await app.listen(8088);
}
bootstrap();
```

接下来，在 app.controller.ts 文件的 getHello 方法中接收参数 id，同时调用 Service 服务的 getHello()方法并传递参数 id，代码如下：

```
@Controller()
export class AppController {
  constructor(private readonly appService: AppService) {}

  @Get(':id')
  getHello(@Param('id') id: string): Person {
    return this.appService.getHello(id);
  }
}
```

最终 getHello()方法会返回新的 JSON 数据：

```
@Injectable()
export class AppService {
  getHello(id: string): Person {
    return {
      name: 'mouse',
      age: 22,
      hello: '你好啊，mouse',
      desc: '我是客户端请求的 id:${id}'
    };
  }
}
```

2.3.4 运行结果

前后端交互逻辑完成之后，在客户端单击按钮发送请求，可以从网络面板看到正常返回的数据并在页面上显示，如图 2-29 所示。

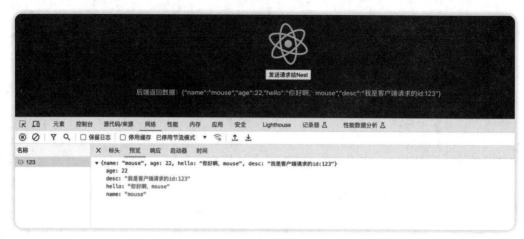

图 2-29　前后端交互结果图

至此，一个简易的前后端交互流程就完成了。

2.3.5　模块化开发

本小节将把请求方式改为 POST，并优化项目结构。通过模块化来管理请求 API。以 user 模块为例，执行 "nest g resource user" 命令来生成 user 模块。接下来，在 user.controller.ts 中修改请求路径，示例代码如下：

```
@Controller('user')
export class UserController {
  constructor(private readonly userService: UserService) {}

  @Post('/create-user')
  create(@Body() createUserDto: CreateUserDto) {
    return this.userService.create(createUserDto);
  }
}
```

同时，在 user.service.ts 中打印并返回提示信息：

```
@Injectable()
export class UserService {
  create(createUserDto: CreateUserDto) {
    console.log('createUserDto', createUserDto);
    return {
      status: true,
      msg: '创建成功'
    };
  }
}
```

接下来，同步修改 web-app-front 前端项目的 App.js，新增 createUser 请求接口和 handleCreateUser 用户事件，代码如下：

```
function getHello() {
  return axios.get('/api/123');
```

```
}
// 新增 createUser 请求接口函数
function createUser(data) {
  return axios.post('/api/user/create-user', data)
}

function App() {
  const [ nestData, setNestData] = useState({})

  const handleGetData = async() => {
    const { data } = await getHello()
    setNestData(data)
  }
  // 新增创建用户页面事件
  const handleCreateUser = async() => {
    const params = {
      name: 'mouse',
      age: '22',
      address: '广州'
    }
    const { data } = await createUser(params)
    setNestData(data)
  }

  return (
    <div className="App">
      <header className="App-header">
        <img src={logo} className="App-logo" alt="logo" />
        <button onClick={handleGetData}>发送请求给 Nest</button>
        <br />
        <button onClick={handleCreateUser}>创建用户</button>
        <br />
        后端返回数据：{ JSON.stringify(nestData) }
      </header>
    </div>
  );
}
```

单击页面上的"创建用户"按钮，后端会正常接收到请求参数，并向客户端发送响应，如图 2-30 所示。

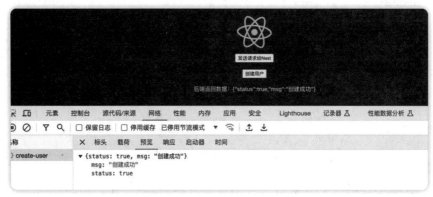

图 2-30　创建用户返回结果

值得注意的是，在请求接口中，/user/*后面的路由对应于 User 控制器下的各类接口方法。这样的设计具有以下优点：它提供了清晰的路由层次结构，并且可以方便地实现对控制器层面的权限管理和拦截器操作。这些内容将在后续章节中详细讲解。

通过本节的学习，一个完整的前后端分离的 Nest 应用已经构建完成。现在，你已经具备了创建自己的 Nest 应用的能力。赶快去实践，将所学知识应用到实际项目中吧！

2.4　Nest 的 AOP 架构理念

Nest 采用了多种优秀的设计模式和软件设计思想，其中之一就是面向切面编程（AOP）。AOP 的引入解决了传统面向对象编程中的一些问题，比如代码重复和业务逻辑的混杂。作为 Nest 最核心的设计理念之一，本节将深入学习 AOP 的工作机制及其应用。

2.4.1　MVC 架构概述

Nest 属于 MVC（Model-View-Controller，模型-视图-控制器）架构体系，实际上，大多数后端框架都是基于这一架构设计的。在 MVC 架构中：

- Model 层负责业务逻辑处理，包括数据的获取、存储、验证以及数据库操作。
- Controller 层通常用于处理用户的输入，调度 Service 服务，以及进行 API 的路由管理。
- View 层在传统的服务器端渲染中，可能使用如 ejs、hbs 等模板引擎。在前后端分离的体系中，通常指的是客户端框架（如 Vue 或 React）负责的部分。

当一个 HTTP 请求到达服务器时，它首先会被 Controller 层接收。Controller 层会根据请求调用 Model 层中的相应模块来处理业务逻辑，并将处理结果返回给 View 层以进行展示。整个基础流程的示意图如图 2-31 所示。

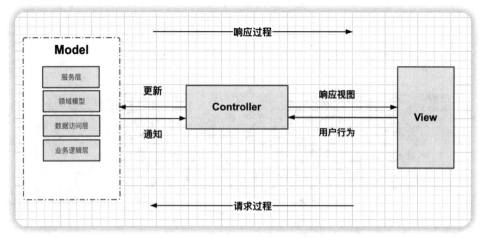

图 2-31　HTTP 请求基本流程

在 MVC 架构的基础上，Nest 还引入了 AOP 的思想，从而具备了面向切面编程的能力。我们经常听到后端开发者提到 AOP 切面，那么究竟什么是面向切面编程呢？接下来将给出答案。

2.4.2 AOP 解决的问题

以一个 HTTP 请求为例，客户端发送请求时首先会经过 Controller（控制器）、Service（服务）、DB（数据访问或操作）等模块。如果想要在这些模块中加入一些操作，例如数据验证、权限校验或日志统计，应该怎么办呢？

首先，我们可能会想到在 Controller 中加入参数校验逻辑或者权限校验，如果不通过验证，就直接返回错误。这样看起来似乎没什么问题。但是，如果有多个功能模块都需要进行校验，并且权限校验的逻辑相同，那么是不是意味着需要在多个 Controller 中重复加入这段逻辑？显然，这样做会导致公共逻辑与业务逻辑耦合。有没有办法可以做到统一管理呢？

答案是有的，办法如图 2-32 所示。

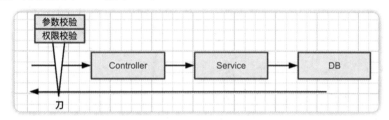

图 2-32 切面工作原理图

由图 2-32 可见，在 Controller 的前后都可以"切一刀"，用来统一处理公共逻辑，这样就不会侵入 Controller、Service 等业务代码。

事实上，在 Nest 中，请求流程可以换一种角度来看，如图 2-33 所示。

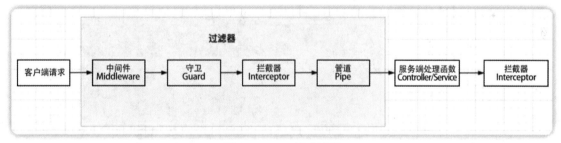

图 2-33 Nest 请求流程

中间的灰色区域属于 AOP 切面部分，包含 Middleware（中间件）、Guard（守卫）、Interceptor（拦截器）、Pipe（管道）和 Filter（过滤器）。它们都是 AOP 思想的具体实现。

2.4.3 AOP 在 Nest 中的应用

前面我们了解了 AOP 思想的优势，接下来看看 AOP 在 Nest 中的应用。这里将以 store-web 项目作为演示，代码已经放到 GitHub 对应章节的目录下。

1. 中间件

Nest 的中间件默认是基于 Express 的，它在请求流程中的位置如图 2-34 所示。

图 2-34　中间件的位置

中间件可以在路由处理程序之前或之后插入执行任务，它们分为全局中间件和局部中间件两种类型。

全局中间件通过 use 方法调用，与 Express 中的使用方式类似。所有进入应用的请求都会经过全局中间件，通常用于执行日志统计、监控、安全性处理等任务。例如，在 main.ts 文件中使用全局中间件的方式如下，其中 LoggerMiddleware 是一个用于日志统计的中间件。示例代码如下：

```
async function bootstrap() {
  const app = await NestFactory.create(AppModule);
  // 中间件
  app.use(new LoggerMiddleware().use)
  // 启动服务
  await app.listen(8088);
}
bootstrap();
```

局部中间件通常应用于特定的控制器或单个路由上，以实现更细粒度的逻辑控制。例如，可以将 LoggerMiddleware 绑定到/person/create-person 路由，并指定其适用的 HTTP 请求方法。以下是如何在控制器中应用局部中间件的示例代码：

```
@Injectable()
// person 模块
export class PersonModule implements NestModule{
  configure(consumer: MiddlewareConsumer) {
      consumer.apply(LoggerMiddleware).forRoutes({
         // 指定路由路径
         path: '/person/create-Person',
         // 指定 HTTP 请求方式
      method: RequestMethod.Post
        })
    }
}
```

对于中间件的更深入的知识，我们将在后续章节中进行详细介绍。

2. 守卫

守卫的职责很明确，通常用于权限、角色等授权操作。守卫所在的位置与中间件类似，可以对请求进行拦截和过滤，如图 2-35 所示。

图 2-35　守卫的位置

守卫在调用路由程序之前返回 true 或者 false 来判断是否通行，分为全局守卫和局部守卫。

守卫必须实现 CanActivate 接口中的 canActivate()方法，代码如下：

```
@Injectable()
export class PersonGuard implements CanActivate {
  canActivate(
    context: ExecutionContext,
  ): boolean | Promise<boolean> | Observable<boolean> {
    console.log('进入守卫');
    // 通常根据 ExecutionContext 信息来判断权限，返回 true/false，表示放行或者禁止通行
    return true;
  }
}
```

全局守卫在 main.ts 中通过 useGlobalGuards 来调用，每个路由程序都会经过它进行权限验证才能够通行。

```
async function bootstrap() {
  const app = await NestFactory.create(AppModule);
  // 守卫
  app.useGlobalGuards(new PersonGuard())
  // 启动服务
  await app.listen(8088);
}
```

同样，与中间件类似，作为局部守卫，可以缩小控制范围，从而实现更加精细的权限控制。控制器中的示例代码如下：

```
@Controller('person')
// 声明守卫
@UseGuards(new PersonGuard())
// 控制器
export class PersonController {}
```

3. 拦截器

拦截器不同于中间件和守卫，它在路由请求之前和之后都可以进行逻辑处理，能够充分操作 request 和 response 对象，如图 2-36 所示。拦截器通常用于记录请求日志、转换或格式化响应数据等。

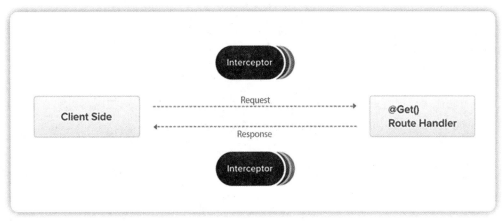

图 2-36　拦截器的位置

为了更好地说明拦截器的作用，下面的代码定义了一个用于统计接口超时的拦截器：

```
import { CallHandler, ExecutionContext, Injectable, NestInterceptor } from "@nestjs/common";
import { log } from "console";
import { Observable, tap, timeout } from "rxjs";

// 统计接口超时的拦截器
@Injectable()
export class TimeoutInterceptor implements NestInterceptor{
  intercept(
    context: ExecutionContext,
    next: CallHandler
  ): Observable<any>{
    log('进入拦截器', context.getClass());
    const now = Date.now();
    return next.handle().pipe(
      tap(() => {
        log('Timeout: ', Date.now() - now)
      }),
      timeout({
        each: 10
      })
    );
  }
}
```

在上述代码中，拦截器必须实现 NestInterceptor 接口中的 intercept 方法。该方法包含与守卫相同的 ExecutionContext 上下文对象作为第一个参数。第二个参数是 CallHandler，它代表每个路由处理程序的方法。只有当拦截器调用了 handle 方法后，控制权才会交给路由处理程序。在 Nest 中，拦截器通常与 RxJS 异步处理库一起使用，以执行一些异步逻辑，例如上例中的超时（timeout）统计。

类似于守卫，拦截器可以设置为控制器作用域、方法作用域或全局作用域。

（1）控制器作用域允许拦截器只作用于某个控制器。当程序执行到控制器时触发拦截器逻辑，通过@UseInterceptors 装饰器将 TimeoutInterceptor 绑定到控制器类，代码如下：

```
@Controller('person')
// 为控制器绑定超时拦截器
@UseInterceptors(new TimeoutInterceptor())

export class PersonController {}
```

（2）方法作用域把拦截器的作用范围限制在某个方法上，比全局或控制器级别的范围更加精确。当程序执行到该方法时，触发拦截器的逻辑。绑定方式如下：

```
@Get()
// 为单独的方法绑定超时拦截器
@UseInterceptors(new TimeoutInterceptor())
findAll() {
  return this.personService.findAll();
}
```

（3）全局作用域允许拦截器应用到整个应用中。在 main.ts 文件中，可以通过 app.useGlobalInterceptors()方法进行绑定，代码如下：

```
async function bootstrap() {
  const app = await NestFactory.create(AppModule);
  // 全局超时拦截器
  app.useGlobalInterceptors(new TimeoutInterceptor());
  // 启动服务
  await app.listen(8088);
}
bootstrap();
```

4. 管道

管道用于处理通用逻辑，其中两个典型的用例是处理请求参数的验证（validation）和转换（transformation）。在执行路由方法之前，会首先执行管道逻辑，并将经过管道转换后的参数传递给路由方法。它的位置如图 2-37 所示。

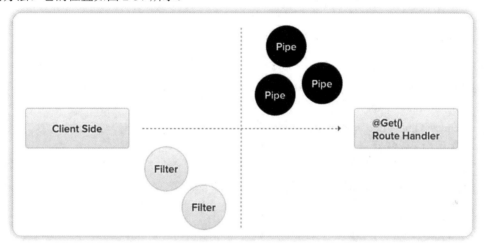

图 2-37　管道的位置

尽管 Nest 框架中已经内置了一些管道（Pipes），例如 ParseIntPipe，但有时我们可能需要实现特定的转换功能，比如将数字转换为八进制。在这种情况下，我们需要创建自定义管道。以下是一个自定义管道的示例代码：

```
import { ArgumentMetadata, BadRequestException, Injectable, PipeTransform } from "@nestjs/common";

@Injectable()
export class ParseIntPipe implements PipeTransform<string, number> {
  transform(value: string, metadata: ArgumentMetadata): number {
    console.log('进入自定义管道');
    // 转为八进制
    const val = parseInt(value, 8);
    if (isNaN(val)) {
      throw new BadRequestException('Validation failed');
    }
    return val;
```

```
    }
}
```

自定义管道需要实现 PipeTransform 接口的 transform()方法。其中，value 表示需要处理的方法参数，而 metadata 则是描述该参数的元数据，用于标识其属于哪种类型，如'body'、'query'、'param' 和 'custom'。

除了自定义管道和内置管道外，通常我们还会结合第三方验证库，例如 class-validator。在后面的章节中将会详细介绍这部分内容。

如果管道验证器验证失败，则需要抛出异常并回应给客户端。这时就需要使用异常过滤器来处理。

5. 过滤器

Nest 中最为常见的是 HTTP 异常过滤器，通常用于在后端服务发生异常时向客户端报告异常的类型。目前内置的 HTTP 异常包含：

- BadRequestException
- UnauthorizedException
- NotFoundException
- ForbiddenException
- NotAcceptableException
- RequestTimeoutException
- ConflictException
- GoneException
- HttpVersionNotSupportedException
- PayloadTooLargeException
- UnsupportedMediaTypeException
- ...

它们都继承自 HttpException 类。当然，我们也可以自定义异常过滤器，并向前端返回统一的数据格式：

```
import { ArgumentsHost, Catch, ExceptionFilter, HttpException } from "@nestjs/common";
import { Request, Response } from "express";

@Catch(HttpException)
export class HttpExceptionFilter implements ExceptionFilter {
    catch(exception: HttpException, host: ArgumentsHost) {
        // 指定传输协议
        const ctx = host.switchToHttp();
        // 获取请求响应对象
        const response = ctx.getResponse<Response>();
        const request = ctx.getRequest<Request>();
        // 获取状态码和异常消息
        const status = exception.getStatus();
        const message = exception.message;
        let resMessage: string | Record<string, any> = exception.getResponse();
```

```
        console.log('进入异常过滤器');

        if (typeof resMessage === 'object') {
            resMessage = resMessage.message
        }
        // 统一组装返回格式
        response.status(status).json({
            message: resMessage || message,
            success: false,
            path: request.url,
            status
        });
    }
}
```

自定义异常过滤器需要实现 ExceptionFilter 接口的 catch()方法来拦截异常。ArgumentsHost 能够获取不同平台的传输协议上下文，用于访问 request 和 response 对象。

另外，@Catch()装饰器用于声明要拦截的异常类型，这里使用的是 HttpException。异常过滤器可应用于控制器作用域、方法作用域和全局作用域，我们都会尝试一遍。

（1）将过滤器绑定到控制器，代码如下：

```
@Controller('person')
// 绑定到控制器
@UseFilters(new HttpExceptionFilter())
export class PersonController {}
```

（2）将过滤器绑定到某个路由方法中，代码如下：

```
@Get()
// 绑定到路由方法
@UseFilters(new HttpExceptionFilter())
findAll() {
  return this.personService.findAll();
}
```

（3）将过滤器绑定到全局中，代码如下：

```
async function bootstrap() {
  const app = await NestFactory.create(AppModule);
  // 全局异常过滤器
  app.useGlobalFilters(new HttpExceptionFilter());
  // 启动服务
  await app.listen(8088);
}
bootstrap();
```

以上就是 Nest 实现 AOP 架构思想的几种方式。不同的切面解决不同场景下的通用逻辑抽离问题，从而实现了更加灵活和可维护的应用程序。

2.5　IoC 思想解决了什么问题

在 2.4 节中，我们介绍了 AOP（面向切面编程）架构理念的基本概念及其在 Nest 中的应用方式。本节将介绍 Nest 中的另一种核心设计思想——IoC（控制反转）思想。IoC 思想贯穿于整个 Nest 应用的开发过程。接下来，让我们一起学习 IoC 如何解决问题，并学习在 Nest 中如何应用 IoC。

2.5.1　IoC 核心思想概述

在后端体系中，通常包含以下几个组件。

- Controller（控制器）：负责接收客户端的请求。
- Service（服务）：处理业务逻辑。
- Dao（数据访问对象）或 Repository（仓库）：负责对数据执行增删改查（CRUD）操作。
- DataSource（数据源）：根据配置信息连接和管理数据库。

这意味着在开发过程中，你需要按照合适的顺序创建这些组件。例如：

```
const dataSource = new DataSource(config)
const dao = new Dao(dataSource)
const Service = new Service(dao)
const Controller = new Controller(Service)
```

在后端架构中，当多个 Controller 模块调用同一个 Service 类时，我们希望确保它们使用的是同一个实例，即维持单例模式。在大型应用中，手动管理这种依赖关系可能会变得复杂，这是传统后端开发经常面临的一个挑战。

幸运的是，IoC（Inverse Of Control，控制反转）提供了一种解决方案。IoC 容器在应用初始化时，会查找每个类上声明的依赖，并按顺序创建相应的实例，然后管理这些实例。当需要使用某个依赖时，IoC 容器会提供相应的对象实例。

依赖注入（DI）是实现 IoC 的一种常见方式。为什么称之为"控制反转"呢？让我们通过一个生活化的比喻来说明：

想象一下，通常我们做饭之前需要准备各种食材，有时还要在市场上讨价还价，回家后要清洗、切割食材，烹饪时还要考虑食材的下锅顺序等，这个过程相当烦琐。那么，有没有更简单的方法呢？

答案是肯定的。如果我们的目的是为了解决饥饿问题，为什么不选择去餐厅就餐呢？我们只需告诉服务员："请给我来一份番茄炒蛋饭。"

此时，后厨接到通知，控制权已经转移。厨师们会根据菜单要求，有序地处理后厨的所有事务，并最终为我们提供一份美味的"番茄炒蛋饭"。

在这个比喻中，后厨相当于 IoC 容器，菜单相当于在类上声明的依赖，服务员则相当于依赖注入的过程。这样，IoC 容器会根据类上声明的依赖来创建和管理对象。

通过 IoC，我们从主动创建和维护对象转变为被动等待依赖注入，实现了从主动下厨到等待服务员上菜的转变，这就是 IoC 控制反转的精髓。

2.5.2　IoC 在 Nest 中的应用

本小节将介绍在 Nest 中如何实现 IoC。首先，通过运行 nest n ioc-test 命令新建一个项目，并在创建过程中选择 pnpm 作为包管理器，如图 2-38 所示。

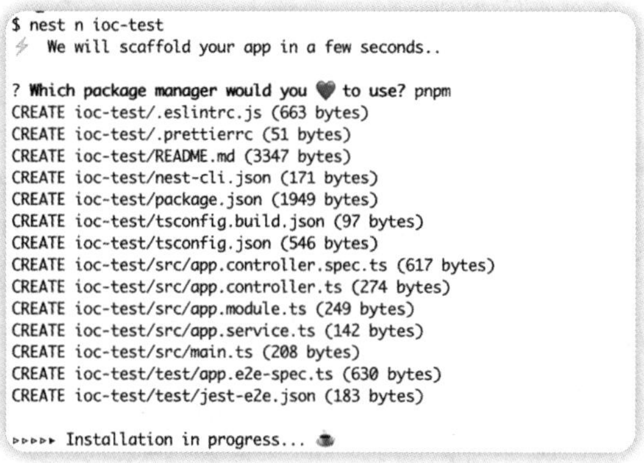

图 2-38　创建 ioc-test 项目

进入该目录，运行 pnpm start:dev 命令启动服务。服务启动成功后的界面如图 2-39 所示。

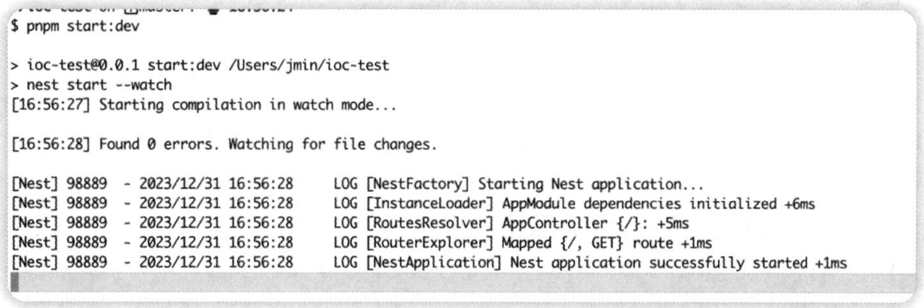

图 2-39　启动服务

下面来看在 Nest 中如何组织代码。打开 app.controller.ts 文件，核心代码如下：

```
@Controller()
export class AppController {
  constructor(private readonly appService: AppService) {}

  @Get()
  getHello(): string {
    return this.appService.getHello();
  }
}
```

可见，AppController 类通过@Controller 装饰器来修饰，表示它可以进行依赖注入，由 Nest 内置的 IoC 容器接管。

接着打开 app.service.ts，代码如下：

```
import { Injectable } from '@nestjs/common';

@Injectable()
```

```
export class AppService {
  getHello(): string {
    return 'Hello World!';
  }
}
```

AppService 类通过@Injectable 进行装饰，表示这个类（class）可以被注入，同时也可以注入其他对象中。我们再创建一个 user 中间件，如图 2-40 所示。

```
$ nest g middleware user
CREATE src/user/user.middleware.spec.ts (180 bytes)
CREATE src/user/user.middleware.ts (196 bytes)
```

图 2-40　创建 user 中间件

打开 user.middleware.ts 文件，代码如下：

```
import { Injectable, NestMiddleware } from '@nestjs/common';

@Injectable()
export class UserMiddleware implements NestMiddleware {
  use(req: any, res: any, next: () => void) {
    next();
  }
}
```

我们发现 UserMiddleware 也是通过@Injectable 装饰器来标记的。这可能会让读者产生一些疑问：

为什么控制器是单独使用@Controller 来装饰的？

除了已知的服务（Service）和中间件（Middleware）使用 @Injectable 装饰，还有哪些组件也使用它？

对于第一个问题，控制器（Controller）只用于处理请求，不作为依赖对象被其他对象组件注入。我们可以把控制器看作是消费者，而服务（Service）和中间件（Middleware）则是提供者。

通常使用@Injectable 进行装饰的还包括过滤器（Filter）、拦截器（Interceptor）、提供者（Provider）、网关（Gateway）等。在没有特殊情况下，Nest 中所有需要依赖注入的模块都可以使用@Injectable 进行标记，以便 IoC 容器进行收集和管理。

回到代码中，这些组件会在 AppModule 中进行引入，如图 2-41 所示。

```
1  import { Module } from '@nestjs/common';
2  import { AppController } from './app.controller';
3  import { AppService } from './app.service';
4
5  @Module({
6    imports: [],
7    controllers: [AppController],
8    providers: [AppService],
9  })
10 export class AppModule {}
```

图 2-41　依赖注入

@Module 装饰器在 Nest 中用于定义模块，这些模块包含需要注入的组件，例如控制器（Controllers）。控制器仅能作为消费者被注入，而提供者（Providers），如服务（Services），既可以作为依赖被注入，也可以注入其他依赖对象中。

除此之外，在 Nest 中实现模块化管理非常简单。imports 属性用于引入其他模块，这有助于实现功能逻辑的分组和重用。例如，我们可以创建一个 UserModule，如图 2-42 所示。

```
$ nest g resource user
? What transport layer do you use? REST API
? Would you like to generate CRUD entry points? Yes
CREATE src/user/user.controller.spec.ts (556 bytes)
CREATE src/user/user.controller.ts (883 bytes)
CREATE src/user/user.module.ts (240 bytes)
CREATE src/user/user.service.spec.ts (446 bytes)
CREATE src/user/user.service.ts (607 bytes)
CREATE src/user/dto/create-user.dto.ts (30 bytes)
CREATE src/user/dto/update-user.dto.ts (169 bytes)
CREATE src/user/entities/user.entity.ts (21 bytes)
UPDATE src/app.module.ts (357 bytes)
```

图 2-42　创建 user 模块

可见，在 AppModule 中自动引入了 UserModule，如图 2-43 所示。

```
1  import { Module } from '@nestjs/common';
2  import { AppController } from './app.controller';
3  import { AppService } from './app.service';
4  import { UserModule } from './user/user.module';
5
6  @Module({
7    imports: [UserModule],
8    controllers: [AppController],
9    providers: [AppService],
10 })
11 export class AppModule {}
```

图 2-43　依赖自动引入

另外，如果还有 aaaModule、bbbModule 等模块，它们也会自动在 AppModule 中引入，最终交给 Nest 工厂方法完成应用的创建，如图 2-44 所示。

```
1  import { NestFactory } from '@nestjs/core';
2  import { AppModule } from './app.module';
3
4  async function bootstrap() {
5    const app = await NestFactory.create(AppModule);
6    await app.listen(3000);
7  }
8  bootstrap();
```

图 2-44　应用主入口

有了这些步骤，在初始化过程中，Nest 就可以轻松通过模块之间的依赖关系找到 UserModule 模块。根据@module 中声明的依赖关系，UserController 只需在构造函数中声明对 UserService 的依赖，而不需要显式创建实例就可以调用 UserService 的实例方法，如图 2-45 所示。

```
import { UpdateUserDto } from './dto/update-user.dto';

@Controller('user')
export class UserController {
  constructor(private readonly userService: UserService) {}

  @Post()
  create(@Body() createUserDto: CreateUserDto) {
    return this.userService.create(createUserDto);
  }

```

图 2-45　调用服务方法

这就是 Nest 的依赖注入与模块机制。在后续章节中，我们将频繁使用这些机制，并享受 IoC 带来的便捷性。

2.6　学会调试 Nest 应用

调试能力是衡量一个开发者水平的重要指标。在编写应用时，首先应该了解如何调试应用程序。通过调试，我们可以清晰地看到程序在运行过程中的具体执行状态。

本节将通过调试请求过程和异常信息来演示如何在 VS Code 中调试 Nest。

2.6.1　Chrome DevTools 调试

首先，运行 nest n nest-debug 命令创建一个项目，如图 2-46 所示。

```
$ nest n nest-debug
⚡  We will scaffold your app in a few seconds..

? Which package manager would you ❤ to use? pnpm
CREATE nest-debug/.eslintrc.js (663 bytes)
CREATE nest-debug/.prettierrc (51 bytes)
CREATE nest-debug/README.md (3347 bytes)
CREATE nest-debug/nest-cli.json (171 bytes)
CREATE nest-debug/package.json (1951 bytes)
CREATE nest-debug/tsconfig.build.json (97 bytes)
CREATE nest-debug/tsconfig.json (546 bytes)
CREATE nest-debug/src/app.controller.spec.ts (617 bytes)
CREATE nest-debug/src/app.controller.ts (274 bytes)
CREATE nest-debug/src/app.module.ts (249 bytes)
CREATE nest-debug/src/app.service.ts (142 bytes)
CREATE nest-debug/src/main.ts (208 bytes)
CREATE nest-debug/test/app.e2e-spec.ts (630 bytes)
CREATE nest-debug/test/jest-e2e.json (183 bytes)

▸▸▸▸▸ Installation in progress... ☕
```

图 2-46　创建 nest-debug 项目

在项目中运行 nest start--watch–debug 命令，可以启动应用并开启热重载以及调试模式，运行成功后，效果如图 2-47 所示。

```
$ nest start --watch --debug
[17:44:14] Starting compilation in watch mode...

[17:44:15] Found 0 errors. Watching for file changes.

Debugger listening on ws://127.0.0.1:9229/520c67ad-51b9-4836-ba80-21d9025e5144
For help, see: https://nodejs.org/en/docs/inspector
[Nest] 776  - 2023/12/31 17:44:15     LOG [NestFactory] Starting Nest application...
[Nest] 776  - 2023/12/31 17:44:15     LOG [InstanceLoader] AppModule dependencies initialized +6ms
[Nest] 776  - 2023/12/31 17:44:15     LOG [RoutesResolver] AppController {/}: +5ms
[Nest] 776  - 2023/12/31 17:44:15     LOG [RouterExplorer] Mapped {/, GET} route +1ms
[Nest] 776  - 2023/12/31 17:44:15     LOG [NestApplication] Nest application successfully started +1ms
```

图 2-47　控制台上的运行结果

访问 http://localhost:3000/，可以看到正常返回的内容，如图 2-48 所示。

图 2-48　浏览器上的运行结果

除此之外，Nest 启动了一个如图 2-47 所示的端口为 9229 的 WebSocket 服务。我们只需要连接这个端口就可以进行调试了。在 Chrome 中输入 chrome://inspect 后按 Enter 键，显示已经连接上远程调试目标，连接的正是 nest-debug 这个项目，如图 2-49 所示。

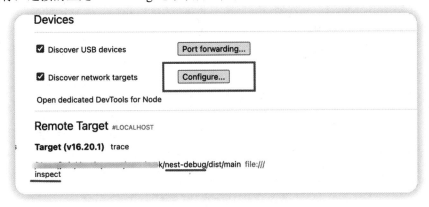

图 2-49　连接调试目标

其中，Configure 是用来配置端口号的。我们可以设置调试服务的端口，Nest 中默认是 9229，如图 2-50 所示。

图 2-50　默认 9229 端口

当然，可以在启动服务时修改端口号，比如输入 nest start --watch --debug 8083，此时启动的就是 8083 端口，如图 2-51 所示。然后在 Configure 中配置即可连接。

```
$ nest start --watch --debug 8083
[17:47:19] Starting compilation in watch mode...

[17:47:19] Found 0 errors. Watching for file changes.

Debugger listening on ws://127.0.0.1:8083/24217f81-35ae-48e3-b74a-3b81053f44a5
For help, see: https://nodejs.org/en/docs/inspector
[Nest] 834  - 2023/12/31 17:47:20     LOG [NestFactory] Starting Nest application...
[Nest] 834  - 2023/12/31 17:47:20     LOG [InstanceLoader] AppModule dependencies initialized +6ms
[Nest] 834  - 2023/12/31 17:47:20     LOG [RoutesResolver] AppController {/}: +5ms
[Nest] 834  - 2023/12/31 17:47:20     LOG [RouterExplorer] Mapped {/, GET} route +1ms
[Nest] 834  - 2023/12/31 17:47:20     LOG [NestApplication] Nest application successfully started +1ms
```

图 2-51　自定义调试端口

在图 2-49 中单击 inspect，导入项目文件夹，在 app.Controller.ts 的 getHello()方法中添加一个断点。刷新页面之后，可以看到代码自动在断点处停下，如图 2-52 所示。

图 2-52　断点演示效果

这样就完成了使用 Chrome 浏览器调试 Nest 应用的过程。然而，这种方法可能不太方便。我们通常会使用 VS Code 等集成开发环境（IDE）进行开发。那么，能否直接通过这些工具进行调试呢？答案是肯定的。接下来将详细讲解。

2.6.2　VS Code 调试

在终端窗口通过命令"nest start --watch –debug"启动调试服务，如图 2-53 所示。

```
[18:00:10] Starting compilation in watch mode...

[18:00:11] Found 0 errors. Watching for file changes.

Debugger listening on ws://127.0.0.1:9229/6e9600c9-60ae-4815-89fc-b6f62af843f5
For help, see: https://nodejs.org/en/docs/inspector
Debugger attached.
[Nest] 1519  - 2023/12/31 18:00:12     LOG [NestFactory] Starting Nest application...
[Nest] 1519  - 2023/12/31 18:00:12     LOG [InstanceLoader] AppModule dependencies initialized +9ms
[Nest] 1519  - 2023/12/31 18:00:12     LOG [RoutesResolver] AppController {/}: +8ms
[Nest] 1519  - 2023/12/31 18:00:12     LOG [RouterExplorer] Mapped {/, GET} route +1ms
[Nest] 1519  - 2023/12/31 18:00:12     LOG [NestApplication] Nest application successfully started +1ms
```

图 2-53　启动调试服务

接着在需要调试的地方打个断点，如图 2-54 所示。

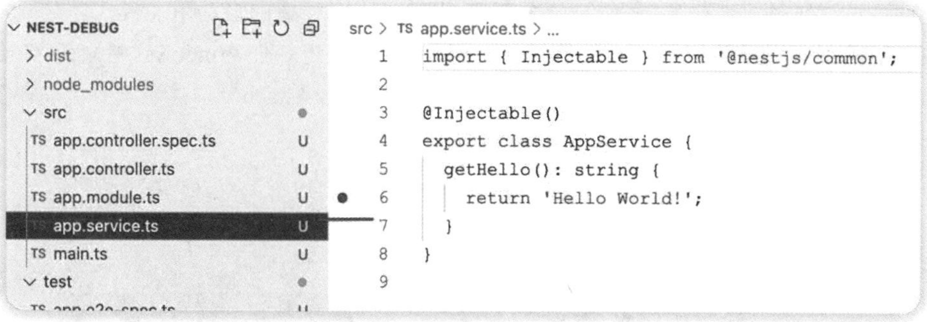

图 2-54　在指定行打断点

然后在 http://localhost:3000/地址刷新浏览器，此时程序会直接在断点处停下，如图 2-55 所示。

图 2-55　断点效果

这种方式是否比之前的方法更简单呢？确实如此。无论是 Chrome DevTools 还是 VS Code 等第三方集成开发环境（IDE），它们都通过 CDP（Chrome DevTools Protocol）协议实现了调试服务的功能。

有些读者可能会提出这样的问题：我不想每次都手动输入 --debug 参数，能否在执行 "yarn start:dev" 命令后自动进入调试模式？

当然可以。你只需在 package.json 文件中添加一个脚本（script）命令即可实现这一点。在 Nest 项目中，这个脚本默认就已经存在了，如图 2-56 所示。

图 2-56　package.json 中的 scripts 配置

确实，使用 Nest CLI 的--debug 参数可以方便地启动调试模式。但如果是其他 Node.js 脚本，情况就不同了。这正是笔者想要提到的：VS Code 提供了一个便捷的自动附加模式。

要使用这个功能，你可以通过按快捷键 Command+Shift+P（在 Windows 中）或 Ctrl+Shift+P（在 Linux 中），然后在弹出的命令面板中输入"Toggle Auto Attach"，接着选择"总是"选项，以自动附加调试器到 Node.js 进程，如图 2-57 所示。

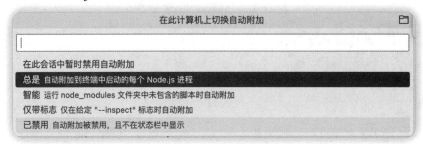

图 2-57　VS Code 自动附加

重新打开控制台并运行"yarn start:dev"命令，此时 VS Code 在启动应用的同时，也会自动创建一个调试器（debugger）服务。该调试服务的默认端口号为 61174，如图 2-58 所示。

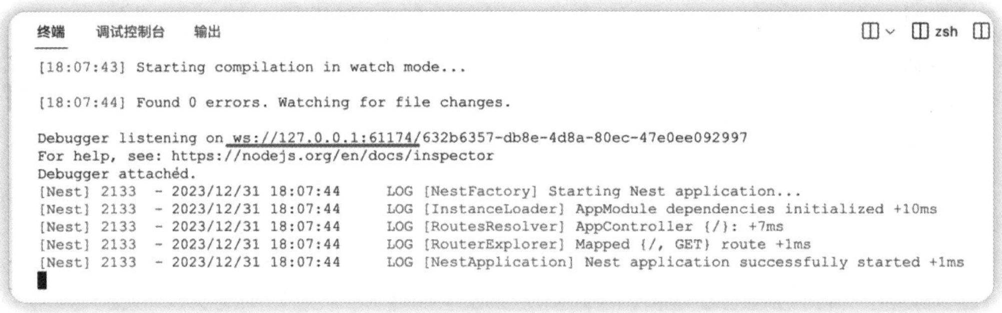

图 2-58　附加模式启动效果

重新在浏览器中访问 http://localhost:3000/，依然可以进入断点调试状态，如图 2-59 所示。

图 2-59　调试效果

以上两种方式会在每次执行命令时创建一个新的 WebSocket 调试服务。然而，对于一些开发者来说，这种方法可能并不适合他们的需求。接下来，在 2.6.3 节中，我们将探讨一些扩展的调试技巧。

2.6.3　扩展调试技巧

前面讲解了基于--debug 参数来启动调试服务，如图 2-60 所示。从 CLI 源码中可以看到，本质上还是执行"node xxxx –inspect"命令来运行的。

```
.command('start [app]')
.option('-c, --config [path]', 'Path to nest-cli configuration file.')
.option('-p, --path [path]', 'Path to tsconfig file.')
.option('-w, --watch', 'Run in watch mode (live-reload).')
.option('-b, --builder [name]', 'Builder to be used (tsc, webpack, swc).')
.option('--watchAssets', 'Watch non-ts (e.g., .graphql) files mode.')
.option('-d, --debug [hostport] ', 'Run in debug mode (with --inspect flag).'
.option('--webpack', 'Use webpack for compilation (deprecated option, use -b
.option('--webpackPath [path]', 'Path to webpack configuration.')
.option('--type-check', 'Enable type checking (when SWC is used).')
```

图 2-60 --debug 参数的定义

因此，我们也可以选择在 VS Code 中使用这种方式进行调试。首先，在调试面板中单击创建 launch.json 文件的选项，如图 2-61 所示。

图 2-61 创建 launch.json 调试文件

然后选择 Node.js 作为调试器，如图 2-62 所示。

图 2-62 选择 Node.js 调试器

接着根据需要配置相关的调试信息。在 launch.json 文件中，将 program 字段的值修改为你的应用程序的入口点，例如 main.ts：

```
{
  // 悬停以查看现有属性的描述
  "version": "0.2.0",
  "configurations": [
    {
      "type": "node",
```

```
            "request": "launch",
            "name": "启动调试",
            "skipFiles": [
              "<node_internals>/**"
            ],
            "program": "${workspaceFolder}/src/main.ts",
            "outFiles": [
              "${workspaceFolder}/dist/**/*.js"
            ]
        }
    ]
}
```

配置完成之后，在 app.service.ts 的 getHello()方法中添加断点，单击左上角的"启动调试"按钮，如图 2-63 所示。

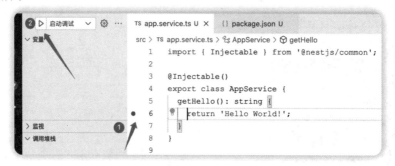

图 2-63　打断点并启动调试

可以看到在"调用堆栈"中新增了一个启动程序，这就是 VS Code 调试器。我们依旧在 Chrome 中访问 http://localhost:3000/，此时程序自动在断点处停住，并且在左侧可以看到调用栈信息和变量情况，如图 2-64 所示。

图 2-64　调试调用栈情况

这种方法的优势在于它避免了每次启动都需要创建一个新的 WebSocket 调试服务，从而节省了运行时的内存消耗。调试程序仅在需要时启动，这正是我们最常用的调试方式。它适用于前端应用，如 Vue、React 以及 Node.js 等多种应用的调试。

第 3 章

Nest 核心概念介绍

本章将介绍 Nest 核心概念及其应用场景。首先，探讨贯穿 Nest 开发过程的装饰器，以及实现功能模块化的模块概念。接着，介绍管理路由的控制器，以及实现复杂业务逻辑并与数据层交互的服务。最后，介绍 AOP 架构下的中间件机制和拦截器功能。读者可结合实际代码示例，全面了解这些概念在实际开发中的应用。

3.1 贯穿全书的装饰器

本节首先介绍装饰器（Decorator）的基本概念和常见的 5 种装饰器类型，包括类装饰器、方法装饰器、属性装饰器、参数装饰器和访问器装饰器。接着介绍 Nest 中常用的装饰器，我们先初步认识它们。

3.1.1 基本概念

顾名思义，装饰器（Java 中有个类似的概念叫注解）是用来装饰和扩展对象功能的。它能够在不改变原有对象结构的前提下，增加额外的功能，以满足更多的实际需求。

举个例子：一间房子里放了一张床，就满足了基本的居住需求。在此基础上，如果新增了沙发、红酒杯、电视机，那么就不仅能够满足休息的需求，还能提供娱乐和休闲的体验。在这个比喻中，沙发、红酒杯、电视机就像是装饰器，它们可以在不影响基本功能的前提下，根据需要随意添加或移除，从而实现功能的扩展和松耦合。

3.1.2 装饰器的种类

上例中，沙发、红酒杯、电视机属于不同的装饰器，它们实现了不同的功能。同样地，常用的装饰器也分为以下几种：

（1）类装饰器。
（2）方法装饰器。
（3）属性装饰器。
（4）参数装饰器。
（5）访问器装饰器。

尽管装饰器仍处于提案（stage-3）阶段，但它们已经广泛应用于 TypeScript 和 Babel 编译中。以方法装饰器为例，在 stage-3 提案中的实现如下：

```typescript
/**
 * 方法装饰器
 */
const Log: MethodDecorator = (
  target: Function,
  context: {
    kind: "method";
    name: string | symbol;
    access: { get(): unknown };
    static: boolean;
    private: boolean;
    addInitializer(initializer: () => void): void;
  }
) => {
  // 此处编写你的逻辑
};
```

上述代码中定义了一个 Log 方法装饰器，用于修饰类方法，其中 target 是被修饰的对象，context 提供上下文信息。在基于 TypeScript 构建的 Nest 项目中，采用的是 stage-1 提案的装饰器版本，具体如下：

```typescript
/**
 * 方法装饰器
 */
const log: MethodDecorator = (
  target: Function,
  propertyKey: string | Symbol,
  descriptor: PropertyDescriptor
) => {
  // 此处编写你的逻辑
};
```

上述代码中，Log 方法装饰器接收 3 个参数：target 是被修饰的值，propertyKey 是被修饰的属性键（方法名），description 是被装饰的方法的属性描述符对象。

下面通过简单的示例分别进行介绍。

1. 类的装饰

装饰器可以用来装饰整个类，以此增强类的功能，示例代码如下：

```
/**
 * 类装饰器
 */
const doc: ClassDecorator = (target: Function) => {
  console.log("---------------类装饰器-----------------");
  target.prototype.name = "mouse";
  console.log(target);
  console.log("---------------类装饰器-----------------");
};

@doc
class App {
  constructor() {}
}

const app: Record<string, any> = new App();
console.log("app name: " + app.name);
```

在上面的代码中，@doc 用来装饰 App 类，同时往原型链上添加一个属性 name，target 是被修饰的类。

我们通过 CodePen 来编译 ES6 代码。CodePen 是一个社区驱动的在线代码编辑器，支持编辑 HTML、CSS 和 JavaScript。CodePen 还支持 Babel，这使得它能够将 ES6 代码转换成 ES5 并进行实时预览。

由于本节案例中我们使用 TypeScript 编码，因此需要在 CodePen 中找到 Settings 下的 JS 选项，并将 JavaScript Preprocessor（JavaScript 预处理器）设置为 TypeScript，如图 3-1 所示。

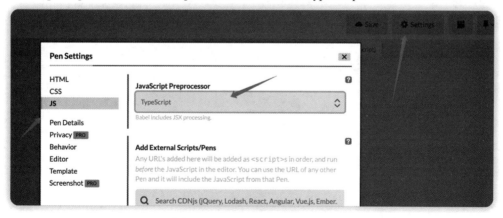

图 3-1　设置 CodePen 预处理器

保存后，类装饰器执行结果如图 3-2 所示。

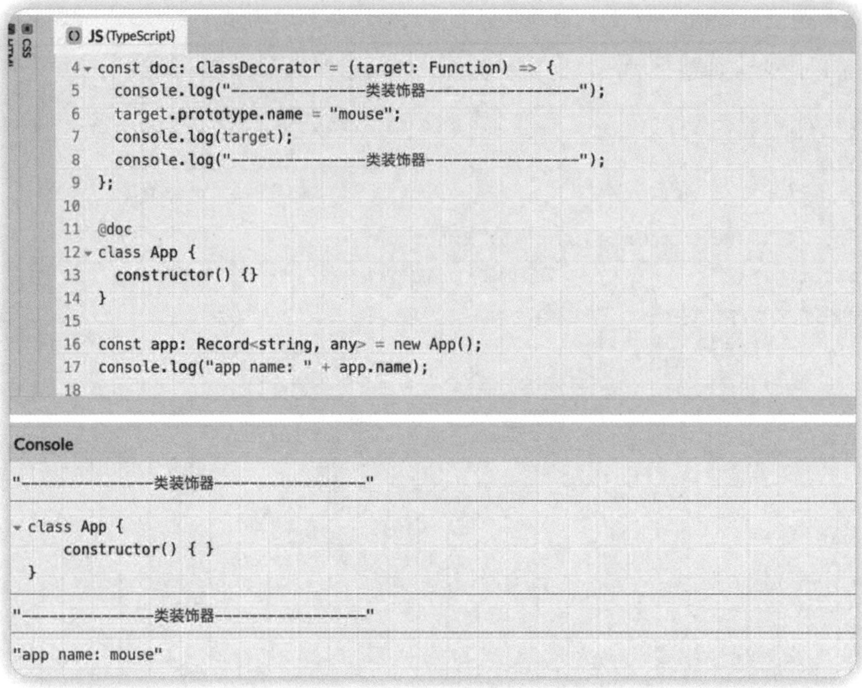

```typescript
const doc: ClassDecorator = (target: Function) => {
  console.log("-----------------类装饰器-----------------");
  target.prototype.name = "mouse";
  console.log(target);
  console.log("-----------------类装饰器-----------------");
};

@doc
class App {
  constructor() {}
}

const app: Record<string, any> = new App();
console.log("app name: " + app.name);
```

```
Console
"-----------------类装饰器-----------------"
class App {
    constructor() { }
}
"-----------------类装饰器-----------------"
"app name: mouse"
```

图 3-2 类装饰器的执行结果

上面打印了代码预期的结果，成功地给实例添加了属性 name，并获取了 App 的构造函数，这样我们就可以动态地修改类的属性和行为了。

如果觉得一个 target 参数不够用，怎么办呢？可以通过工厂函数来增强装饰器的能力。代码如下：

```typescript
/**
 * 类装饰器
 */
const doc: ClassDecorator = (module) => {
  return (target: Function) => {
    console.log("-----------------类装饰器-----------------");
    target.prototype.module = module;
    console.log(target);
    console.log("-----------------类装饰器-----------------");
  };
};

@doc("user")
class App {
  constructor() {}
}

const app: Record<string, any> = new App();
console.log("app module: " + app.module);
```

通过工厂函数增强装饰器的能力后，@doc 可以接收外部参数，使得 App 被实例化时能够获取更多的信息来组织代码，具体效果如图 3-3 所示。

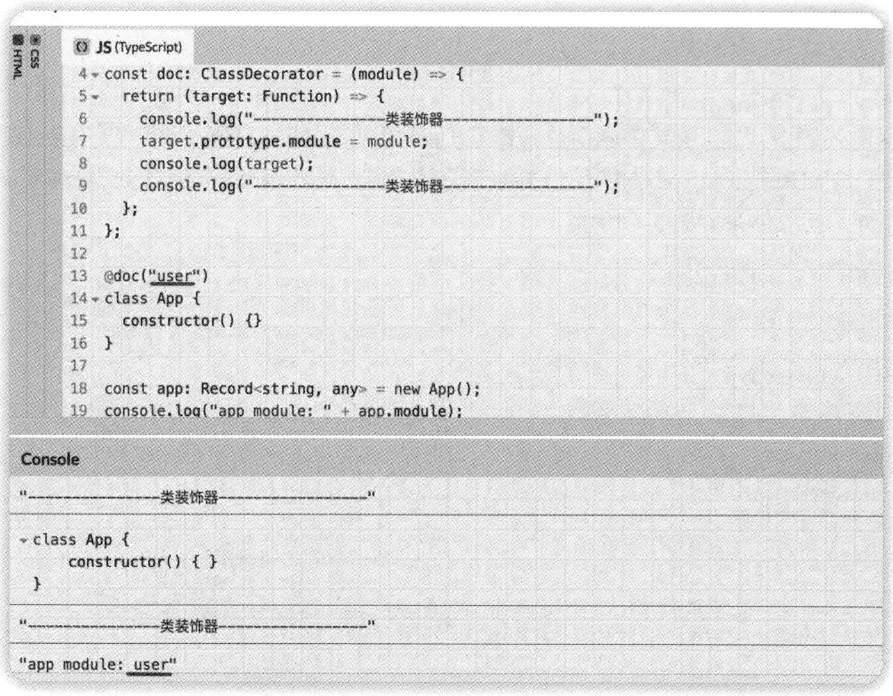

第 3 章　Nest 核心概念介绍

```typescript
const doc: ClassDecorator = (module) => {
  return (target: Function) => {
    console.log("-----------------类装饰器-----------------");
    target.prototype.module = module;
    console.log(target);
    console.log("-----------------类装饰器-----------------");
  };
};

@doc("user")
class App {
  constructor() {}
}

const app: Record<string, any> = new App();
console.log("app module: " + app.module);
```

Console:
```
"-----------------类装饰器-----------------"
class App {
    constructor() { }
}
"-----------------类装饰器-----------------"
"app module: user"
```

图 3-3　装饰器工厂

至此，你应该已经理解了 Nest 中的装饰器的实现方式，它在依赖注入中扮演着重要的角色。然而，请不要急于下结论，让我们继续深入探讨。

2. 方法的装饰

装饰器可以用来装饰类的方法，代码如下：

```typescript
/**
 * 方法装饰器
 */
const Log: MethodDecorator = (
  target: Object,
  propertyKey: string | Symbol,
  descriptor: PropertyDescriptor
) => {
  console.log("----------------方法装饰器-----------------");
  console.log(target);
  console.log(propertyKey);
  console.log(descriptor);
  console.log("----------------方法装饰器-----------------");
};

class User {
  @Log
  getName() {
```

```
    return "mouse";
  }
}
```

在上面的代码中，@Log 装饰器函数修饰了 User 类的 getName 方法。其中，target 是被"装饰"的类的原型，即 User.prototype。由于此时类还没有被实例化，因此只能装饰原型对象。这与类装饰器有所不同。运行结果如图 3-4 所示。

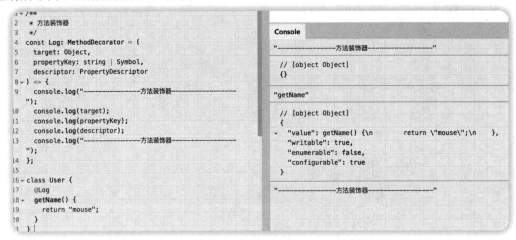

图 3-4　方法装饰器的运行结果

由于 User 类的原型尚未附加属性方法，因此是空对象{}。

3. 属性的装饰

装饰器也可以用来装饰类的属性，代码如下：

```
/**
 * 属性装饰器
 */
const prop: PropertyDecorator = (
  target: Object,
  propertyKey: string | Symbol
) => {
  console.log("------属性装饰器-------");
  console.log(target);
  console.log(propertyKey);
  console.log("------属性装饰器-------");
};

class User {
  @prop
  name: string = "mouse";
}
```

在上面的代码中，@prop 用于装饰 User 类的 name 属性。此时，target 是被装饰的类的原型对象，propertyKey 是被修饰的属性名。具体的运行结果如图 3-5 所示。

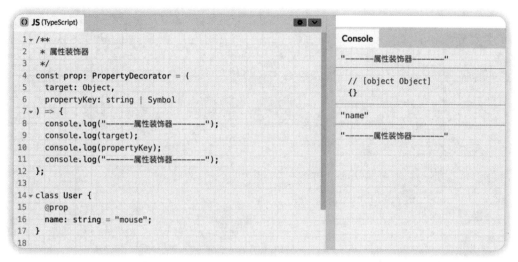

图 3-5 属性装饰器的运行结果

4. 参数的装饰

除了装饰类、类方法和类属性外，装饰器还能装饰类方法的参数，示例代码如下：

```
/**
 * 参数装饰器
 */

const Param: ParameterDecorator = (
  target: Object,
  propertyKey: string | Symbol | undefined,
  index: number
) => {
  console.log("-------------参数装饰器------------------");
  console.log(target);
  console.log(propertyKey);
  console.log(index);
  console.log("-------------参数装饰器------------------");
};
class User {
  getName(@Param name: string) {
    return name;
  }
}
```

在上述代码中，@Param 用于装饰类方法 getName 的参数，其中 target 是被装饰类的原型，propertyKey 是被装饰的类方法，index 是装饰方法参数的索引位置。具体的运行结果如图 3-6 所示。

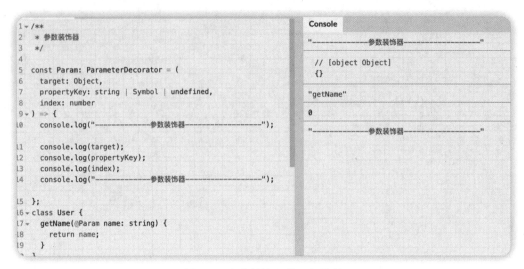

图 3-6　参数装饰器的运行结果

5. 访问器的装饰

装饰器还可以用来修饰类的访问器，即属性的 getter 方法。示例代码如下：

```
/**
 * 访问器装饰器
 */
const Immutable: ParameterDecorator = (
  target: any,
  propertyKey: string | Symbol,
  descriptor: PropertyDescriptor
) => {
  console.log("-------------访问器装饰器------------------");
  console.log(target);
  console.log(propertyKey);
  console.log(descriptor);
  console.log("-------------访问器装饰器------------------");
};
class User {
  private _name = "mouse";
  @Immutable
  get name() {
    return this._name;
  }
}
```

访问器装饰器与类方法装饰器类似，唯一的区别在于它们的描述符（descriptor）中某些键（key）不同。

方法装饰器的描述器键包括：

- value
- writable
- enumerable

- configurable

访问器装饰器的描述器的 key 为：

- get
- set
- enumerable
- configurable

运行结果如图 3-7 所示。

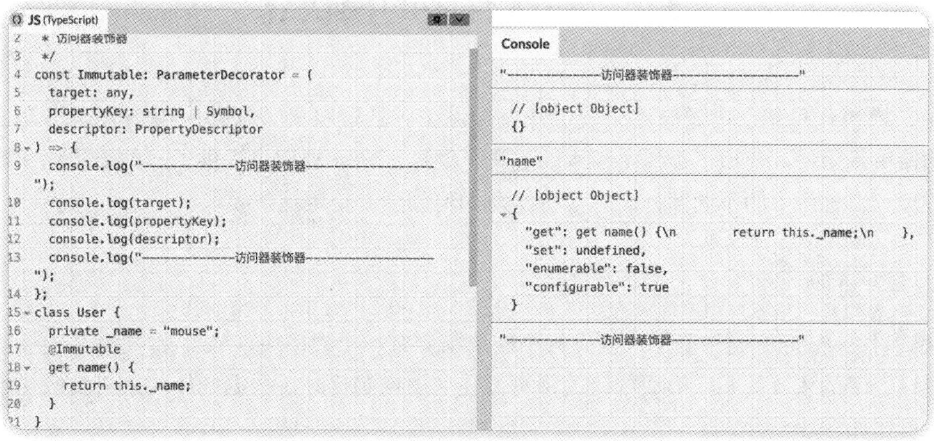

图 3-7 访问器装饰器的运行结果

了解了这几个装饰器的概念后，接下来学习 Nest 中的装饰器，你应该能够轻松理解。

3.1.3 Nest 中的装饰器

在 Nest 中实现了前面列举的前 4 种装饰器，它们包含但不限于：

- 类装饰器：@Controller、@Injectable、@Module、@UseInterceptors。
- 方法装饰器：@Get、@Post、@UseInterceptors。
- 属性装饰器：@IsNotEmpty、@IsString、@IsNumber。
- 参数装饰器：@Body、@Param、@Query。

下面分别对它们进行解释。

- @Controller()：用于装饰控制器类，使之能够管理应用中的路由程序，并通过设置路由路径前缀来模块化管理路由。
- @Injectable()：装饰后成为服务提供者，可以被其他对象进行依赖注入。
- @Module()：模块装饰器，用于在 Nest 中划分功能模块并限制依赖注入的范围。
- @UseInterceptors()：用于绑定拦截器，将拦截器的作用范围限制在控制器类范围中（当然也可以作用在类方法上）。
- @Get、@Post：用于定义路由方法的 HTTP 请求方式。

- @IsNotEmpty、@IsString、@IsNumber：用于在参数的验证场景中校验 HTTP 请求的参数是否符合预期。
- @Body、@Param、@Query：用于接收 HTTP 请求发送的数据，不同的请求方式对应不同的参数接收方式。

Nest 中的装饰器主要分为这几类。理解了它们将有助于我们在使用过程中更加得心应手。关于这些装饰器的具体使用方法，我们将在后续章节中逐一进行详细介绍。

3.2 井然有序的模块化

当前，模块化思想已广泛应用于软件开发领域。它通过将大型软件应用划分为相互独立且功能明确的模块或组件，实现了类似搭积木的过程。同样，Nest 框架也提供了一种结构化和模块化的方法来管理应用程序中的不同部分。本节将学习如何创建并使用这些模块。

3.2.1 基本概念

模块通过@Module 装饰器来声明。每个应用都会有一个根模块，Nest 框架会从根模块开始收集各个模块之间的依赖关系，形成依赖关系树。在应用初始化时，根据依赖关系树实例化不同的模块对象，具体如图 3-8 所示。

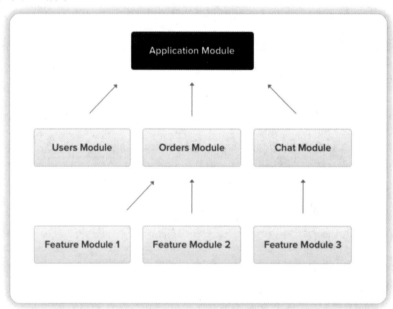

图 3-8　模块依赖关系

在模块树中，每个模块都有自己独立的作用域。它们之间的代码是相互隔离的，各自拥有自己的控制器（Controllers）、服务提供者（Providers）、中间件（Middlewares）和其他组件。当然，它们也可以通过一定的方式进行相互共享。

3.2.2 创建模块

下面通过创建一个名为 nest-module 的项目来演示。运行"nest n nest-module -p pnpm"命令，具体的运行结果如图 3-9 所示。

```
$ nest n nest-module -p pnpm
⚡ We will scaffold your app in a few seconds..

CREATE nest-module/.eslintrc.js (663 bytes)
CREATE nest-module/.prettierrc (51 bytes)
CREATE nest-module/README.md (3347 bytes)
CREATE nest-module/nest-cli.json (171 bytes)
CREATE nest-module/package.json (1952 bytes)
CREATE nest-module/tsconfig.build.json (97 bytes)
CREATE nest-module/tsconfig.json (546 bytes)
CREATE nest-module/src/app.controller.spec.ts (617 bytes)
CREATE nest-module/src/app.controller.ts (274 bytes)
CREATE nest-module/src/app.module.ts (249 bytes)
CREATE nest-module/src/app.service.ts (142 bytes)
CREATE nest-module/src/main.ts (208 bytes)
CREATE nest-module/test/app.e2e-spec.ts (630 bytes)
CREATE nest-module/test/jest-e2e.json (183 bytes)

✓ Installation in progress... 🐢

🚀 Successfully created project nest-module
```

图 3-9　创建项目

首先来看 app.module.ts 文件，代码如下：

```typescript
import { Module } from '@nestjs/common';
import { AppController } from './app.controller';
import { AppService } from './app.service';

@Module({
  imports: [],
  controllers: [AppController],
  providers: [AppService],
})
export class AppModule {}
```

其中，AppModule 是默认的根模块。类装饰器@Module 的参数中，controllers 用于注入该模块的控制器集合，而 providers 用于注入该模块的服务提供者，这些服务提供者将在该模块中共享。

Imports 用于导入应用中的其他模块，默认为空。下面以创建新的 User 和 Order 模块为例，运行如下命令：

```
// 生成 User 模块
nest g resource User --no-spec
// 生成 Order 模块
nest g resource Order --no-spec
```

--no-spec 表示不生成单元测试文件，运行结果如图 3-10 所示。

```
$ nest g resource User --no-spec
nest g resource Order --no-spec
? What transport layer do you use? REST API
? Would you like to generate CRUD entry points? No
CREATE src/user/user.controller.ts (204 bytes)
CREATE src/user/user.module.ts (240 bytes)
CREATE src/user/user.service.ts (88 bytes)
UPDATE src/app.module.ts (310 bytes)
? What transport layer do you use? REST API
? Would you like to generate CRUD entry points? No
CREATE src/order/order.controller.ts (210 bytes)
CREATE src/order/order.module.ts (247 bytes)
CREATE src/order/order.service.ts (89 bytes)
UPDATE src/app.module.ts (375 bytes)
```

图 3-10 生成 User 和 Order 模块

回到 AppModule 根模块，可以看到 UserModule 和 OrderModule 被自动导入根模块，成为 AppModule 的子模块。具体代码如下：

```
import { Module } from '@nestjs/common';
import { AppController } from './app.controller';
import { AppService } from './app.service';
import { UserModule } from './user/user.module';
import { OrderModule } from './order/order.module';

@Module({
  // 模块被自动导入了
  imports: [UserModule, OrderModule],
  controllers: [AppController],
  providers: [AppService],
})
export class AppModule {}
```

3.2.3 共享模块

既然有 imports，那么必然也有 exports。假设存在这样的需求，Order 模块需要依赖 User 模块中的 UserService。这时可以将 UserService 添加到 UserModule 的 exports 中，使之成为共享服务。这样，Order 模块只需导入 UserModule 即可访问 UserService。下面通过一个案例来演示。

在 User 模块中导出 User 服务：

```
@Module({
  controllers: [UserController],
  providers: [UserService],
  // 导出 UserService 服务
  exports: [UserService]
})
export class UserModule {}
```

接着在 Order 模块中导入 User 模块，代码如下：

```
@Module({
  // 导入 UserModule
  imports: [UserModule],
  controllers: [OrderController],
```

```
  providers: [OrderService]
})
export class OrderModule {}
```

此时，在 Order 模块的任何地方，都可以共享 UserService 服务了。下面来演示一下。

在 order.controller.ts 中定义路由方法 getOrder，然后调用 order.service.ts 中的 getOrderDesc 方法：

```
// order.controller.ts
import { Controller, Get } from '@nestjs/common';
import { OrderService } from './order.service';

@Controller('order')
export class OrderController {
  constructor(private readonly orderService: OrderService) {}

  @Get()
  getOrderDesc(): string {
    return this.orderService.getOrderDesc()
  }
}
```

在 order.service.ts 中通过属性注入 UserService 依赖，同时调用 UserService 的 getUserHello 方法，最终返回一个问候语字符串，代码如下：

```
// order.service.ts
import { Inject, Injectable } from '@nestjs/common';
import { UserService } from 'src/user/user.service';

@Injectable()
export class OrderService {
  // 依赖注入之属性注入共享的服务
  @Inject(UserService)
  private userService: UserService;
  getOrderDesc(): string {
    let name = this.userService.getUserName()
    return '订单 ID：xxxx，下单人：${name}'
  }
}

// user.service.ts
import { Injectable } from '@nestjs/common';

@Injectable()
export class UserService {
  getUserName(): string {
    return 'mouse'
  }
}
```

除使用属性注入依赖外，还可以使用构造函数注入：

```
import { Inject, Injectable } from '@nestjs/common';
import { UserService } from 'src/user/user.service';
```

```
@Injectable()
export class OrderService {
  // 依赖注入之构造函数注入
  constructor(private userService: UserService) {}
  getOrderDesc(): string {
    let name = this.userService.getUserName()
    return '订单ID: xxxx, 下单人: ${name}'
  }
}
```

最后，在浏览器中访问 http://localhost:3000/order，页面成功返回了内容，这表明实现了模块间的数据交互，如图 3-11 所示。

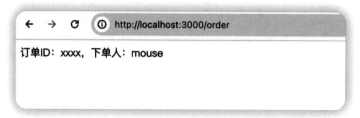

图 3-11　代码运行结果

3.2.4　全局模块

如果某个模块在多个地方被引用，为了简化管理，可以使用@Global 装饰器将其声明为全局模块。这样，便可以直接注入通过 exports 导出的 providers，而无须在每个模块的 imports 中重复声明它们。代码如下：

```
// user.module.ts
// 声明为全局模块
@Global()
@Module({
  controllers: [UserController],
  providers: [UserService],
  exports: [UserService]
})
export class UserModule {}

// order.module.ts
@Module({
  // 这里无须再导入
  // imports: [UserModule],
  controllers: [OrderController],
  providers: [OrderService]
})
export class OrderModule {}
```

需要注意的是，使用全局模块之前，需要确保你的模块确实需要全局使用，以避免不必要地增加模块之间的耦合性。

3.2.5 动态模块

前面介绍的都是静态模块的绑定和使用。Nest 中还提供了动态加载模块的功能，使得应用可以在运行时创建模块，通常用于动态读取配置或根据权限判断来加载模块。

创建动态模块的第一步是在需要使用动态模块的地方，例如某个服务或控制器中，使用 forFeature 方法来动态加载模块。代码如下：

```
// dynamic.module.ts
import { Module, DynamicModule } from '@nestjs/common';

@Module({})
export class DynamicModule {
  static forFeature(entities: Function[]): DynamicModule {
    // 在此方法中根据需要创建模块
    return {
      module: DynamicModule,
      providers: [],
      exports: [],
    };
  }
}
```

动态加载模块时，调用 loadModule 方法即可加载相应的模块。代码如下：

```
// some.service.ts
import { Injectable } from '@nestjs/common';
import { DynamicModule } from './dynamic.module'

@Injectable()
export class SomeService {
  constructor(private readonly dynamicModule: DynamicModule) {}
  // 在某个方法中动态加载模块
  loadModule(entities: Function[]) {
    return this.dynamicModule.forFeature(entities);
  }
}
```

除 loadModule 方法之外，还有 register 和 forRoot 方法可用于加载动态模块。其中，register 方法通常用于与外部模块或第三方库集成时，它能将外部模块动态加载到 Nest.js 模块中。

```
import { Module } from '@nestjs/common';

@Module({})
export class DynamicModule {
  static register(options: SomeOptions) {
    // 在此方法中注册外部模块
    return {
      module: DynamicModule,
      providers: [],
      exports: [],
    };
  }
}
```

而 forRoot 方法在 Nest 中通常用于注册根模块,例如配置根模块的全局服务或中间件等。代码如下:

```
import { Module } from '@nestjs/common';

@Module({})
export class CoreModule {
  static forRoot(options: SomeOptions) {
    // 在此方法中配置根模块的全局服务或特性
    return {
      module: CoreModule,
      providers: [],
      exports: [],
    };
  }
}
```

以上介绍了加载静态和动态模块的几种方式。动态模块的加载方式在后续章节中会频繁出现,特别是与第三方服务集成时。

3.3 控制器与服务的默契配合

在了解模块的基础知识之后,本节将深入学习模块中的核心部分:控制器和服务。首先介绍它们各自在模块中承担的职责。通过请求流程解释控制器如何管理应用路由和处理请求参数,以及服务是如何进行数据处理的。此外,还将介绍服务与服务提供者之间的区别,以帮助读者理清思路,从而更好地理解依赖注入的核心概念。

3.3.1 基本概念

Nest 提供了分层结构,其中 Controller 和 Service 分别扮演着不同但紧密相关的角色。它们各自承担的任务也不相同,你必须清楚地了解它们的职责,才能确保你的应用既清晰又可维护。

1. Controller 的职责

(1)处理请求和路由:接收 HTTP 请求并确定路由,一个控制器中通常存在多个路由,不同的路由执行不同的操作。

(2)解析和验证输入数据:控制器通常负责解析并验证请求的数据。例如,通过使用管道或数据传输对象(Data Transfer Object,DTO),控制器可以定义和实施数据验证、格式化或转换操作。

(3)调用 Service:Controller 层用于调度 Service 层执行业务逻辑处理,是 Service 层的入口。

2. Service 的职责

(1)处理业务逻辑:包含对数据的处理、外部系统的调用和复杂业务逻辑处理。

(2)数据持久化:与数据库、Redis 缓存等存储库进行交互,执行持久化操作。通常在

Service 中使用模型或实体来进行数据的增删改查操作。

（3）由此可见，Controller 负责与 HTTP 交互相关事项，如路由管理和参数处理等，而 Service 则负责业务数据处理。

下面我们通过实际案例来演示说明。

3.3.2 Controller 管理请求路由

创建一个新的 Nest 项目，运行"nest n nest-controller-service -p pnpm"命令，结果如图 3-12 所示。

```
$ nest n nest-controller-service -p pnpm
⚡  We will scaffold your app in a few seconds..

CREATE nest-controller-service/.eslintrc.js (663 bytes)
CREATE nest-controller-service/.prettierrc (51 bytes)
CREATE nest-controller-service/README.md (3347 bytes)
CREATE nest-controller-service/nest-cli.json (171 bytes)
CREATE nest-controller-service/package.json (1964 bytes)
CREATE nest-controller-service/tsconfig.build.json (97 bytes)
CREATE nest-controller-service/tsconfig.json (546 bytes)
CREATE nest-controller-service/src/app.controller.spec.ts (617 bytes)
CREATE nest-controller-service/src/app.controller.ts (274 bytes)
CREATE nest-controller-service/src/app.module.ts (249 bytes)
CREATE nest-controller-service/src/app.service.ts (142 bytes)
CREATE nest-controller-service/src/main.ts (208 bytes)
CREATE nest-controller-service/test/app.e2e-spec.ts (630 bytes)
CREATE nest-controller-service/test/jest-e2e.json (183 bytes)

✔ Installation in progress... ☕

🚀  Successfully created project nest-controller-service
```

图 3-12　创建项目

接着通过 CLI 快速创建 User 模块的 CRUD 控制器，执行"nest g resource User"命令，选择 REST API 和 CRUD 入口，如图 3-13 所示。

```
$ nest g resource User
? What transport layer do you use? REST API
? Would you like to generate CRUD entry points? Yes
CREATE src/user/user.controller.spec.ts (556 bytes)
CREATE src/user/user.controller.ts (883 bytes)
CREATE src/user/user.module.ts (240 bytes)
CREATE src/user/user.service.spec.ts (446 bytes)
CREATE src/user/user.service.ts (607 bytes)
CREATE src/user/dto/create-user.dto.ts (30 bytes)
CREATE src/user/dto/update-user.dto.ts (169 bytes)
CREATE src/user/entities/user.entity.ts (21 bytes)
UPDATE package.json (1997 bytes)
UPDATE src/app.module.ts (308 bytes)
✔ Packages installed successfully.
```

图 3-13　生成 User 模块

创建完成后，打开 user.controller.ts 文件，核心代码如下：

```
// 定义 user 路径前缀
@Controller('user')
```

```
export class UserController {
  constructor(private readonly userService: UserService) {}

  @Post()
  create(@Body() createUserDto: CreateUserDto) {
    return this.userService.create(createUserDto);
  }

  @Get()
  findAll() {
    return this.userService.findAll();
  }
}
```

在 Nest 应用程序中，User 控制器会自动生成对应的路由 API，并在 Controller()装饰器中指定一个路径前缀，例如 user。这样做的好处是可以实现路由的分组，同时最大限度地减少代码重复。在大型应用系统中，这种路由分组方式使得开发者能够通过路径快速识别路由所属的控制器。例如，通过/orders/*路径可以清楚地知道这些路由是由 OrderController 管理的，而/products/*路由则属于 ProductController。

Create 方法通过@Post()装饰器声明了请求方式，告诉 HTTP 客户端以/user 路径和 POST 方法进行请求访问。如果使用/user 路径和 GET 方法请求 Nest 应用，则会映射到 findAll 这个路由处理程序中。

如果一个控制器中存在多个 POST 请求或 GET 请求，可以通过向装饰器传递参数来区分它们。例如，设置@Post("create-user")会将请求路径组合为/user/create-user，并将该路径映射到 create 方法。示例代码如下：

```
// 定义 user 路径前缀
@Controller('user')
export class UserController {
  constructor(private readonly userService: UserService) {}
  // 定义方法路径
  @Post('create-user')
  create(@Body() createUserDto: CreateUserDto) {
    return this.userService.create(createUserDto);
  }
}
```

另外，Nest 支持所有标准的 HTTP 方法，如@Get()、@Post()、@Put()、@Delete()、@Patch()、@Options()和@Head()等，读者可以根据需要自由组合路径。

3.3.3 Controller 处理请求参数与请求体

在 Express 中，通常需要通过 Request、Response 请求对象来获取 header、query、body 等属性。而 Nest 基于 Express 作为底层的 HTTP 框架，自然也提供了类似的对象，如@Req、@Res，但一般情况下不需要那么麻烦。Nest 提供了开箱即用的装饰器@Headers、@Body、@Param、@Query 等，示例代码如下：

```
@Get(':id')
findOne(@Param('id') id: string) {
```

```
    return this.userService.findOne(+id);
}

@Patch(':id')
update(@Param('id') id: string, @Body() updateUserDto: UpdateUserDto) {
    return this.userService.update(+id, updateUserDto);
}

@Delete(':id')
remove(@Param('id') id: string) {
    return this.userService.remove(+id);
}
```

首先，我们来探讨路由中携带参数的情况。使用@Get(':id')装饰器可以构建出像/user/123 这样的请求 URL，而@Param('id')用于从 URL 中提取参数 id，其值为 123。如果是查询参数形式的请求 URL，如/user?id=123，则可以通过@Query('id')来获取参数 id，其值为 123。

接下来是 POST 请求的处理。我们通常通过@Body()来获取请求体，并利用最常见的数据传输对象（Data Transfer Object，DTO）进行数据验证。例如，如上文代码所示的 updateUserDto，它将用于验证数据。更详细的 DTO 介绍将在本章后续部分进行。

完成一系列操作后，我们将调用服务层（Service）进行数据持久化。服务层的方法通常与控制器（Controller）中的路由方法名一一对应。例如，在 use.service.ts 中定义的方法，其代码如下：

```
import { Injectable } from '@nestjs/common';
import { CreateUserDto } from './dto/create-user.dto';
import { UpdateUserDto } from './dto/update-user.dto';

@Injectable()
export class UserService {
    create(createUserDto: CreateUserDto) {
        return 'This action adds a new user';
    }

    findAll() {
        return `This action returns all user`;
    }

    findOne(id: number) {
        return `This action returns a #${id} user`;
    }

    update(id: number, updateUserDto: UpdateUserDto) {
        return `This action updates a #${id} user`;
    }

    remove(id: number) {
        return `This action removes a #${id} user`;
    }
}
```

3.3.4 Service 处理数据层

至此，我们已经了解到，当客户端发出一个请求时，首先由控制器（Controller）处理，然后控制器会调度服务层（Service）的特定方法来执行业务逻辑，例如与数据库交互以执行数据的增加、删除、更新和查询操作。那么，在服务层中，数据又是如何被处理的呢？

先来看下面的伪代码：

```
import { Injectable } from '@nestjs/common';
import { CreateXxxxDto } from './dto/create-xxx.dto';
import { UpdateXxxDto } from './dto/update-xxx.dto';
import { InjectRepository } from '@nestjs/typeorm';
import { Repository } from 'typeorm';
import { Xxx } from './entities/xxx.entity';

@Injectable()
export class XxxService {
  // 注入 Xxx 实体
  @InjectRepository(Xxx) private xxxRepository: Repository<User>
  // 新增
  async create(createXxxDto: CreateXxxxDto) {
    createXxxDto.createTime = createXxxDto.updateTime = new Date()
    return await this.xxxRepository.save(createXxxDto);
  }
  // 查询全部
  async findAll() {
    return await this.xxxRepository.find();
  }
  // 查询单个
  async findOne(id: number) {
    return this.xxxRepository.findBy;
  }
  // 修改
  async update(id: number, updateXxxDto: UpdateXxxDto) {
    updateXxxDto.updateTime = new Date()
    return await this.xxxRepository.update(id, updateXxxDto);
  }
  // 删除
  async remove(id: number) {
    return await this.xxxRepository.delete(id);
  }
}
```

由此可见，服务层（Service）会调用 Repository 实例方法对实体数据进行 CRUD 操作，而实体对象数据的改变将会同步映射到数据库中对应字段的更新。这就是对象关系映射（Object-Relational Mapping，ORM）的作用，后面会更加详细地介绍 ORM 的概念。

在 Nest 框架中，客户端发出的请求会经历一个分层处理过程：首先由控制器（Controller）接收请求，然后传递给服务层（Service），接着服务层可能会调用仓库层（Repository）进行数据操作，最终将处理结果返回给客户端。这一流程正是 Nest 分层架构的典型体现。

3.3.5 服务与服务提供者

Service 服务是 Nest 分层架构中的一个组成部分。通过使用@Injectable 装饰器声明，它可以被控制器（Controller）调度使用，或者被其他服务共享。这时，一些读者可能会有疑问：服务与服务提供者之间是什么关系？实际上，所有通过@Injectable 装饰器装饰的类都是服务提供者。这是一个抽象概念，它不仅包括服务类，还可能包括中间件、拦截器、管道等，它们可以是工厂函数或常量。这些服务提供者被依赖注入系统注入其他组件中（例如控制器、服务等），以提供特定的功能、数据或操作。

3.4 耳熟能详的中间件

后端体系中有很多中间件，比如消息中间件如 Kafka 和 MQ，缓存中间件如 Redis，数据库中间件如 Hibernate 和 MyBatis 等。在 Node.js 生态体系中，Express 和 Koa 等第三方框架都实现了中间件功能。同样地，Nest 默认基于 Express 框架，自然也支持中间件，但实际上它们并不完全相同。本节将探讨 Nest 的中间件。

Nest 实现面向切面编程（AOP）思想的方式之一是中间件。根据实现方式，可以分为函数式中间件和类中间件；根据作用域，又可以分为局部中间件和全局中间件。

创建一个 Nest 项目，运行"nest n nest-middleware -p pnpm"命令，如图 3-14 所示。

```
$ nest n nest-middleware -p pnpm
⚡ We will scaffold your app in a few seconds..

CREATE nest-middleware/.eslintrc.js (663 bytes)
CREATE nest-middleware/.prettierrc (51 bytes)
CREATE nest-middleware/README.md (3347 bytes)
CREATE nest-middleware/nest-cli.json (171 bytes)
CREATE nest-middleware/package.json (1956 bytes)
CREATE nest-middleware/tsconfig.build.json (97 bytes)
CREATE nest-middleware/tsconfig.json (546 bytes)
CREATE nest-middleware/src/app.controller.spec.ts (617 bytes)
CREATE nest-middleware/src/app.controller.ts (274 bytes)
CREATE nest-middleware/src/app.module.ts (249 bytes)
CREATE nest-middleware/src/app.service.ts (142 bytes)
CREATE nest-middleware/src/main.ts (208 bytes)
CREATE nest-middleware/test/app.e2e-spec.ts (630 bytes)
CREATE nest-middleware/test/jest-e2e.json (183 bytes)

✔ Installation in progress... ☕

🚀 Successfully created project nest-middleware
```

图 3-14 创建项目

创建成功后，接下来演示这几种中间件的使用及它们之间的区别。

3.4.1 类中间件

类中间件通过使用@Injectable()装饰器来声明，并需要实现 NestMiddleware 接口的 use 方法。在默认情况下，如果使用 Express 作为 HTTP 框架，中间件可以操作 Express 提供的 Request 和 Response 对象。而当使用 Fastify 作为 HTTP 框架时，对应的请求和响应对象分别是 FastifyRequest

和 FastifyReply 对象。

运行"nest g middleware logger"命令生成一个中间件，logger.middleware.ts 代码如下：

```typescript
import { Injectable, NestMiddleware } from '@nestjs/common';
import { Request, Response, NextFunction } from 'express';

@Injectable()
export class LoggerMiddleware implements NestMiddleware {
  use(req: Request, res: Response, next: NextFunction) {
    console.log('before 中间件');
    next();
    console.log('after 中间件');
  }
}
```

接着运行"nest g resource user"命令创建 User 模块，并在 user.module.ts 文件中使用中间件，示例代码如下：

```typescript
import { MiddlewareConsumer, Module, NestModule } from '@nestjs/common';
import { UserService } from './user.service';
import { UserController } from './user.controller';
import { LoggerMiddleware } from 'src/logger/logger.middleware';

@Module({
  controllers: [UserController],
  providers: [UserService]
})
export class UserModule implements NestModule {
  configure(consumer: MiddlewareConsumer) {
      // 针对此模块的所有路由绑定中间件
      consumer.apply(LoggerMiddleware).forRoutes('*')
  }
}
```

可见，@Module 装饰器中并没有直接提供注册中间件的位置。使用中间件的模块必须实现 NestModule 接口的 configure 方法，并通过 consumer 辅助类对中间件进行配置。forRoutes 方法用于指定应用中间的路由。该中间件可以应用到 UserController 中的所有路由程序中，当然，我们也可以指定更精确的路由路径或请求方式，例如：

```typescript
import { MiddlewareConsumer, Module, NestModule, RequestMethod } from '@nestjs/common';
import { UserService } from './user.service';
import { UserController } from './user.controller';
import { LoggerMiddleware } from 'src/logger/logger.middleware';

@Module({
  controllers: [UserController],
  providers: [UserService]
})
export class UserModule implements NestModule {
  configure(consumer: MiddlewareConsumer) {
      // consumer.apply(LoggerMiddleware).forRoutes('*')
      // 指定应用中间件的路由和请求方式
      consumer.apply(LoggerMiddleware).forRoutes({
```

```
      path: '/user',
      method: RequestMethod.GET
    })
  }
}
```

我们指定中间件应用在 /person 路径，并指定了 GET 请求方式，完成后在浏览器访问 localhost:3000，控制台打印效果如图 3-15 所示。

```
[Nest] 53714  - 2024/01/03 15:18:43    LOG [RouterExplorer] Mapped {/user/:id, GET} route +0ms
[Nest] 53714  - 2024/01/03 15:18:43    LOG [RouterExplorer] Mapped {/user/:id, PATCH} route +0ms
[Nest] 53714  - 2024/01/03 15:18:43    LOG [RouterExplorer] Mapped {/user/:id, DELETE} route +0ms
[Nest] 53714  - 2024/01/03 15:18:43    LOG [NestApplication] Nest application successfully started +1ms
before中间件
after中间件
```

图 3-15　中间件的运行结果

在 Nest 中，类中间件的作用不仅限于处理 HTTP 请求和响应，更重要的是它能够实现依赖注入。这意味着我们可以在中间件中注入特定的依赖项，并调用这些依赖项内部的方法，例如 UserService 服务。示例代码如下：

```
import { Injectable, NestMiddleware } from '@nestjs/common';
import { Inject } from '@nestjs/common/decorators';
import { Request, Response, NextFunction } from 'express';
import { UserService } from 'src/user/user.service';

@Injectable()
export class LoggerMiddleware implements NestMiddleware {
  // 注入 UserService 依赖
  @Inject(UserService)
  private userService: UserService;
  use(req: Request, res: Response, next: NextFunction) {
    console.log('before 中间件');
    console.log('中间件作用的方法执行结果: ', this.userService.findAll());
    next();
    console.log('after 中间件');
  }
}
```

刷新浏览器，运行结果如图 3-16 所示。

```
[Nest] 53533  - 2024/01/03 15:04:21    LOG [RouterExplorer] Mapped {/user, GET} route +0ms
[Nest] 53533  - 2024/01/03 15:04:21    LOG [RouterExplorer] Mapped {/user/:id, GET} route +0ms
[Nest] 53533  - 2024/01/03 15:04:21    LOG [RouterExplorer] Mapped {/user/:id, PATCH} route +1ms
[Nest] 53533  - 2024/01/03 15:04:21    LOG [RouterExplorer] Mapped {/user/:id, DELETE} route +0ms
[Nest] 53533  - 2024/01/03 15:04:21    LOG [NestApplication] Nest application successfully started +1ms
before中间件
中间件作用的方法执行结果:  This action returns all user
after中间件
```

图 3-16　中间件注入外部依赖

这就是 Nest 中间件的依赖注入。当然，如果不需要依赖，可以用轻量的函数式中间件。这就好比在 React/Vue 中，当我们的组件不需要状态管理时，优先选择函数式组件是类似的道理。

3.4.2 函数式中间件

Nest 中提供了函数式中间件，又称为功能中间件。当中间件不需要成员、没有额外的方法和依赖时，应优先考虑使用它。

下面以 LoggerMiddleware 为例：

```
import { Request, Response, NextFunction } from 'express';

export function LoggerMiddleware(
  req: Request,
  res: Response,
  next: NextFunction,
) {
  console.log('中间件 before');
  next();
  console.log('中间件 after');
}
```

简单吧？与定义普通的函数没什么区别。

3.4.3 局部中间件

在前面的例子中，我们将中间件应用到局部作用域中，例如通过 forRoutes 指定特定的路由范围。通过使用通配符（*）可以将中间件应用到整个模块中，除此之外，还可以指定应用的控制器，例如在 app.module.ts 中这样使用：

```
export class AppModule implements NestModule {
  configure(consumer: MiddlewareConsumer) {
    // 将中间件绑定到指定控制器中
    consumer.apply(LoggerMiddleware).forRoutes(UserController)
  }
}
```

如果有多个中间件，apply 方法支持绑定多个中间件列表。通过执行 next 函数可以调用下一个中间件函数。示例代码如下：

```
export class AppModule implements NestModule {
  configure(consumer: MiddlewareConsumer) {
    // 绑定多个中间件
    consumer
      .apply(cors(), helmet(), LoggerMiddleware)
      .forRoutes(UserController);
  }
}
```

这些局部中间件统称为路由级别的中间件。

3.4.4 全局中间件

全局中间件通过 app.use 方法来注册。示例代码如下：

```
import { NestFactory } from '@nestjs/core';
import { AppModule } from './app.module';
```

```
import { LoggerMiddleware } from './logger/logger.middleware';

async function bootstrap() {
  const app = await NestFactory.create(AppModule);
  // 中间件
  app.use(new LoggerMiddleware().use)
  // 启动服务
  await app.listen(8088);
}
bootstrap();
```

与 Express 相比，Nest 框架中的中间件提供了更丰富的功能。在 Express 中，通常不能为中间件指定路由或控制器，并且不支持依赖注入。例如，如果在中间件中强制注入 UserService，可能会导致错误。因此，在 Express 中，通常推荐在应用模块（AppModule）中通过 configure 方法来注册中间件。

Nest 框架中的中间件分为类中间件和函数式中间件，它们各自有不同的使用场景和作用域。类中间件支持依赖注入，可以通过 forRoutes 方法绑定到一个或多个具体的路由处理器或控制器。此外，也可以通过 app.use 方法将中间件注册为全局中间件。

在 Nest 应用的初始化过程中，中间件的执行顺序是首先运行全局中间件，然后执行局部（模块级）中间件。

3.5 拦截器与 RxJS 知多少

拦截器是 AOP 编程思想的第二种实现方式，第一种是 3.4 节介绍的中间件。本节首先介绍拦截器的基本概念和常见的应用场景，然后通过代码演示如何实现一个接口超时的拦截器。这个拦截器需要与异步事件处理库 RxJS 配合来使用，最后还会介绍 RxJS 常用的 API，并且练习如何使用它。

3.5.1 基本概念

拦截器的请求和响应流程遵循先进后出的顺序。请求首先通过全局拦截器，然后是控制器级别的拦截器，最后是路由级别的拦截器进行处理。响应流程则相反，即从路由级别的拦截器开始，经过控制器级别的拦截器，最终到达全局拦截器。这样的设计允许在请求处理的任何阶段，包括由管道、控制器或服务抛出的错误，都能够通过拦截器进行捕获和处理。

每个拦截器通过@Injectable()来声明，并且需要实现 NestInterceptor 接口的 intercept 方法，接收两个参数：ExecutionContext 上下文对象和 CallHandler 处理程序。

ExecutionContext 能够访问当前请求的详细信息，包括路由信息、HTTP 方法、请求体以及响应体数据。它主要应用于以下几个场景：

（1）记录请求和响应的日志，用于追踪、监控和调试。
（2）进行身份验证和权限检查。
（3）根据请求头或路由信息来设置缓存策略。
（4）修改或转换响应数据，例如对响应进行包装、格式化、加密操作。

CallHandler 实现了 handle 方法，必须在拦截器中调用 handle 方法才能执行路由处理方法。示例代码如下：

```
import {
  CallHandler,
  ExecutionContext,
  Injectable,
  NestInterceptor,
} from '@nestjs/common';

@Injectable()
export class TimeoutInterceptor implements NestInterceptor {
  intercept(context: ExecutionContext, next: CallHandler): Observable<any> {
    // 调用 handle
    return next.handle();
  }
}
```

通常情况下，intercept 方法返回 RxJS 的 Observable 对象。那么，什么是 RxJS 呢？

RxJS 是一个用于处理异步数据流的 JavaScript 库。你可以把它理解为一个"管道"，它可以帮你更方便地处理各种事件和数据流。比如，当你需要处理用户的单击事件、网络请求返回的数据、定时器触发的事件等，RxJS 能够帮助你更加优雅地管理这些复杂的异步操作。

然而，我们不必担心这会增加额外的认知负担，可以简单地将其视为前端的 Lodash 工具集。接下来，我们通过新建项目来演示说明这一点。

3.5.2　创建项目

运行"nest n nest-interceptor -p pnpm"命令创建项目，如图 3-17 所示。

```
$ nest n nest-interceptor -p pnpm
⚡  We will scaffold your app in a few seconds..

CREATE nest-interceptor/.eslintrc.js (663 bytes)
CREATE nest-interceptor/.prettierrc (51 bytes)
CREATE nest-interceptor/README.md (3347 bytes)
CREATE nest-interceptor/nest-cli.json (171 bytes)
CREATE nest-interceptor/package.json (1957 bytes)
CREATE nest-interceptor/tsconfig.build.json (97 bytes)
CREATE nest-interceptor/tsconfig.json (546 bytes)
CREATE nest-interceptor/src/app.controller.spec.ts (617 bytes)
CREATE nest-interceptor/src/app.controller.ts (274 bytes)
CREATE nest-interceptor/src/app.module.ts (249 bytes)
CREATE nest-interceptor/src/app.service.ts (142 bytes)
CREATE nest-interceptor/src/main.ts (208 bytes)
CREATE nest-interceptor/test/app.e2e-spec.ts (630 bytes)
CREATE nest-interceptor/test/jest-e2e.json (183 bytes)

✔ Installation in progress... 🐢

🚀  Successfully created project nest-interceptor
```

图 3-17　创建项目

3.5.3　拦截器的基本使用方法

在 src 目录下新建一个名为 interceptor 的子目录，专门用于存放应用拦截器。接下来，我们将实现一个用于统计接口超时的拦截器。示例代码如下：

```typescript
import {
  CallHandler,
  ExecutionContext,
  Injectable,
  NestInterceptor,
} from '@nestjs/common';
import { log } from 'console';
import { Observable, tap, timeout } from 'rxjs';

@Injectable()
export class TimeoutInterceptor implements NestInterceptor {
  intercept(context: ExecutionContext, next: CallHandler): Observable<any> {
    console.log('进入拦截器', context.getClass());
    const now = Date.now();
    // 调用 handle
    return next.handle().pipe(
      tap(() => {
        // 统计耗时
        log('Timeout: ', Date.now() - now);
      }),
      // 超时时间
      timeout(1000),
    );
  }
}
```

在上面的代码中，tap 和 timeout 操作符不会改变数据或干扰响应周期，而是用于执行额外的逻辑处理。

将这些操作符绑定到 app.controller.ts 文件中的 getHello 方法上，具体示例如图 3-18 所示。

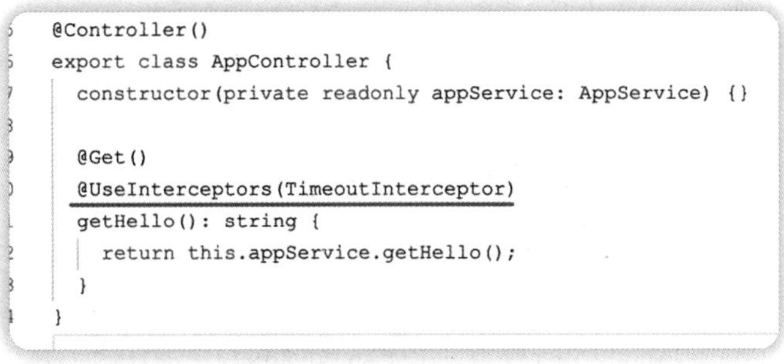

```typescript
@Controller()
export class AppController {
  constructor(private readonly appService: AppService) {}

  @Get()
  @UseInterceptors(TimeoutInterceptor)
  getHello(): string {
    return this.appService.getHello();
  }
}
```

图 3-18　绑定拦截器

接着在浏览器输入 localhost:3000，运行结果如图 3-19 所示。

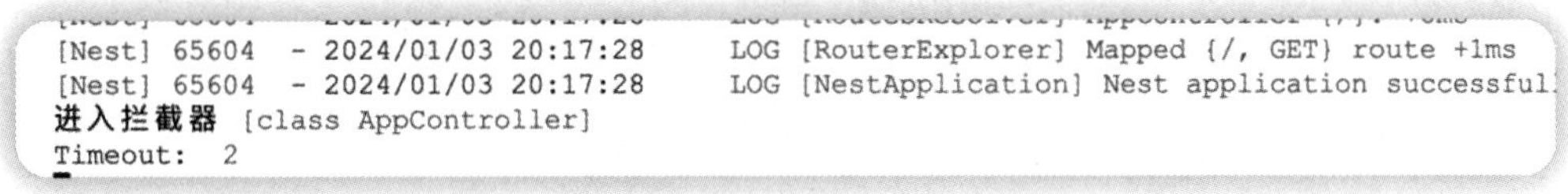

```
[Nest] 65604   - 2024/01/03 20:17:28     LOG [RouterExplorer] Mapped {/, GET} route +1ms
[Nest] 65604   - 2024/01/03 20:17:28     LOG [NestApplication] Nest application successful
进入拦截器 [class AppController]
Timeout: 2
```

图 3-19　拦截器的运行结果

除此之外，还有其他操作符，如 map、filter、delay、from、toArray、catchError 等。我们接下

来定义一个稍微复杂一些的拦截器，用于转换并过滤指定用户的名称。示例代码如下：

```typescript
import {
  Injectable,
  NestInterceptor,
  ExecutionContext,
  CallHandler,
} from '@nestjs/common';
import { Observable, of } from 'rxjs';
import { tap, map, catchError, filter, toArray } from 'rxjs/operators';

@Injectable()
export class AuthInterceptor implements NestInterceptor {
  intercept(context: ExecutionContext, next: CallHandler): Observable<any> {
    // 在请求处理之前进行日志记录
    console.log('Request received...');
    return next.handle().pipe(
      // 使用 map 操作符将每个用户名转换为大写
      map((name) => name.toUpperCase()),

      // 使用 filter 操作符过滤出名字长度大于 2 的用户
      filter((name) => name.length > 2),

      // 使用 tap 操作符进行简单的日志记录
      tap((arr) => console.log(`Filtered Name: ${arr}`)),

      // 转为数组
      toArray(),

      // 使用 catchError 操作符捕获并处理任何错误
      catchError((error) => {
        console.error('Error occurred:', error);
        // 返回新的 Observable
        return of('Fallback Value');
      }),
    );
  }
}
```

在上面的代码中，通过 map 操作符把结果转为大写，然后使用 filter 操纵符过滤长度大于 2 的名称，接着打印日志并最后转为数组。如果出现异常，使用 catchError 来捕获并返回新的数据。

将这些操作符绑定到 app.controller.ts 文件中的 getNameList 方法上，具体示例如图 3-20 所示。

```
6  @Get('name')
7  @UseInterceptors(AuthInterceptor)
8  getNameList(): Observable<any> {
9    return from(['小二', '老王', 'mouse', 'nestjs'])
0  }
```

图 3-20　绑定拦截器

在浏览器中输入 localhost:3000/name，运行结果如图 3-21 所示。

图 3-21　拦截器的运行结果

如果接口在执行过程中抛出异常，catchError 操作符将介入并返回数据，作为错误处理的最后手段，如图 3-22 所示。

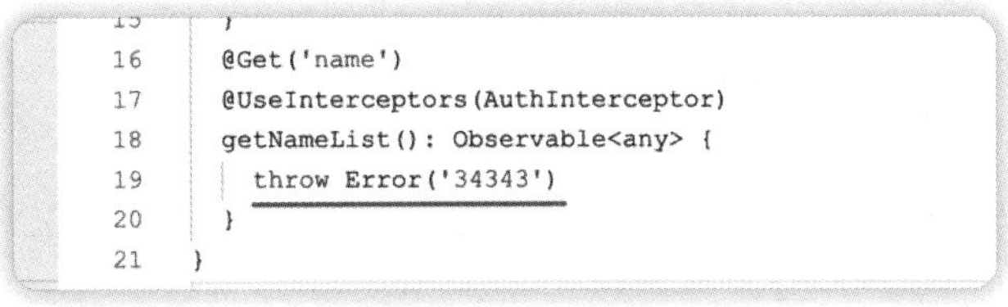

图 3-22　接口抛出异常

刷新浏览器，返回"Fallback Value"，效果如图 3-23 所示。

图 3-23　返回数据

除此之外，在前面的 2.4 节中，我们学习了 HTTP 框架拥有许多内置的异常过滤器。这些异常过滤器不仅可以捕获和处理异常，还可以作为"Fallback Value"（备用返回值）使用。

3.6　数据之源守护者：管道

前面讲过，控制器层（Controllers）的职责之一是解析和验证请求数据。这项工作是由什么来完成的呢？正是本节将要介绍的管道。本节首先介绍管道的基本概念和用法，包括 Nest 框架中内置的 9 种开箱即用的管道验证器。接下来，将演示这些验证器如何应用于处理 GET 和 POST 请求参数。最后，通过实现自定义管道，将进一步加深对这一概念的理解。

3.6.1　基本概念

在后端开发中，数据库表的字段类型在创建时就已经被明确定义，任何不符合预期类型的数据保存操作都会导致错误。为了确保传入的数据满足预期的格式和规范，Nest 框架会在客户端发起请求时，将请求数据传递给管道进行预处理。这些预处理操作包括数据验证、转换或过滤等，以确保数据的准确性。处理后的数据随后会被传递给路由处理程序。

如果在验证过程中遇到任何异常，Nest 框架可以使用异常过滤器来统一捕获和处理这些异常。

这包括记录错误日志、返回统一格式的错误响应等操作。

通常情况下，我们会将管道绑定在方法参数上，这样管道就会与特定的路由方法关联起来，这种方式称为参数级别管道。除此之外，Nest 还允许将管道绑定在全局作用域，使其适用于每个控制器和路由方法。

3.6.2　内置管道

Nest 内置了 9 种开箱即用的管道验证器，它们分别是：

- ParseIntPipe
- ParseFloatPipe
- ParseBoolPipe
- ParseUUIDPipe
- ParseEnumPipe
- DefaultValuePipe
- ValidationPipe
- ParseArrayPipe
- ParseFilePipe

接下来创建一个 Nest 项目，用来分别演示 GET 和 POST 请求中常用的几个管道。执行"nest n nest-pipe -p pnpm"命令，结果如图 3-24 所示。

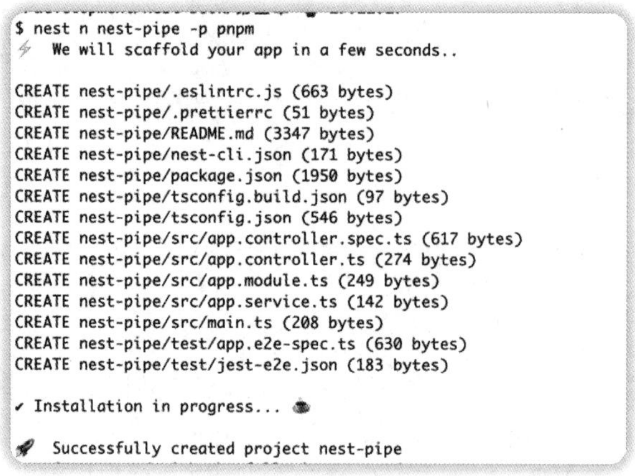

图 3-24　创建项目

执行"pnpm start:dev"命令以启动服务，然后在浏览器中访问 localhost:3000。如果运行成功，结果如图 3-25 所示。

图 3-25　成功运行结果

1. ParseIntPipe

ParseIntPipe 接收的参数必须能够被解析为整数（parseInt），否则将会抛出异常。示例代码如下：

```
@Get('int/:id')
getHello(@Param('id', ParseIntPipe) id: number): string {
  log('id:', id);
  return this.appService.getHello();
}
```

@Param 装饰器用于接收 URL 中的参数，例如在浏览器中访问 localhost:3000/int/123，即可成功获取 id 参数，如图 3-26 所示。

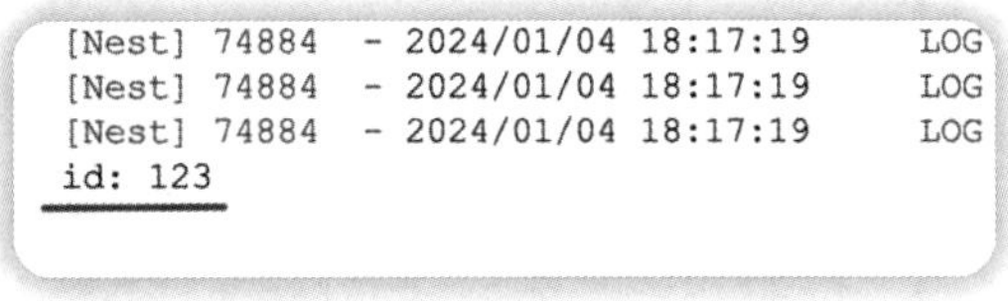

图 3-26 ParseIntPipe 验证参数

把地址改为 localhost:3000/xxx 并刷新，此时请求不能通过管道验证，会抛出异常，如图 3-27 所示。

图 3-27 验证异常效果

这个错误响应是 Nest 默认定义的，也可以自定义友好的错误提示信息，验证器允许传递下面两个参数。

（1）errorHttpStatusCode：验证器失败时抛出的 HTTP 状态码，默认为 400（错误请求）。

（2）exceptionFactory：工厂函数，用于接收错误消息并返回相应的错误对象。

修改后的代码如下：

```
@Get(':id')
getHello(
  @Param(
    'id',
    new ParseIntPipe({
      exceptionFactory: () => {
        // 抛出 HTTP 异常
        throw new HttpException('参数 id 类型错误', HttpStatus.BAD_REQUEST);
      },
    }),
  )
  id: number,
```

```
): string {
  log('id:', id);
  return this.appService.getHello();
}
```

刷新浏览器，返回结果如图 3-28 所示。

图 3-28　自定义异常信息

同样地，接下来要介绍的验证器也允许我们通过这两个参数来自定义错误提示信息。

2. ParseFloatPipe

接下来，我们来看 ParseFloatPipe。此管道的作用是将接收到的数据转换为浮点数（Float）类型。示例代码如下：

```
@Get('float/:id')
getHello2(@Param('id', ParseFloatPipe) id: Number): string {
  log('float id:', id);
  return this.appService.getHello()
}
```

在浏览器中输入 localhost:3000/float/123.45，运行结果如图 3-29 所示。

```
[Nest] 76469  - 2024/01/04 22:30:17
[Nest] 76469  - 2024/01/04 22:30:17
[Nest] 76469  - 2024/01/04 22:30:17
float id: 123.45
```

图 3-29　ParseFloatPipe 验证参数

3. ParseBoolPipe

ParseBoolPipe 负责将参数转换为布尔（Boolean）类型。我们可以通过@Query()装饰器来接收查询参数，并利用此管道进行转换。示例代码如下：

```
@Get('bool')
getHello3(@Query('flag', ParseBoolPipe) flag: boolean): string {
  log('bool flag:', flag)
  return this.appService.getHello()
}
```

浏览器中输入 localhost:3000/bool?flag=true，运行结果如图 3-30 所示。

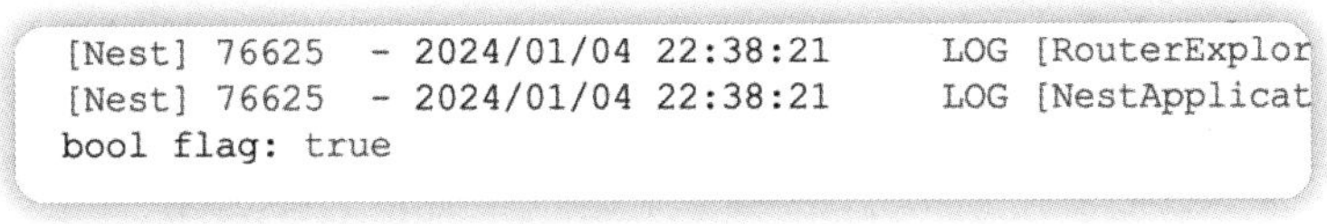

```
[Nest] 76625  - 2024/01/04 22:38:21     LOG [RouterExplor
[Nest] 76625  - 2024/01/04 22:38:21     LOG [NestApplicat
bool flag: true
```

图 3-30　ParseBoolPipe 验证参数

4. ParseUUIDPipe

uuid 是一串随机生成且唯一的字符，广泛应用于需要唯一性或保护隐私的场景，例如数据库主键、订单号、用户标识等。

uuid 有多个版本，包括 v1、v3、v4 和 v5。其中，v1 和 v4 是较为常用的版本。v1 版本通过结合时间戳和节点信息来保证时间顺序和唯一性，而 v4 版本则生成完全随机的 uuid。v3 和 v5 版本基于哈希算法，能够确保对于相同的输入，生成的 uuid 是一致的。

在接收 UUID 参数时，可以使用 ParseUUIDPipe 管道来验证其格式正确性。示例代码如下：

```
@Get('uuid')
getHello4(@Query('id', ParseUUIDPipe) id: string): string {
  log('uuid id:', id)
  return this.appService.getHello()
}
```

在浏览器中输入 localhost:3000/uuid?id=1262c798-aacf-4c41-ad73-7dd29a708a68，运行结果如图 3-31 所示。

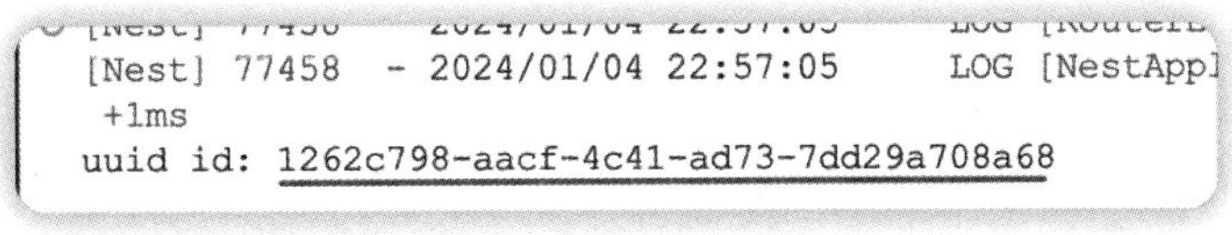

```
[Nest] 77458  - 2024/01/04 22:57:05     LOG [NestApp]
 +1ms
uuid id: 1262c798-aacf-4c41-ad73-7dd29a708a68
```

图 3-31　ParseUUIDPipe 验证参数

5. ParseEnumPipe

ParseEnumPipe 是用于验证枚举值的管道。如果接收到的参数不在预定的枚举值内，验证将失败。例如，我们可以定义一组枚举值来描述某种状态。示例代码如下：

```
enum StatusEnum {
  ACTIVE = 'active',
  INACTIVE = 'inactive',
  PENDING = 'pending',
}
```

同时，在路由方法中验证参数是否合法：

```
@Get('enum')
getHello5(
  @Query('status', new ParseEnumPipe(StatusEnum)) status: string
): string {
  log('enum status:', status);
  return this.appService.getHello();
}
```

在浏览器中输入 localhost:3000/enum?status=active，运行结果如图 3-32 所示。

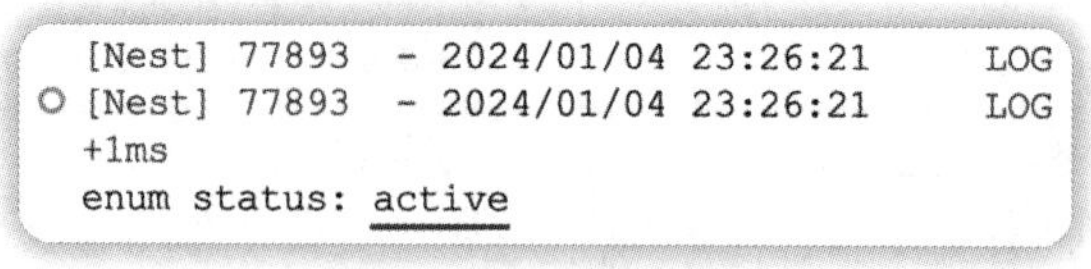

```
[Nest] 77893    - 2024/01/04 23:26:21    LOG
[Nest] 77893    - 2024/01/04 23:26:21    LOG
+1ms
enum status: active
```

图 3-32　ParseEnumPipe 验证参数

此时,如果传递一个不在枚举范围内的参数,则会触发异常,如图 3-33 所示。

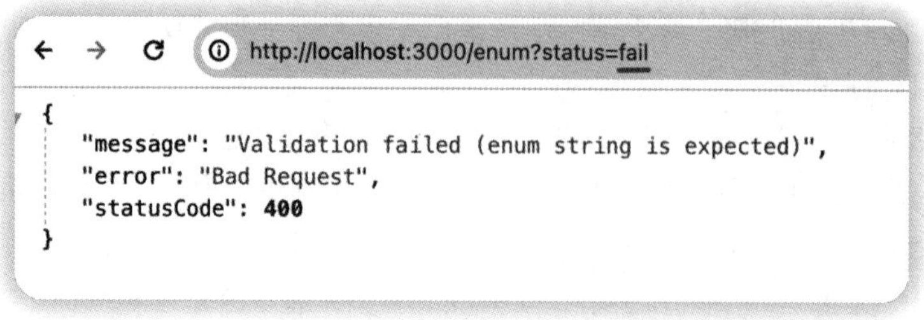

```
http://localhost:3000/enum?status=fail
{
  "message": "Validation failed (enum string is expected)",
  "error": "Bad Request",
  "statusCode": 400
}
```

图 3-33　enum 参数验证异常

在实际开发中,验证枚举参数的场景并不常见。然而,它能够将前后端的枚举状态统一起来维护,这对于基于 MVC 架构体系的项目来说是一个不错的选择。

6. DefaultValuePipe

DefaultValuePipe 的作用非常直接:当请求中未传递参数时,它允许我们为该参数指定一个默认值。示例代码如下:

```
@Get('default')
getHello6(
  @Query('value', new DefaultValuePipe('jmin')) value: string
): string {
  log('default value:', value);
  return this.appService.getHello();
}
```

在浏览器中输入 localhost:3000/default,运行结果如图 3-34 所示。

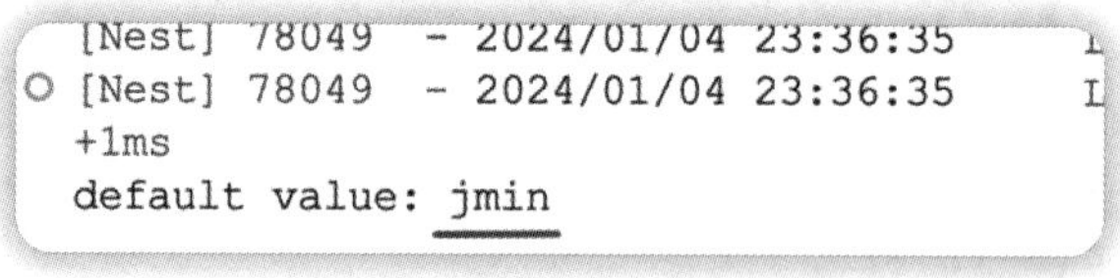

```
[Nest] 78049    - 2024/01/04 23:36:35    L
[Nest] 78049    - 2024/01/04 23:36:35    L
+1ms
default value: jmin
```

图 3-34　DefaultValuePipe 验证参数

需要注意的是,如果输入地址为 localhost:3000/default?value=,则不会触发默认赋值。Nest 会认为你传递的参数为空字符串,运行结果如图 3-35 所示。

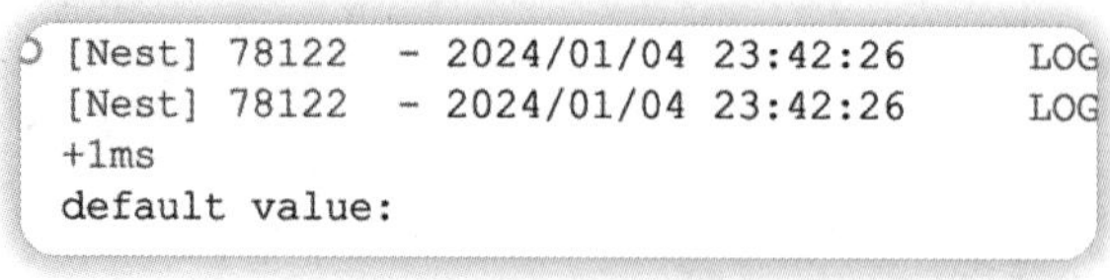

```
[Nest] 78122    - 2024/01/04 23:42:26    LOG
[Nest] 78122    - 2024/01/04 23:42:26    LOG
+1ms
default value:
```

图 3-35　传递空字符串参数

7. ParseArrayPipe

ParseArrayPipe 用于将传递的字符串参数转换为数组类型。例如，在需要传递多个用户 ID 的场景中，客户端可能会发送形如/array/1,2,3 的请求。通过使用 ParseArrayPipe，这样的请求可以被转换为[1,2,3]这样的数组形式。

由于 ParseArrayPipe 依赖于 class-validator 和 class-transformer 这两个 npm 包，在开始使用它之前，需要先进行安装。可以通过执行 "pnpm add class-validator class-transformer –save" 命令来添加这些依赖。安装完成后，可以继续编写相关的代码：

```
@Get('array')
getHello7(@Query('ids', ParseArrayPipe) ids: Number[]): string {
  log('array ids:', ids);
  return this.appService.getHello();
}
```

在浏览器中输入 localhost:3000/array?ids=1,2,3，运行结果如图 3-36 所示。

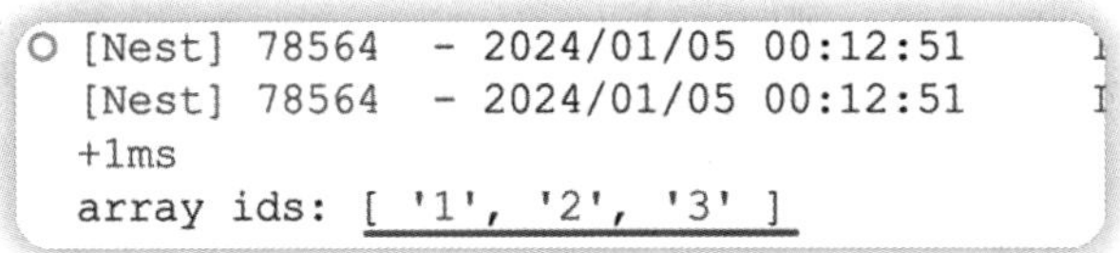

```
[Nest] 78564    - 2024/01/05 00:12:51
[Nest] 78564    - 2024/01/05 00:12:51
+1ms
array ids: [ '1', '2', '3' ]
```

图 3-36　ParseArrayPipe 验证参数

ParseArrayPipe 默认使用逗号（,）作为分隔符来分隔字符串参数。然而，在 Nest 框架中，我们可以灵活地自定义分隔符以满足不同的需求。例如：

```
@Get('array')
getHello7(
  @Query('ids', new ParseArrayPipe({
    separator: '-'
  })) ids: Number[]
): string {
  log('array ids:', ids);
  return this.appService.getHello();
}
```

这样就可以通过 localhost:3000/array?ids=4-5-6 这种方式来传参了，运行结果如图 3-37 所示。

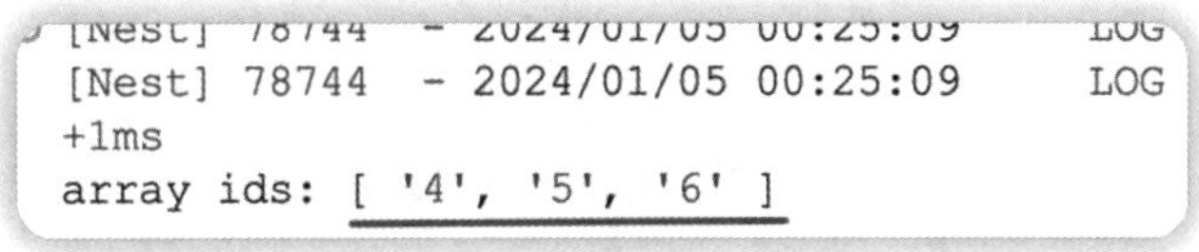

```
[Nest] 78744    - 2024/01/05 00:25:09    LOG
[Nest] 78744    - 2024/01/05 00:25:09    LOG
+1ms
array ids: [ '4', '5', '6' ]
```

图 3-37　使用自定义参数分隔符后的运行结果

8. ValidationPipe

前面介绍的都是 GET 请求的参数验证，如果是 POST 请求发送的请求体，又该如何校验呢？这就要用到 ValidationPipe 了。POST 请求体可以通过@Body 装饰器来接收，需要配合数据传输对象（Data Transfer Object，DTO）来使用。新建 src/dto/create-user.dto.ts 文件。示例代码如下：

```
import { IsInt, IsNotEmpty, IsString } from "class-validator";
```

```
export class CreateUserDto {
    @IsNotEmpty()
    @IsString()
    name: string

    @IsInt({ message: "age 必须是数字类型"})
    age: number

    @IsString({ message: "sex 必须是字符串类型"})
    sex: string

    phone: string

    email: string
}
```

ValidationPipe 的使用需要依赖 class-validator 和 class-transformer 这两个包。class-validator 提供了一系列用于验证字段的装饰器，例如@IsNotEmpty、@IsString 和@IsInt。这些装饰器允许我们对字段进行验证，并可以自定义错误提示信息（message）。之后，我们可以在路由方法中绑定这些验证器来执行数据验证：

```
@Post('create-user')
createUser(@Body(ValidationPipe) createUserDto: CreateUserDto): Record<string, any> {
    return createUserDto
}
```

接下来，可以通过 Postman 或者 AJAX 来调用接口。实际上，VS Code 也提供了一种更加快捷的方式来完成 mock 请求和响应的流程。具体步骤如下：

（1）安装 VS Code 插件 REST Client，如图 3-38 所示。

图 3-38　安装插件

（2）在项目根目录下创建.http 文件，设置请求 URL、Content-type、请求参数/体，多个接口用###分隔。示例代码如下：

```
GET http://localhost:3000/array?ids=1-2-3
###

POST http://localhost:3000/create-user
Content-Type: application/json

{
  "name": "",
  "age": 22,
```

```
  "sex": "男",
  "phone": "",
  "email": "mouse@example.com",
}
```

(3) 单击 Send Request 发送请求, 如图 3-39 所示。

图 3-39　发送请求

(4) 发送请求后, 在 VS Code 中可以看到响应体, 如图 3-40 所示。

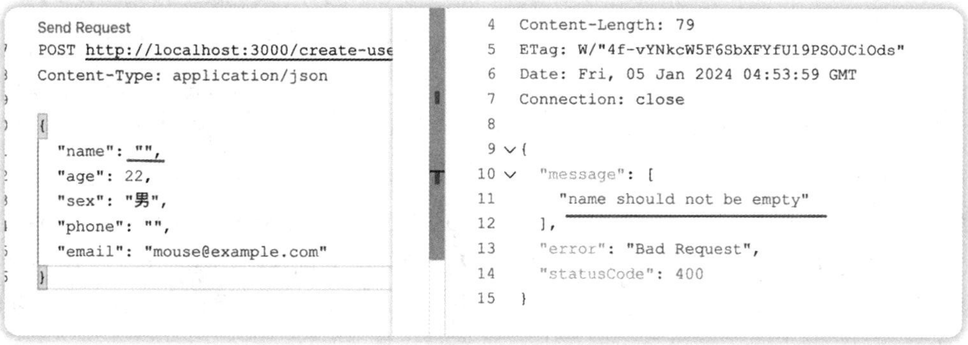

图 3-40　查看响应结果

图 3-40 中提示的 name 不能为空, 这就是 ValidationPipe 的默认验证结果。我们将 age 改为 String 类型, 重新发送请求, 可以看到返回的自定义提示信息, 如图 3-41 所示。

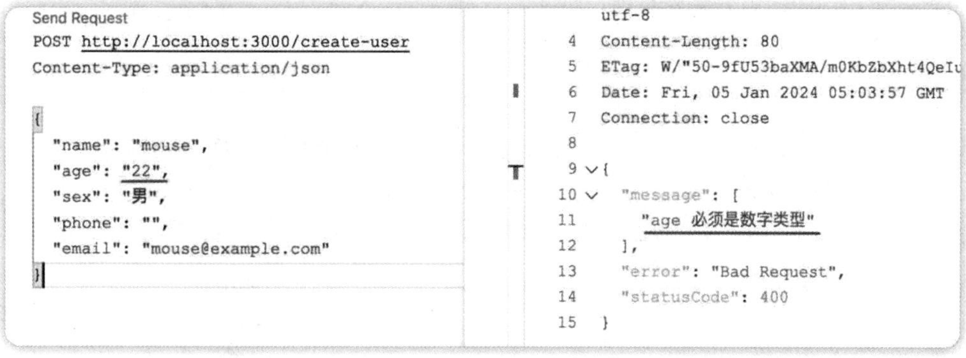

图 3-41　自定义验证提示

在项目开发过程中, 通常会有多个 POST 请求需要进行验证。如果在每个路由方法中单独绑定验证器，可能会显得相当烦琐。为了简化这一过程，我们可以将管道验证器配置为全局生效。例如：

```
import { NestFactory } from '@nestjs/core';
import { AppModule } from './app.module';
import { ValidationPipe } from '@nestjs/common';

async function bootstrap() {
  const app = await NestFactory.create(AppModule);
  // 绑定全局验证器
  app.useGlobalPipes(new ValidationPipe())
  await app.listen(3000);
}
bootstrap();
```

输入正确的参数，重新单击 Send Request，运行结果如图 3-42 所示。

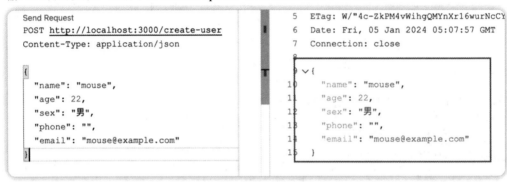

图 3-42　全局验证器的运行结果

综上所述，在 Nest 框架中，我们可以通过 DTO 来管理和校验 POST 请求参数。DTO 作为一个强大的工具，能够帮助我们确保数据的准确性和安全性。在实际应用中，为了保持代码的整洁和可维护性，通常会将 DTO 定义单独抽取到一个公共的模块或服务中进行管理。这样做可以使得整个应用的分层结构更加清晰，同时也便于跨服务或跨模块复用 DTO 定义。

9. ParseFilePipe

ParseFilePipe 是专用于文件上传验证的管道，能够校验上传文件的类型、大小等信息。关于 ParseFilePipe 更详细的使用方法和特性，将在下一节进行深入介绍。

3.6.3　自定义管道

至此，我们已经探讨了 Nest 框架提供的数据验证器，这些工具在大多数业务场景中已经足够使用。然而，如果你面临特定的定制化需求，Nest 也支持通过自定义管道来满足这些需求。

为了加深对前面讨论的 ValidationPipe 的理解，我们将通过一个自定义实现的示例来演示其工作原理。示例代码如下：

```
import { PipeTransform, Injectable, ArgumentMetadata, BadRequestException } from '@nestjs/common';
import { validate } from 'class-validator';
import { plainToInstance } from 'class-transformer';
```

```
@Injectable()
export class ValidationPipe implements PipeTransform<any> {
  async transform(value: any, { metatype }: ArgumentMetadata) {
    // 如果是原生的 JavaScript 类型，则跳过验证
    if (!metatype || !this.toValidate(metatype)) {
      return value;
    }
    // 将普通对象转为类型化对象才能够验证
    const object = plainToInstance(metatype, value);
    // 调用验证方法
    const errors = await validate(object);
    if (errors.length > 0) {
      throw new BadRequestException('Validation failed');
    }
    return value;
  }

  private toValidate(metatype: Function): boolean {
    const types: Function[] = [String, Boolean, Number, Array, Object];
    return !types.includes(metatype);
  }
}
```

自定义管道需要实现 PipeTransform 接口的 transform 方法，其中 value 是我们需要处理的方法参数，而 metatype 是描述这个参数的元数据，用于标识当前处理的参数是原始 JavaScript 类型还是 DTO 类。在这里，我们需要验证 DTO 类型的参数，因此需要通过 toValidate 方法把原生 JavaScript 数据类型过滤掉，这就是验证器的原理。

在实际开发中，我们可以直接使用 Nest 中开箱即用的 ValidationPipe，它提供了丰富的选项配置，允许我们根据具体需求进行定制化设置。

3.7 Nest 实现文件上传

文件上传几乎是每个应用开发中必备的功能，包括单文件上传、多文件上传以及大文件的分片上传等多种场景。本节将通过 Nest 框架来学习如何实现最常见的单文件和多文件上传。

首先，我们将介绍 Nest 框架中内置的上传中间件 Multer。在前端部分，将使用 React 结合 Ant Design UI 组件库来模拟一个接近实际开发环境的模式，创建一个简单的上传页面。该页面将允许用户将文件上传到 Nest 后端项目中的 uploads 目录。

3.7.1 初识 Multer

Multer 是一个 Node.js 中间件，主要用于处理文件上传，接收以 multipart/form-data 格式发送的数据，在 HTTP 中通过 POST 请求来实现。它提供了单文件 .single(filename)、多文件 .array(filename[, maxCount])、混合文件 .fileds(fileds) 等多种形式的上传接口，并且支持自定义文件存储引擎和自定义文件存储位置，可以更好地控制上传内容。

Nest 在此基础上做了一层封装，通过装饰器模式让上传文件变得更加简单和灵活。下面我们

来创建一个前后端分离项目来进行演示。

先来创建 Nest 项目，执行"nest n nest-file-upload -p pnpm"命令，指定 pnpm 为项目包管理器。项目创建成功后如图 3-43 所示。

```
$ nest n nest-file-upload -p pnpm
⚡ We will scaffold your app in a few seconds..
CREATE nest-file-upload/.eslintrc.js (663 bytes)
CREATE nest-file-upload/.prettierrc (51 bytes)
CREATE nest-file-upload/README.md (3347 bytes)
CREATE nest-file-upload/nest-cli.json (171 bytes)
CREATE nest-file-upload/package.json (1957 bytes)
CREATE nest-file-upload/tsconfig.build.json (97 bytes)
CREATE nest-file-upload/tsconfig.json (546 bytes)
CREATE nest-file-upload/src/app.controller.spec.ts (617 bytes)
CREATE nest-file-upload/src/app.controller.ts (274 bytes)
CREATE nest-file-upload/src/app.module.ts (249 bytes)
CREATE nest-file-upload/src/app.service.ts (142 bytes)
CREATE nest-file-upload/src/main.ts (208 bytes)
CREATE nest-file-upload/test/app.e2e-spec.ts (630 bytes)
CREATE nest-file-upload/test/jest-e2e.json (183 bytes)

✓ Installation in progress... 🍵

🚀 Successfully created project nest-file-upload
```

图 3-43　创建后端项目

在 main.ts 主入口文件中将端口改为 8088，接下来前端通过代理这个端口请求接口，代码如下：

```
async function bootstrap() {
  const app = await NestFactory.create(AppModule);
  await app.listen(8088);
}
```

执行"pnpm start:dev"命令启动项目，如图 3-44 所示。

```
[22:46:01] File change detected. Starting incremental compilation...

[22:46:01] Found 0 errors. Watching for file changes.

[Nest] 50152  - 2024/01/11 22:46:01     LOG [NestFactory] Starting Nest application...
[Nest] 50152  - 2024/01/11 22:46:01     LOG [InstanceLoader] MulterModule dependencies initial
[Nest] 50152  - 2024/01/11 22:46:01     LOG [InstanceLoader] AppModule dependencies initialize
[Nest] 50152  - 2024/01/11 22:46:01     LOG [RoutesResolver] AppController {/}: +6ms
[Nest] 50152  - 2024/01/11 22:46:01     LOG [RouterExplorer] Mapped {/, GET} route +1ms
[Nest] 50152  - 2024/01/11 22:46:01     LOG [RouterExplorer] Mapped {/upload, POST} route +0ms
[Nest] 50152  - 2024/01/11 22:46:01     LOG [RouterExplorer] Mapped {/uploads, POST} route +0
```

图 3-44　启动后端项目

接下来创建前端 React 项目。首先，执行"create-react-app file-upload-front"命令初始化项目，一旦项目创建成功后，我们将继续依赖安装，需要用到 antd、@antd-design/icons 和 axios 这三个包。我们可以通过执行"pnpm add antd@4.x @ant-design/icons axios -S"命令来添加这些依赖，结果如图 3-45 所示。

```
Packages: +64
++++++++++++++++++++++++++++++++++++++++++++++++++++++++++++++++
Progress: resolved 1329, reused 1329, downloaded 0, added 64, done

dependencies:
+ @ant-design/icons 5.2.6
+ antd 4.24.15 (5.12.8 is available)
+ axios 1.6.5
```

图 3-45　安装依赖

最后，完善代理配置。在 React 项目中，反向代理通过 setupProxy.js 文件实现。在 src 目录下创建这个文件，配置代码如下：

```
// 引入 http-proxy-middleware，react 脚手架已经安装
const proxy = require("http-proxy-middleware");

module.exports = function (app) {
  app.use(
    // 遇见/api前缀的请求，就会触发该代理配置
    proxy.createProxyMiddleware("/api", {
      // 请求转发给谁
      target: "http://localhost:8088",
      // 控制服务器收到的请求头中 Host 的值
      changeOrigin: true,
      // 重写请求路径
      pathRewrite: { "^/api": "" },
    })
  );
};
```

配置完成后，执行"pnpm start"命令启动项目，效果如图 3-46 所示。

```
Compiled successfully!

You can now view file-upload-front in the browser.

  Local:            http://localhost:3000
  On Your Network:  http://192.168.0.100:3000

Note that the development build is not optimized.
To create a production build, use npm run build.

webpack compiled successfully
```

图 3-46　启动前端项目

3.7.2　单文件上传

在 Nest 框架中，上传单个文件的功能由 FileInterceptor()拦截器和@UploadedFile()装饰器共同实现。当 FileInterceptor 绑定到控制器方法时，它负责拦截请求中包含的文件，并将这些文件保存到预设的位置。而@UploadedFile 装饰器用于从请求中提取已上传的文件，以便在控制器方法中进行进一步的处理。

为了演示这一功能，我们可以在 app.controller.ts 文件中添加一个用于文件上传的接口。示例代码如下：

```
// 上传单个文件
@Post('/upload')
@UseInterceptors(FileInterceptor('file'))
uploadFile(@UploadedFile() file: Express.Multer.File, @Body() body) {
  console.log(file);
  return {
    message: '上传成功',
    file: file.filename
  }
}
```

此时 TypeScript 会提示类型错误，如图 3-47 所示。

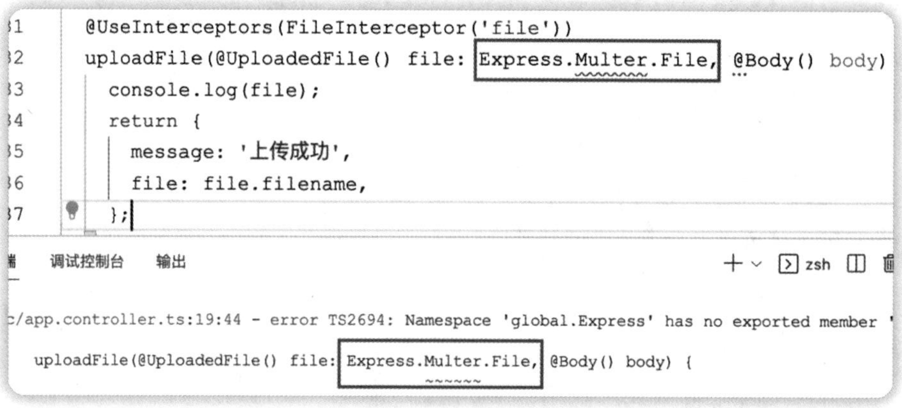

图 3-47　类型错误提示

执行"pnpm add @types/multer -D"命令，为 Multer 添加类型安全。然后重新执行"pnpm start:dev"命令启动服务。

我们还需要设置文件上传后的保存位置。Multer 中间件提供了文件存储引擎，在 Nest 中的 app.module.ts 中可以这样配置：

```
import { Module } from '@nestjs/common';
import { AppController } from './app.controller';
import { AppService } from './app.service';
import { MulterModule } from '@nestjs/platform-express';
import { diskStorage } from 'multer';
import { extname } from 'path';

@Module({
  imports: [
    MulterModule.register({
      // 定义存储引擎
      storage: diskStorage({
        // 定义文件存储的目录
        destination: './uploads',
        filename: (req, file, cb) => {
          // 创建随机文件名
          const randomName = Array(32)
```

```
            .fill(null)
            .map(() => Math.round(Math.random() * 16).toString(16))
            .join('');
          return cb(null, `${randomName}${extname(file.originalname)}`);
        },
      }),
    }),
  ],
  controllers: [AppController],
  providers: [AppService],
})
export class AppModule {}
```

在上述代码实现中,我们指定将上传的文件保存在 uploads 目录下。如果该目录不存在,Nest 框架将会自动创建它。为了确保文件名的一致性和避免潜在的命名冲突,接下来将对上传的文件进行重命名处理,然后将其保存到指定的目录中。

接下来完善前端代码,修改 App.js。示例代码如下:

```
import "./App.css";
import { UploadOutlined } from "@ant-design/icons";
import { Button, message, Upload } from "antd";
import { useState } from "react";
import axios from "axios";
// 上传单个文件
function fileUpload(data) {
  return axios.post("/api/upload", data, {
    headers: {
      "Content-type": "multipart/form-data",
    },
  });
}

const App = () => {
  const [fileList, setFileList] = useState([]);
  // 移除文件
  const handleRemove = (file) => {
    const files = fileList.filter((item) => item.uid !== file.uid);
    setFileList(files);
  };
  // beforeUpload 钩子
  const handleBeforeUpload = async (file) => {
    const formData = new FormData();
    formData.append("file", file);
    await fileUpload(formData);
    setFileList([...fileList, file]);
    message.success("上传成功");
  };

  return (
    <div className="App">
      <Upload
        fileList={fileList}
        onRemove={handleRemove}
```

```
        beforeUpload={handleBeforeUpload}
      >
        <Button icon={<UploadOutlined />}>选择文件</Button>
      </Upload>
    </div>
  );
};
export default App;
```

单击"选择文件"按钮上传文件,上传成功后将返回正确的结果,效果如图 3-48 所示。

图 3-48　单文件上传成功

上传成功后,在 uploads 文件夹中也可以看到刚刚上传的文件,如图 3-49 所示。

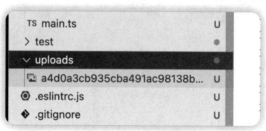

图 3-49　上传到指定文件夹

3.7.3　多文件上传

在成功实现单个文件上传的基础上,接下来我们探讨上传多个文件的场景。Nest 框架为此提供了两个专用的 API:FilesInterceptor()拦截器和@UploadedFiles()装饰器。与处理单个文件的 API 相比,它们在命名上仅多了一个 s,这表明它们用于处理多个文件的上传。为了演示这一功能,我们可以在 app.controller.ts 文件中添加相应的接口方法。示例代码如下:

```
// 上传多个文件(单表单域字段)
@Post('/uploads')
@UseInterceptors(FilesInterceptor('files', 3)) // 限制文件数量
uploadFiles(@UploadedFiles() files: Array<Express.Multer.File>, @Body() body) {
  console.log(files)
```

```
    return {
      files
    }
}
```

在以上代码中，files 用于接收表单域的字段，并且限制了最多上传 3 个文件，@Body 用于接收除 files 之外的其他参数。

在之前的基础上，修改前端代码如下：

```
import "./App.css";
import { UploadOutlined } from "@ant-design/icons";
import { Button, message, Upload } from "antd";
import { useState } from "react";
import axios from "axios";
// 上传多个文件（单表单域字段）
function filesUpload(data) {
  return axios.post("/api/uploads", data, {
    headers: {
      "Content-type": "multipart/form-data",
    },
  });
}

const App = () => {
  const [fileList, setFileList] = useState([]);
  // 移除文件
  const handleRemove = (file) => {
    const files = fileList.filter((item) => item.uid !== file.uid);
    setFileList(files);
  };
  // 保存选择的文件
  const handleBeforeUpload2 = async (file, files) => {
    setFileList([...fileList, ...files]);
  };
  // 上传多个文件
  const handleUploadAll = async () => {
    const formData = new FormData();
    fileList.forEach((item) => {
      formData.append("files", item);
    });
    await filesUpload(formData);
    message.success("上传成功");
  };

  return (
    <div className="App">
      <Upload
        multiple
        fileList={fileList}
        onRemove={handleRemove}
        beforeUpload={handleBeforeUpload2}
      >
        <Button icon={<UploadOutlined />}>选择文件</Button>
      </Upload>
```

```
      <br />
      <Button onClick={handleUploadAll}>单击上传多文件</Button>
    </div>
  );
};
export default App;
```

上面的代码应该很容易理解。相比单文件上传，这里首先在 beforeUpload 钩子中收集上传的文件，然后通过单击事件发送请求将文件上传到 uploads 接口中。

接下来，选择 3 个文件后，单击"单击上传多文件"按钮，执行结果如图 3-50 所示。

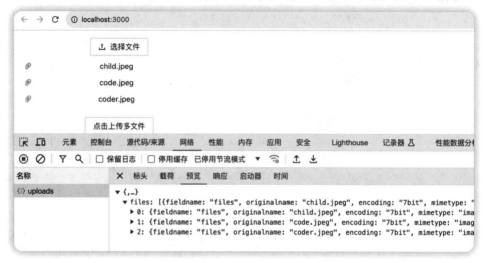

图 3-50　多文件成功上传

这是关于使用单表单域字段 files 上传的情况。另一种情景涉及多文件上传，需要使用不同的表单域字段进行上传。表单字段的设置如下：

```
const formData = new FormData();
formData.append("files1", files[0]);
formData.append("files2", files[1]);
```

在后端，为了接收多个文件字段，我们可以使用 FileFieldsInterceptor 拦截器来定义各个文件的字段名，并设置上传文件的数量限制。这样，我们可以在 app.controller.ts 中添加一个处理多文件上传的接口方法。示例代码如下：

```
// 上传多个文件（多表单域字段）
@Post('/fieldUploads')
@UseInterceptors(
  FileFieldsInterceptor([
    { name: 'files1', maxCount: 3 },
    { name: 'files2', maxCount: 3 },
  ]),
) // maxCount 限制文件数量
uploadFilesField(
  @UploadedFiles()
  files: { files1?: Express.Multer.File; files2?: Express.Multer.File },
  @Body() body,
) {
```

```
    console.log(files);
    return {
      files,
    };
  }
```

对应的前端代码如下：

```
import "./App.css";
import { UploadOutlined } from "@ant-design/icons";
import { Button, message, Upload } from "antd";
import { useState } from "react";
import axios from "axios";
// 上传多个文件（多表单域字段）
function fieldFilesUploads(data) {
  return axios.post("/api/fieldUploads", data, {
    headers: {
      "Content-type": "multipart/form-data",
    },
  });
}

const App = () => {
  const [fileList, setFileList] = useState([]);
  // 移除文件
  const handleRemove = (file) => {
    const files = fileList.filter((item) => item.uid !== file.uid);
    setFileList(files);
  };
  // 保存选择的文件
  const handleBeforeUpload2 = async (file, files) => {
    setFileList([...fileList, ...files]);
  };
  // 上传多个文件
  const handleUploadAll = async () => {
    const formData = new FormData();
    fileList.forEach((item) => {
      formData.append("files1", item);
      formData.append("files2", item);
    });
    await fieldFilesUploads(formData);
    message.success("上传成功");
  };

  return (
    <div className="App">
      <Upload
        multiple
        fileList={fileList}
        onRemove={handleRemove}
        beforeUpload={handleBeforeUpload2}
      >
        <Button icon={<UploadOutlined />}>选择文件</Button>
      </Upload>
      <br />
```

```
        <Button onClick={handleUploadAll}>单击上传多文件</Button>
      </div>
    );
};
export default App;
```

为了演示方便，上面的代码在 handleUploadAll 方法中设置了多个表单字段 files1、files2 来模拟多个表单域的情况。上传文件的效果如图 3-51 所示。

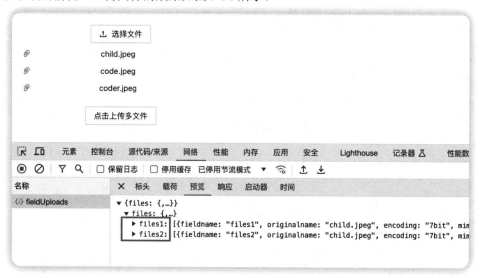

图 3-51　多文件上传成功（多字段）

3.7.4　上传任意文件

在某些情况下，如果不确定具体哪个字段用于表示文件数据，可以使用 AnyFilesInterceptor 拦截器来处理这种情况。此拦截器允许从请求的任何字段中拦截文件上传。示例代码如下：

```
// 上传任意文件
@Post('/anyUploads')
@UseInterceptors(
  AnyFilesInterceptor({
    // 限制文件数量
    limits: { files: 3 },
  }),
)
uploadAnyFiles(
  @UploadedFiles() files: Array<Express.Multer.File>,
  @Body() body,
) {
  console.log(files);
  return {
    files,
  };
}
```

对应的前端代码改动如下：

```
// 上传多个文件
const handleUploadAll = async () => {
  const formData = new FormData();
  fileList.forEach((item) => {
    formData.append("files1", item);
    formData.append("files2", item);
  });
  // 改为上传任何文件接口
  await anyFilesUpload(formData);
  message.success("上传成功");
};
```

上传效果如图 3-52 所示。

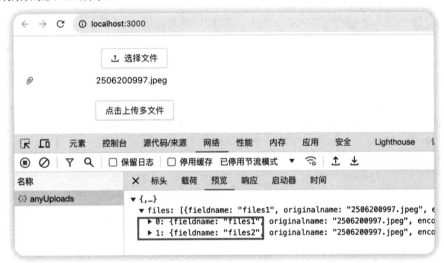

图 3-52　文件上传成功

至此，我们介绍了 Nest 框架中几种典型的文件上传方法。然而，为了构建一个健壮的文件上传功能，我们还需要进一步完善上传逻辑。仅仅限制文件数量是不够的，我们还需要加入文件验证机制，这包括对文件大小、类型等属性进行校验，以确保上传的文件符合特定的要求和标准。

3.7.5　文件验证

前面提到，文件验证通常通过管道来完成，Nest 中内置了用于文件验证的 ParseFilePipe 管道。接下来，在 anyUploads 接口中添加文件大小和文件类型验证逻辑。示例代码如下：

```
// 上传任何文件
@Post('/anyUploads')
@UseInterceptors(
  AnyFilesInterceptor({
    // 限制文件数量
    limits: { files: 3 },
  }),
)
uploadAnyFiles(
  @UploadedFiles(new ParseFilePipe({
    validators: [
```

```
      // 文件大小限制
      new MaxFileSizeValidator({maxSize: 1024 * 1000}),
      // 文件类型限制
      new FileTypeValidator({fileType: 'image/png'}),
    ]
  })) files: Array<Express.Multer.File>,
  @Body() body,
) {
  console.log(files);
  return {
    files,
  };
}
```

接下来上传一幅大小超过 1024KB 的图片，其运行结果如图 3-53 所示。

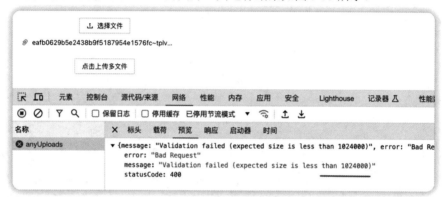

图 3-53　文件大小限制

如果类型错误，应怎么办呢？可以尝试上传一个类型不符合要求的文件，其运行结果如图 3-54 所示。

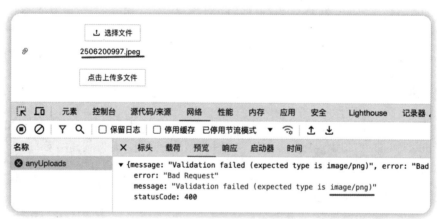

图 3-54　文件类型限制

在 Nest 框架中，文件验证主要依赖于这两个核心组件：文件大小验证和文件类型验证。结合我们之前讨论的自定义管道验证器的实现方法，你应该已经掌握了足够的知识来应用它们。现在，是时候将这些理论知识付诸实践了。我鼓励读者立刻开始，实现自己的文件验证逻辑！

第 2 部分

进 阶 篇

第 2 部分将深入探索 Nest 框架的进阶部分,学习如何有效集成 MySQL 和 Redis 等数据库技术,掌握后端服务中至关重要的身份验证和授权机制,并将 Nest 应用部署到特定的服务器环境。通过全面理解这些进阶概念,我们将具备开发和部署自己的 Nest 应用服务的能力。

第 4 章

Nest 与数据库

在前面的章节中,我们学习了 Nest 的基础部分。从本章开始,我们将深入探讨 Nest 的进阶功能,包括如何使用 MySQL 数据库进行数据持久化、使用 ORM(对象关系映射)来操作数据库、以及如何通过 Redis 缓存来提升性能。接着,我们将介绍应用中必不可少的身份验证与授权,展示不同的实现方式来管理用户权限。最后,我们会介绍后端常用的运维工具 Docker,展示如何利用它快速创建和部署后端服务。

4.1 快速上手 MySQL

MySQL 是目前广泛应用于各种 Web 应用的数据库之一,扮演着关键的数据存储和管理角色。对于任何资深的后端工程师而言,MySQL 都是一个不可或缺的工具。他们利用 MySQL 接收前端提交的数据,并高效地将其存储到数据库中,同时能够快速检索数据以供前端渲染使用。这些技能对我们后续章节的学习和项目实践至关重要。

本节将引导读者完成 MySQL 的安装过程,确保它能够在读者的个人电脑上正常运行。此外,我们将通过命令行和可视化工具这两种方式,演示 MySQL 的常用命令,帮助读者更好地理解和掌握这一强大的数据库管理系统。

4.1.1 安装和运行

MySQL 可以在各种操作系统上运行,如 Windows、Linux、macOS 等,也可以在容器平台如 Docker 和 K8s 上运行。接下来,我们把 MySQL 安装到个人主机上。在 MySQL 官网上,依次单击 DOWNLOADS→MySQL Community Downloads,选择适合自己主机的版本,然后单击 Download 按钮,如图 4-1 所示。

第 4 章 Nest 与数据库

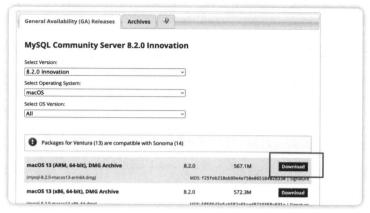

图 4-1　下载指定版本的 MySQL

接着会提示登录或者注册，这里可以直接选择下载，如图 4-2 所示。

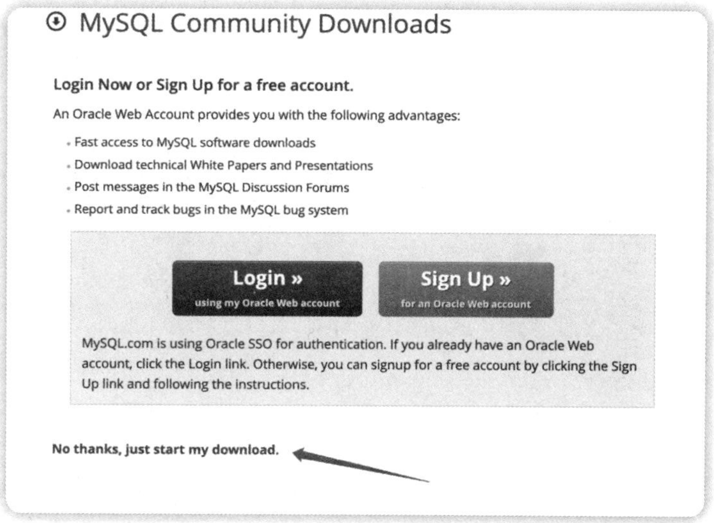

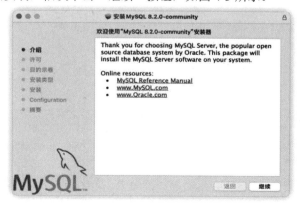

图 4-2　跳过登录注册

下载完成后，开始安装，依次单击"继续"按钮，如图 4-3 所示。

图 4-3　开始安装

到了这一步,需要输入你的数据库密码,如图 4-4 所示。连接数据库时需要用到这个密码,默认账号是 root。

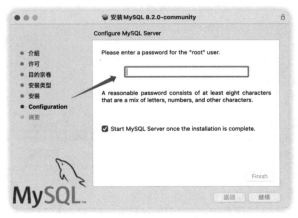

图 4-4 输入数据库密码

输入完成后,单击 Finish 按钮完成安装。MySQL 默认已经启动,我们可以在"系统偏好设置"选项的底部找到它,如图 4-5 所示。

图 4-5 MySQL 安装位置

单击 MySQL 图标,可以看到 MySQL 已经成功运行了,如图 4-6 所示。

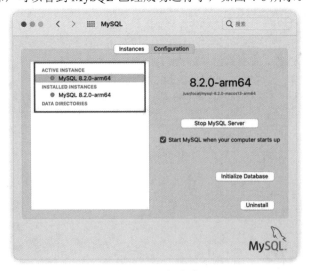

图 4-6 运行成功

如果读者在尝试启动 MySQL 时遇到问题，并且界面显示如图 4-7 所示，经过多次尝试仍无法成功启动，那么可能需要采取进一步的解决措施。建议在这种情况下，卸载当前版本的 MySQL，并重新安装一个与你的操作系统版本兼容的 MySQL 版本。

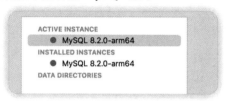

图 4-7 MySQL 版本与系统版本不匹配

接下来，我们分别通过命令行方式和可视化客户端来操作数据库。

4.1.2 MySQL 的常用命令

操作 MySQL 的一种方式是通过命令行工具。要连接到 MySQL 服务器，我们需要在系统的根目录输入命令“/usr/local/mysql/bin/mysql -u root –p”。

在此命令中，-u 参数后面跟着的是 MySQL 的账号名称，例如 root。执行该命令后，系统会提示你输入该账号的密码，如图 4-8 所示。请注意，具体的 MySQL 安装路径/usr/local/mysql/bin/可能会因安装方式和操作系统的不同而有所变化，因此读者可能需要根据自己的实际情况调整该路径。

```
~ 17:10:16
$ /usr/local/mysql/bin/mysql -u root -p
Enter password:
```

图 4-8 登录 MySQL

输入密码后，就可以进行命令行操作了。执行“show databases;”命令，可以查看默认存在的数据库，如图 4-9 所示。

```
Copyright (c) 2000, 2022, Oracle and/or its affiliates.

Oracle is a registered trademark of Oracle Corporation and/or its
affiliates. Other names may be trademarks of their respective
owners.

Type 'help;' or '\h' for help. Type '\c' to clear the current input statement.

mysql> show databases;
+--------------------+
| Database           |
+--------------------+
| information_schema |
| mysql              |
| performance_schema |
| sys                |
+--------------------+
4 rows in set (0.04 sec)

mysql>
```

图 4-9 查看数据库

我们创建了一个名为 nest-mysql 的数据库，执行“create database nest_mysql;”命令并展示，如图 4-10 所示。

```
mysql> create database nest_mysql;
Query OK, 1 row affected (0.00 sec)

mysql> show databases;
+--------------------+
| Database           |
+--------------------+
| information_schema |
| mysql              |
| nest_mysql         |
| performance_schema |
| sys                |
+--------------------+
5 rows in set (0.01 sec)
```

图 4-10 nest_mysql 数据库

接下来，进入这个数据库查看或创建表，执行"use nest_mysql;"命令使用数据库，并通过"show tables;"命令查看有哪些表，如图 4-11 所示。

```
mysql> use nest_mysql;
Database changed
mysql> show tables;
Empty set (0.01 sec)
```

图 4-11 查看数据表

创建 user 表并插入一些基础数据，运行如下命令：

```
create table user(id int(4) primary key, name char(20), age int(3));
```

其中，id、name、age 为表的列，int、char 为字段类型，primary key 是主键，不能重复。

接下来执行"describe user;"或"desc user;"命令，查看表结构，如图 4-12 所示。

```
mysql> create table user(id int(4) primary key, name char(20), age int(3));
Query OK, 0 rows affected, 2 warnings (0.01 sec)

mysql> desc user;
+-------+----------+------+-----+---------+-------+
| Field | Type     | Null | Key | Default | Extra |
+-------+----------+------+-----+---------+-------+
| id    | int      | NO   | PRI | NULL    |       |
| name  | char(20) | YES  |     | NULL    |       |
| age   | int      | YES  |     | NULL    |       |
+-------+----------+------+-----+---------+-------+
3 rows in set (0.00 sec)
```

图 4-12 表结构

图 4-12 中显示的数据类型，MySQL 的数据类型还有很多，诸如：

- 整数类型：TINYINT、SMALLINT、MEDIUMINT、INIT、BIGINT。
- 浮点数字类型：FLOAT、DOUBLE。
- 定点数字类型：DECIMAL、MUMARIC。
- 字符串类型：CHAR、VARCHAT、TEXT、BLOB。
- 日期类型：DATE、TIME、DATETIME、TIMESTAMP。

操作 MySQL 的命令也有很多，诸如：

- 创建数据库：

```
create database <数据库名>;
```

- 进入数据库：

```
use <数据库名>;
```

- 查看数据库：

```
show databases;
```

- 删除数据库：

```
drop database <数据库名>;
```

- 创建表：

```
create table 表名 (
  列名1 数据类型1 auto_increment,
  列名2 数据类型2,
  ...
  列名n 数据类型n,
  primary key (主键列名)
);
```

- 查看表结构：

```
desc <表名>;
```

- 修改表名称：

```
alter table <表名> rename <新表名>;
```

- 删除表：

```
drop table <表名>;
```

- 修改表字段信息：

```
alter table <表名> modify <列名> <数据类型>;
```

- 查询表数据：

```
select * from <表名>;
```

你可能已经注意到，每次通过命令行执行数据库操作可能会显得有些烦琐。尤其是当处理大量数据时，组织 SQL 语句并逐一执行它们可能是一项相当耗时且工作量巨大的任务。为了简化这一过程并提高效率，我们可以使用数据库可视化工具来操作数据库。这些工具通常提供了用户友好的界面，使得数据管理和 SQL 开发变得更加直观和便捷。

4.1.3 可视化操作 MySQL

MySQL 有许多 GUI 客户端可供选择，比如 MySQL 官方提供的 GUI 工具——MySQL Workbench、广泛使用的 Navicat 以及适用于 VS Code 的 MySQL 插件。在这里，我们将使用最熟悉的 VS Code 来演示。

在 VS Code 应用商店中搜索 mysql，从搜索结果中选择一个进行安装即可，如图 4-13 所示。

图 4-13 安装 MySQL 插件

安装完成后，从 VS Code 工具栏中找到该 MySQL 插件，并单击"+"图标以创建新的数据库连接。此时将会显示如图 4-14 所示的配置界面。

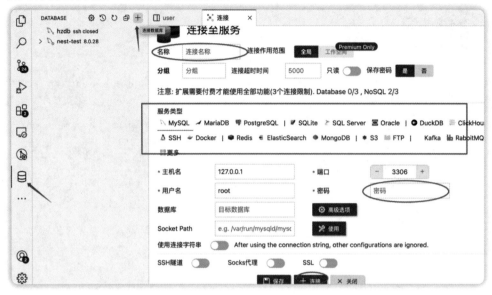

图 4-14　连接 MySQL

从图 4-14 可以看到，我们不仅可以连接 MySQL 数据库，还可以连接 MySQL 之外的各种数据库。在输入连接名称（例如案例中的 nest-test）、账户和密码后，单击下方的"连接"按钮即可完成连接。

一旦连接成功，可以打开之前创建的表，查看其表结构和新增的数据，如图 4-15 所示。

图 4-15　查看表数据

如前所述，MySQL 提供了多种数据类型和操作命令。然而，在当前阶段，我们并不需要记住这些细节。通过 GUI 面板，可以直观地进行数据库操作，因为插件为我们提供了一个包含丰富操作元素的用户界面，这简化了许多常规任务，如图 4-16 所示。

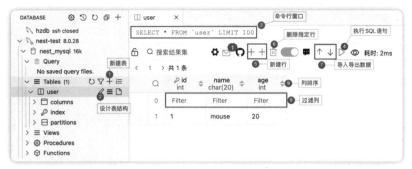

图 4-16　GUI 界面介绍

比如新建一个 product 表，需要编辑默认的 SQL 语句，如图 4-17 所示。

图 4-17　新建表

将其修改为如下的 SQL 语句：

```
CREATE TABLE product(
    id int NOT NULL PRIMARY KEY AUTO_INCREMENT COMMENT 'Primary Key',
    create_time DATETIME COMMENT 'Create Time',
    name VARCHAR(255)
) COMMENT '产品表';
```

单击 Execute 按钮执行语句，如图 4-18 所示，可以看到成功创建 product 表。这种方式相比手动编写 SQL 语句，确实方便得多，是吧？

接下来设计 product 表，新增 updateTime（创建时间）和 price（价格）字段，如图 4-19 所示。

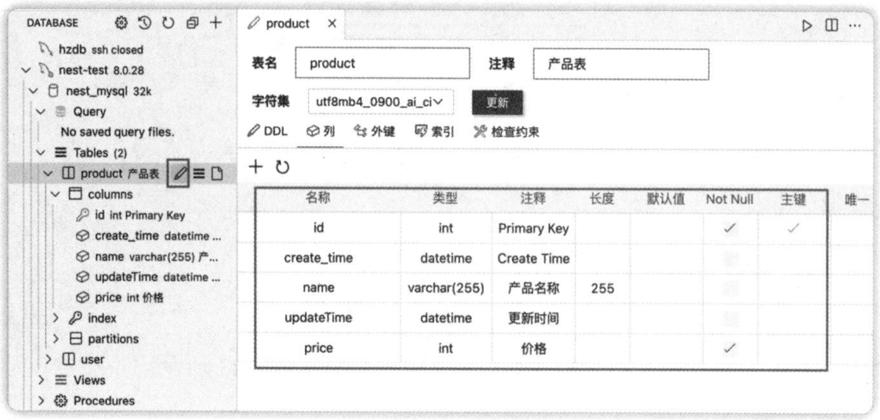

图 4-19　表结构设计

在数据库管理中，执行数据定义语言（Data Definition Language，DDL）语句，用于设置和管理表结构，包括定义外键、创建索引以及设置检查约束等。这些重要的概念将会在后续章节中进行详细介绍。

接下来，为了丰富我们的表结构，可以添加一个新的字段，例如 description（产品描述），如图 4-20 所示。

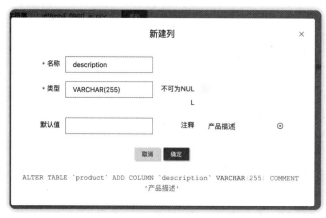

图 4-20　新建列

设计完表结构之后，回到数据表中，就可以进行插入数据、导入数据等操作了。插入数据的过程如图 4-21 所示。

图 4-21　插入数据

在图 4-21 中，id 不用填，会自动自增；price 为非空字段。

我们也可以通过指定 SQL 语句快速生成多条数据，如图 4-22 所示。

图 4-22　执行 SQL 语句生成数据

此外，还有更为高效的手段来辅助我们的开发和测试工作。我们可以执行"run mock"命令自动生成测试数据，这极大地简化了我们的日常测试流程。具体操作如图 4-23 所示。

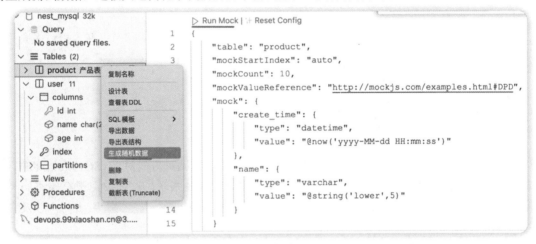

图 4-23　生成随机数据

接下来学习清空表。右击产品表，在弹出的快捷菜单中选择"截断表（Truncate）"，表中的数据就会被清空，如图 4-24 所示。

图 4-24　清空表

最后学习如何删除表。右击产品表，在弹出的快捷菜单中选择"删除"命令来删除表，如图 4-25 所示。

图 4-25　删除表

至此，我们已经学习了常用的数据库单表操作方法。在下一节中，我们将深入探讨多表关联以及复杂 SQL 查询的技巧，这些内容对于提升数据库操作的效率和灵活性至关重要。

4.2　MySQL 表之间的关系

在上一节中，我们学习了如何通过命令行和可视化工具这两种方式对 MySQL 数据库进行单表的增删改查操作。在实际业务场景中，数据库往往会包含多张表，这些表用于存储不同类型的信息，并在它们之间建立特定的联系。常见的关系类型包括一对一、一对多/多对一以及多对多关系。

本节内容将从实际应用出发，通过生活中的例子来具体说明这四种关系。我们会探讨这些关系是如何通过外键在数据表之间建立联系的，以及如何对涉及这些关系的数据库进行有效的操作和管理。

4.2.1　一对一关系

在数据库设计中，有些数据表之间是一对一的关系。以下是一些典型的一对一关系示例：

- 一个人拥有一个独一无二的身份证，而这个身份证也只属于这个人。
- 一个学生分配有一个学号，而这个学号在系统中仅对应这一个学生。
- 每个人的指纹是独一无二的，而每个指纹也只对应一个特定的人。

以身份证为例，我们可以设计两个表来维护用户信息和身份证信息，如图 4-26 所示。

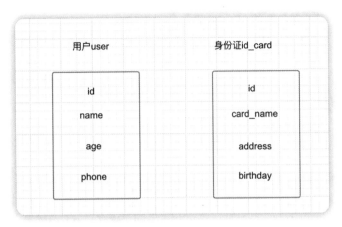

图 4-26　用户与身份证关系表

在此案例中，user 表拥有一个主键 id，该 id 是唯一的，用于标识单个用户。为了实现用户信息与身份证信息之间的一对一关联，我们可以在 id_card 表中引入一个外键字段 user_id，该字段用于存储对应的 user 表的 id 值。这样，通过 user 表的 id，我们可以快速地查询到与之关联的身份证信息，如图 4-27 所示。

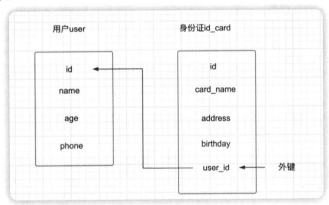

图 4-27　外键关联

在图 4-27 中，user_id 作为外键存在于 id_card 表中，它引用了 user 表的主键。在这种关系中，user 表被称为主表，而 id_card 表则被称为从表。这种设置定义了数据表之间的主从关系。

读者可能会问，如何确定哪个表应该作为主表，哪个表作为从表呢？在设计主从关系时，我们需要综合考虑多个因素，包括数据的唯一性、业务逻辑、数据表的访问频率、数据完整性以及系统的扩展性等因素。在我们的案例中，user 表在业务流程中扮演核心角色，它不仅包含更丰富的用户相关信息，而且由于访问频率较高，数据也更为完整。此外，user 表在业务发展中的扩展性也更强。

关于如何进行关联查询，我们将通过实际建表演示来详细说明。

1. JOIN 查询

在上一节中，我们已经创建了 user 表。接下来，为了生成 id_card 表，需要在数据库管理界面中单击"新建表"按钮。然后，根据你的需求修改系统默认的 SQL 语句。完成这些步骤后，运行修改后的 SQL 语句，即可创建新的 id_card 表，如图 4-28 所示。

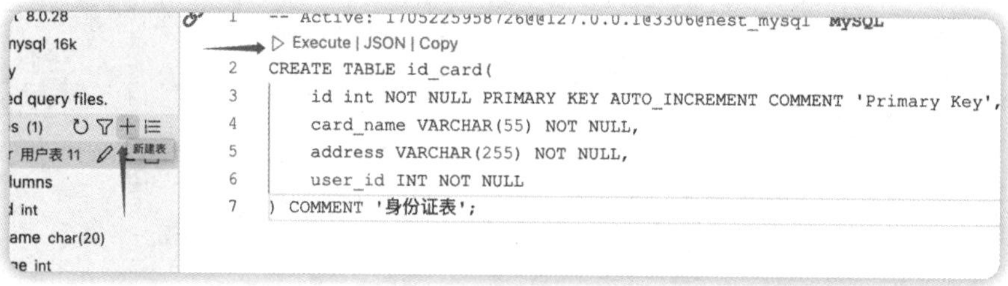

图 4-28 创建 id_card 表

在成功创建 id_card 表之后，接下来我们需要向该表添加一个外键 user_id，它将关联到 user 表的主键 id 上，以建立两个表之间的一对一关系。这样的设计允许 id_card 表通过 user_id 引用 user 表中的相应记录。具体操作如图 4-29 所示。

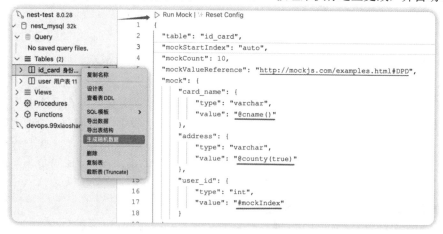

图 4-29 关联外键

需要注意的是，On Update 和 On Delete 选项暂时保持默认的 NO ACTION，这涉及级联操作，我们稍后会详细讨论。

接下来，为了向 id_card 表中添加随机数据，我们需要对 SQL 语句进行相应的修改，具体如图 4-30 所示。修改完成后，单击界面上的"Run Mock"按钮来执行这些更改，并自动填充表数据。

图 4-30 生成 id_card 表数据

在图 4-30 中，@cname 用于表示生成的随机名字，而@county 表示生成的随机地址。这些特定的 mock 值允许自动为数据库表填充示例数据。如果需要了解更多关于这些 mock 值的细节，可以参考 mockValueReference 提供的信息。完成这些步骤后，生成的数据如图 4-31 所示。

图 4-31 id_card 表数据

以同样的方式向 user 表中添加一些数据，具体如图 4-32 所示。

图 4-32 user 表数据

如何关联查询呢？执行下面这条 SQL 语句：

SELECT * FROM user JOIN id_card ON user.id = id_card.user_id;

执行结果如图 4-33 所示。

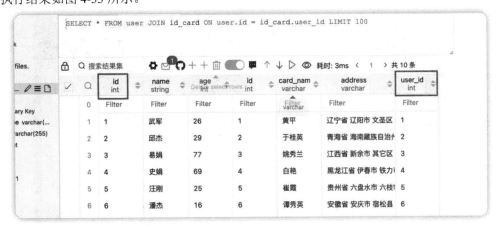

图 4-33 关联查询

可见 user 表和 id_card 表同时被查询出来了，这得益于 JOIN ON 语句，关联方式是 user.id = id_card.user_id，即主表中的 id 与从表中的 user_id 外键相关联。

2. 级联

在数据库设计中，我们经常会遇到需要定义级联操作的情况，特别是涉及主表（也称为父表）与从表（也称为子表）之间的关系时。级联操作定义了当父表中的记录发生变化时，子表应如何响应。具体来说，有以下几种可选项：

- RESTRICT：当父表中的某条记录被删除或更新时，子表中存在与之相关的记录，则会阻止父表的删除或更新操作，即不允许对父表进行删除或更新操作。
- CASCADE：当父表中的某条记录被删除或更新时，子表中与之相关的记录也会被自动删除或更新。这样可以确保数据的一致性，避免出现孤儿记录。
- SET NULL：当父表中的某条记录被删除或更新时，子表中与之相关的外键字段被设置为空值（NULL）。这样可以避免删除或更新父表记录时引发外键约束错误。
- NO ACTION：在 MySQL 中，NO ACTION 与 RESTRICT 具有相同的行为。

接下来，让我们通过实际操作来演示如何应用这些级联选项。首先，如果想要将级联选项设置为 CASCADE，则需要在 VS Code 的 MySQL 插件中进行一些调整。这通常涉及先删除现有的外键约束，如图 4-34 所示。

图 4-34　删除现有的外键约束

完成这一步骤后，我们可以重新定义外键，并设置其级联行为为 CASCADE，以确保在父表记录变更时，子表中的相关记录也会自动更新或删除，如图 4-35 所示。

图 4-35　设置 CASCADE 级联方式

> **注　意**
>
> 细心的读者可能会注意到，在图 4-35 中多了一个前文未提及的 SET DEFAULT 级联选项。这是因为 MySQL 默认情况下不支持该级联类型，而 VS Code 插件支持连接多种类型的数据库，如 Oracle、PostgreSQL 等，这些数据库中支持 SET DEFAULT 选项。该选项代表当父表中的某条记录被删除或更新时，子表中相关的外键字段将被设置为默认值。

此时尝试修改父表中的数据，切换到 user 表中，把 id=1 的记录修改为 id=100，并按 Enter 键确认，如图 4-36 所示。

图 4-36　修改父表中的数据

再切换到 id_card 子表中，发现 id=1 的记录中，user_id 已经被同步更新为 100，如图 4-37 所示。

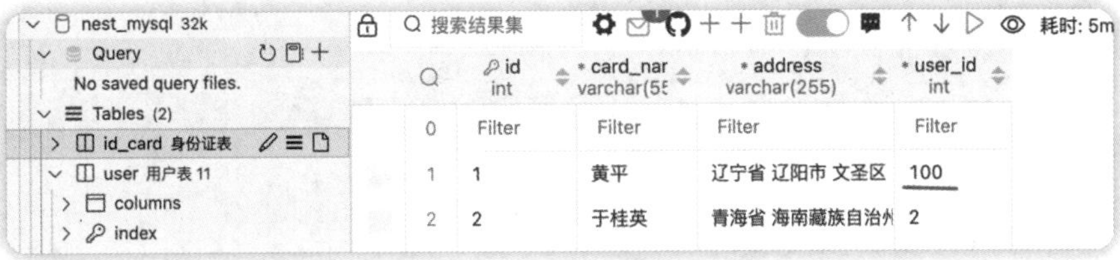

图 4-37　子表数据同步更新

同样地，CASCADE 级联选项也适用于删除操作。例如，在 user 表中，如果我们选择并删除了 id 等于 100 的记录，根据 CASCADE 规则，id_card 子表中与之关联的记录也会自动被删除，如图 4-38 所示。因此，在执行删除操作后，id_card 表中将不再存在 id_card 等于 100 的记录。

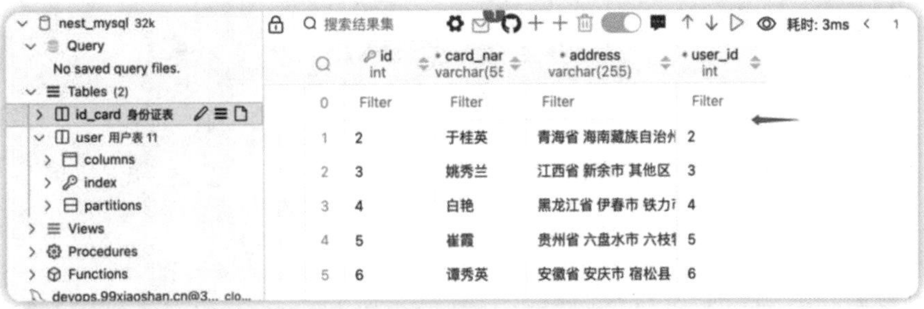

图 4-38　子表数据同步删除

接下来，让我们尝试创建一个新的外键，并将其级联方式设置为 SET NULL。然而，在单击"确定"按钮后，出现了一个错误，如图 4-39 所示。错误提示指出 user_id 字段不能为 NULL，这与我们选择的级联选项相冲突。因为 SET NULL 级联选项的意图是，当父表中的记录被更新或删除时，子表中的相关外键字段应被设置为 NULL。显然，如果 user_id 字段不允许为空，这就违反了外键约束的规则，导致操作无法进行。这就是所谓的外键约束冲突。

图 4-39　外键非空约束

要解决这个外键约束冲突，我们需要调整 user_id 字段的设置。首先，切换到数据库表设计的"列"标签页。接下来，在"列"栏中找到 user_id 字段，并修改其属性以允许为空（NULL）。这样，当父表记录被更新或删除时，子表中的 user_id 字段就可以被设置为 NULL，而不会违反外键约束。更改后的界面如图 4-40 所示。

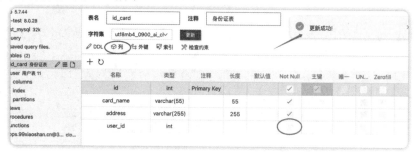

图 4-40　修改外键设置，允许设置为空

在成功删除旧的外键并创建新的外键之后，我们可以继续进行测试。现在，如果我们在主表中更新或删除 id 等于 2 的记录，应该可以观察到 id_card 表中相应的 user_id 外键字段被设置为 NULL。这一效果验证了 SET NULL 级联选项配置正确，并且外键约束得到了适当的处理。操作结果如图 4-41 所示。

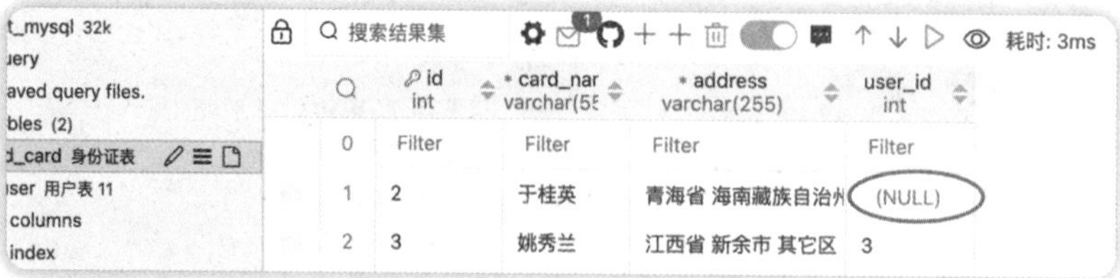

图 4-41　外键设置为 NULL

综上所述，MySQL 提供了多种级联选项，包括 RESTRICT、CASCADE、SET NULL 和 NO ACTION，以适应不同的业务需求和数据模型设计。在本案例中，考虑到当用户记录被删除时，其对应的身份证信息也随之失去实际用途，因此选择 CASCADE 作为级联方式是恰当的。这样，用户记录的删除将自动触发相关身份证信息的删除，确保数据的一致性和相关性得到维护。

4.2.2 一对多/多对一关系

除一对一关系外,生活中还普遍存在一对多或多对一的关系。以下是一些典型的例子:

- 一个公司有多个员工,公司通常只在一家公司工作。
- 一个学校有多名学生,每位学生通常只属于一个学校。
- 一个作者可以写多篇文章,文章也只属于一个作者。

在数据库设计中,一对多和多对一关系是相互关联的。为了在数据库中准确描述这种关系,我们通常会使用外键来实现。以公司与员工的关系为例,员工表中会包含一个指向公司表的外键,以此表明一名员工只属于一个公司,而公司可以有多个员工,如图 4-42 所示。这种设计允许我们维护数据的完整性并清晰地表示实体间的关联。

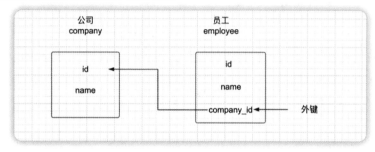

图 4-42　公司表与员工表关系

在数据库中,我们需要创建两个表:company 表和 employee 表。对于 employee 表,我们将设置一个外键 company_id,该外键引用 company 表的主键。此外,我们将此外键的级联方式设置为 CASCADE。这样的设计意味着,如果 company 表中的某条记录被删除,那么 employee 表中与之关联的所有员工记录也会被自动删除,以保证数据的一致性。同样,如果 company 表中的记录被更新(例如公司名称变更),employee 表中相应的外键字段也会自动更新,确保员工合同上的信息保持最新。

完成表的创建和数据插入后,company 表和 employee 表的结构将如图 4-43 所示。

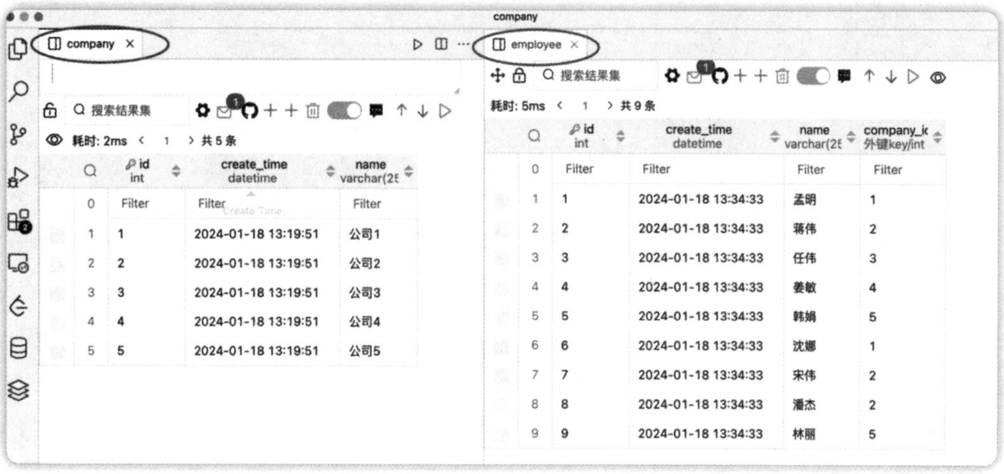

图 4-43　company 表和 employee 表的结构

根据图 4-43 的展示，我们可以看到 company 表和 employee 表都包含 id、create_time 和 name 列。除此之外，employee 表还增加了一个 company_id 列，作为外键关联到 company 表的主键 id 字段。

现在，为了查询特定公司（例如 id 为 2 的公司）的员工信息，我们将执行一个 SQL 查询，该查询将返回关键字段 id、name 和 company_id。以下是执行此查询的 SQL 语句：

```
SELECT
company.id,
company.name,
employee.id as employee_id,
employee.name as employee_name,
company_id
FROM company
JOIN employee
ON company.id = employee.company_id
WHERE company.id = 2
LIMIT 100
```

在涉及多个表的查询中，如果存在字段名称重复的情况，我们可以使用 AS 关键字对字段进行重命名，以确保查询结果的清晰性。通过执行带有 JOIN ... ON 条件的 SQL 语句，我们可以检索出 company 表中 id 等于 2 的公司所关联的员工数据。根据图 4-44 的展示，该查询结果显示，该公司共有 3 名员工。

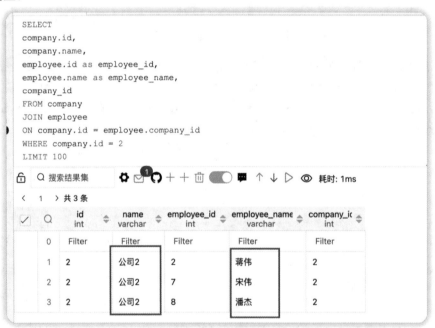

图 4-44　一对多级联查询

接下来，我们将执行删除操作。要删除 company 表中 id 等于 2 的公司记录，需要在数据库管理界面中选中该记录，然后单击"删除"按钮来执行删除操作。这一过程的结果将反映在数据库中，如图 4-45 所示。

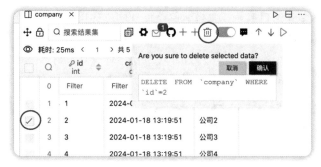

图 4-45　删除指定公司

接下来，单击界面上出现的"确认"按钮，以执行之前的删除查询语句。完成这一操作后，当我们再次查询数据库时，会发现先前与 id 等于 2 的公司记录相关联的所有数据都已被清空，这反映了 CASCADE 级联删除的效果。如图 4-46 所示，查询结果现在显示为空，这正是我们期望中 CASCADE 级联操作的结果。

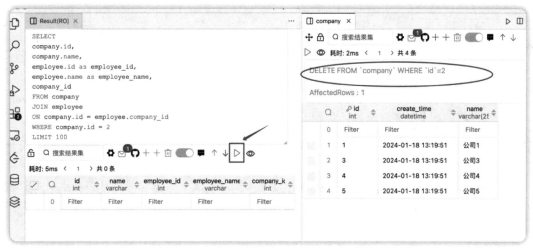

图 4-46　一对多级联删除

通过上述操作，你可能已经注意到，一对多/多对一关系与一对一关系在表操作方面有许多相似之处。确实，这两种关系在基本的数据库操作上可能看起来差不多，但在实际的数据库设计实践中，它们各自扮演着不同的角色。

一对一关系通常用于规范化数据，它有助于保持数据的一致性和减少冗余。而一对多关系则更常用于建立表之间的层次结构，这种结构可以清晰地表示实体间的从属或关联关系。

在本案例中，公司与员工表之间的关系就是一个典型的父子关系示例。company 表可以被视为父表，因为它包含了公司的整体信息；而 employee 表则作为子表，存储了属于各个公司的员工详细信息。通过在 employee 表中使用 company_id 外键，我们可以有效地表达和维护公司与员工之间的层次关系。

4.2.3　多对多关系

在数据库设计中，多对多关系是一种更为复杂的实体关系。以下是一些典型的多对多关系示例：

- 一个订单可以包含多种商品，而这些商品也可能出现在多个不同的订单中。
- 一个学生可以注册多门不同的课程，同时一门课程也可以有多名注册学生。
- 一篇文章可以被标记为多个不同的标签，而每个标签也可能被用来分类多篇文章。
- 一个用户可能拥有多个微信好友，同样，一个好友也可能与多个用户建立联系。

以订单和商品的关系为例，它们之间的多对多关系可以通过一个关联表来实现。这个关联表通常包含两个外键，分别指向订单表和商品表的主键。图 4-47 展示了这种关系是如何在数据库中建模的。

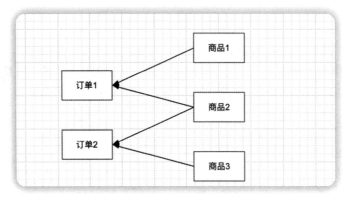

图 4-47　订单与商品的多对多关系

在现实世界中，多对多关系具有以下几个显著特点。

- 复杂性：多对多关系反映了两个或多个实体之间复杂的相互作用，在许多实际应用场景中非常常见。
- 中间表（关联表）：为了在数据库中实现多对多关系，通常需要创建一个中间表，也称为关联表。该表存储了参与关系的两个实体的主键作为外键，并可能包含其他与这种关系直接相关的属性信息。

图 4-48 将用于详细描述这种多对多关系的结构和工作方式，它将帮助我们更清晰地理解涉及的实体如何通过中间表相互关联。

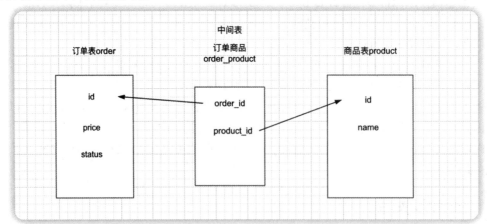

图 4-48　中间表

图 4-48 展示了 order 表和 product 表之间的关系。它们不是通过直接的外键关系来维护彼此之间的关系，而是通过一个名为 order_product 的中间表来维护对应的关系。这个中间表不仅维护了两个表之间的外键关系，还可以存储额外的信息，例如订单中商品的数量、价格和优惠折扣等。

为了演示多对多关系的实际应用，我们将创建这三个表：order、product 和 order_product。在创建 order_product 中间表时，我们将 order_id 和 product_id 都设置为表的主键，形成了一个复合主键。这样的设计是因为 order_product 表中的每条记录都唯一地关联了 order 表和 product 表中的记录，确保了这种组合的唯一性。

创建完成后的表结构和关系如图 4-49 所示，清晰地展示了多对多关系如何在数据库中得到实现和管理。

图 4-49 创建多对多关系表

接下来，我们需要为 order_product 中间表分别添加指向 order 表和 product 表的外键。这些外键将确保中间表能够维护与 order 表和 product 表的关联关系。此外，我们将这些外键的级联选项设置为 CASCADE。这意味着，如果 order 表或 product 表中的记录被删除，那么 order_product 中间表中相应的关联记录也将被自动删除。这种设置保证了数据的一致性，并确保了数据库中的数据关系保持有意义。

完成这些设置后，数据库的结构如图 4-50 所示。

图 4-50 设置中间表外键

三个表都创建完毕后，接下来我们将使用 mock 方式生成随机测试数据。在订单表中，status 取值范围限制在 0 至 2 之间，价格限制在 0 至 100 之间。mock 的配置如下：

```
{
    "table": "order",
```

```
    "mockStartIndex": "auto",
    "mockCount": 10,
    "mockValueReference": "http://mockjs.com/examples.html#DPD",
    "mock": {
        "price": {
            "type": "int",
            "value": "@integer(1,100)"
        },
        "status": {
            "type": "int",
            "value": "@integer(0,2)"
        }
    }
}
```

在商品表中随机生成"商品+序号"格式的数据,配置如下:

```
{
    "table": "product",
    "mockStartIndex": "auto",
    "mockCount": 10,
    "mockValueReference": "http://mockjs.com/examples.html#DPD",
    "mock": {
        "name": {
            "type": "varchar",
            "value": "商品 @increment()"
        }
    }
}
```

中间表中两个主键分别生成 0 至 10 范围内的数据,配置如下:

```
{
    "table": "order_product",
    "mockStartIndex": "auto",
    "mockCount": 10,
    "mockValueReference": "http://mockjs.com/examples.html#DPD",
    "mock": {
        "order_id": {
            "type": "int",
            "value": "@integer(1,10)"
        },
        "product_id": {
            "type": "int",
            "value": "@integer(1,10)"
        }
    }
}
```

在成功生成数据之后,接下来的问题是,我们如何进行多对多关系的查询操作呢?

正如前面所讨论的,我们可以使用 JOIN 查询来实现这一目标。不同于简单的一对一或一对多查询,多对多查询需要连接多个表。以查询特定订单(例如 id 等于 5 的订单)下所有关联的商品为例,我们需要执行一系列 JOIN 操作来关联 order 表、product 表以及 order_product 中间表。SQL 语句如下:

```
SELECT * FROM `order`
JOIN order_product ON order.id = order_id
JOIN product ON product.id = product_id
WHERE order.id = 2
LIMIT 100
```

执行 SQL 查询后，我们得到了结果：订单 id 为 2 的订单中绑定了商品 id 为 103 和 104 的两个商品，如图 4-51 所示。

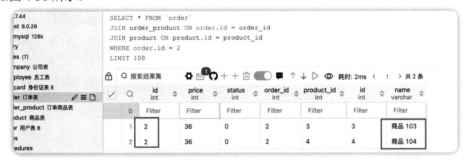

图 4-51　多对多查询结果

接下来选中 id=2 的这条订单记录执行删除操作，如图 4-52 所示。

图 4-52　删除 order 表数据

此时，执行 SQL 语句，可以看到中间表的查询结果为空，如图 4-53 所示，这说明 CASCADE 级联方式已经生效。

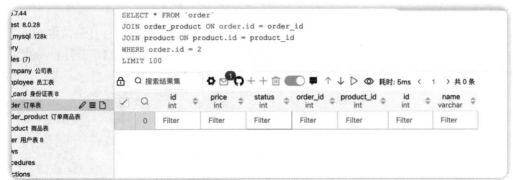

图 4-53　查询结果为空

以上内容阐述了如何通过数据表来实现生活中的多对多关系模型。掌握本节知识后，我们能够对生活中常见的实体关系进行有效的建模，例如朋友圈的互相点赞、社交网络中的好友关系以及权限与角色之间的关系等。这些知识能帮助读者更好地理解和分析业务场景。

4.3 快速上手 TypeORM

在 4.2 节中，我们学习了通过可视化界面和命令行两种方式来操作 MySQL 数据库中的单表及多表的增删改查。你可能会发现，命令行操作不仅需要记住各种 SQL 语法，而且在复杂场景下编写多表查询、子查询等语句也相当烦琐。ORM（Object-Relational Mapping，对象关系映射）框架可以有效地解决这类问题。本节将介绍主流 ORM 框架之一——TypeORM 的基本概念，并探讨它如何实现数据库操作。随后，我们将通过代码演示来学习 TypeORM 的使用方法，这将有助于我们在后续的 NestJS 项目中更有效地应用它。

4.3.1 基本概念

ORM 允许开发者通过创建一个"虚拟对象数据库"来简化数据库操作。这种技术将关系数据库的操作映射为对这些虚拟对象的操作，使得开发者可以直接在代码中进行数据库操作。目前，TypeORM、Sequelize 和 Prisma 是一些主流的 ORM 框架。TypeORM 特别值得注意，因为它是一个成熟的 Node.js ORM 框架，使用 TypeScript 编写，并且能够与 NestJS 框架友好集成。

4.3.2 项目准备

运行"npx typeorm init --name typeorm-test --database mysql2"命令来创建 TypeORM 项目，其中--name 为项目名称，--database 为数据库类型。

> **提示**
> 如果你尚未安装过 TypeORM 包，使用 npx 命令时，系统会先检测是否已安装该包。如果检测到未安装，npx 命令会询问你是否要进行安装。选择"是"后，脚本将自动安装包括"reflect-metadata"在内的相关依赖包。

创建项目成功后，打开终端并输入"cd typeorm-test"命令以进入项目目录。接下来，安装 mysql2 包，可以通过在终端运行"npm install mysql2 --save"命令来完成。这将安装 mysql2 并将其作为依赖项添加到你的 package.json 文件中，如图 4-54 所示。

```
$ npm install mysql2 -S
added 12 packages in 2m
```

图 4-54　安装 mysql2 包

4.3.3 创建模型及实体

要处理数据库，首先需要创建表。在 TypeORM 中，我们通过定义实体来告诉它如何创建一个数据表。以创建用户表为例，其基本模型定义如下：

```
export class User {
    id: number
    name: string
    age: number
    nickname: string
    phone: string
}
```

然而，仅有模型还不够，我们还需要将其定义为一个实体。这可以通过使用@Entity 装饰器来实现，代码如下：

```
import { Entity } from "typeorm"

@Entity()
export class User {
    id: number
    name: string
    age: number
    nickname: string
    phone: string
}
```

4.3.4 定义数据列及类型

创建数据表后，下一步是向表中添加列。在 TypeORM 中，我们使用装饰器来修饰实体类的属性，以定义列的特性。用于装饰列的装饰器包括@Column、@PrimaryColumn 和@PrimaryGeneratedColumn。其中，@PrimaryColumn 用于定义普通主键列，而@PrimaryGeneratedColumn 用于定义自增主键列。装饰器还允许我们指定列的数据类型。示例代码如下：

```
import { Entity, Column, PrimaryGeneratedColumn } from "typeorm"
@Entity()
export class User {
    @PrimaryGeneratedColumn()
    id: number

    @Column({
        length: 100,
    })
    name: string

    @Column()
    age: number

    @Column({
        length: 100
    })
    nickname: string

    @Column()
    phone: string

    @Column("text")
    desc: string
```

```
    @Column("double")
    other: number
}
```

请注意，string 类型默认映射为数据库中的 VARCHAR(255)类型（以 MySQL 为例），而 number 类型默认映射为整数类型。如果我们希望某个字段是文本（TEXT）类型或浮点（DOUBLE）类型，可以通过@Column 装饰器指定。例如，将 description 字段设置为 TEXT 类型，将 other 字段设置为 DOUBLE 类型。完成这些设置后，将上述代码替换到 src/entity/User.ts 文件中，实体的定义就完成了。

4.3.5 连接数据库

为了与数据库进行交互，我们需要设置一个数据源（DataSource）。数据源包含了所有与数据库连接相关的配置信息。通过这些配置，我们可以建立与数据库的初始连接或连接池。示例代码如下：

```
import "reflect-metadata"
import { DataSource } from "typeorm"
import { User } from "./entity/User"

export const AppDataSource = new DataSource({
    type: "mysql",
    host: "localhost",
    port: 3306,
    username: "root",
    password: "你的数据库密码",
    database: "typeorm_mysql",
    entities: [User],
    synchronize: true,
    logging: true,
})
```

在以上代码中，数据库配置属性说明如下。

- **type**：表示数据库类型，例如 mysql、oracle、sqlite、mongodb 等。
- **database**：表示你需要连接的数据库名。
- **entities**：用于定义需要加载的实体，它接收实体类或目录路径，如./**/entity/*.ts。
- **synchronize**：表示在应用启动时是否自动创建和更新数据库结构。
- **logging**：用于启动日志记录，记录实际执行了哪些 SQL 语句。
- **host、port、username、password**：为数据库的基础配置，包括数据库服务器地址、端口、用户名和密码。

创建 DataSource 实例后，需要调用该实例的 initialize 方法来建立数据库连接。一旦数据库连接成功，我们可以使用实体管理器（manager）来执行数据的插入和查询操作。修改 src/index.ts 的代码如下：

```
import { AppDataSource } from "./data-source"
import { User } from "./entity/User"

async function bootstrap() {
```

```
    await AppDataSource.initialize()
    const user = new User()
    user.name = "mouse"
    user.nickname = "奶酪"
    user.age = 25
    user.phone = '13000000000'
    user.desc = '默认描述'
    user.other = 1.0
    // 插入数据
    await AppDataSource.manager.save(user)
    // 查询数据
    const users = await AppDataSource.manager.find(User)
    console.log("查询用户: ", users)
}
bootstrap()
```

运行"npm run start"命令启动服务，可以看到成功插入数据并打印查询结果，如图 4-55 所示。

```
                    `rc`.`CONSTRAINT_NAME` = `kcu`.`CONSTRAINT_NAME`
query: SELECT VERSION() AS `version`
query: SELECT * FROM `INFORMATION_SCHEMA`.`COLUMNS` WHERE `TABLE_SCHEMA` = 'typeorm_mysql' AND `TABLE_NAME` = 'typeorm_metadata
query: COMMIT
query: START TRANSACTION
query: INSERT INTO `user`(`id`, `name`, `age`, `nickname`, `phone`, `desc`, `other`) VALUES (DEFAULT, ?, ?, ?, ?, ?, ?) -- PARA
METERS: ["mouse",25,"奶酪","13000000000","默认描述",1]
query: COMMIT
query: SELECT `User`.`id` AS `User_id`, `User`.`name` AS `User_name`, `User`.`age` AS `User_age`, `User`.`nickname` AS `User_ni
ckname`, `User`.`phone` AS `User_phone`, `User`.`desc` AS `User_desc`, `User`.`other` AS `User_other` FROM `user` `User`
查询用户: [
  User {
    id: 2,
    name: 'mouse',
    age: 25,
    nickname: '奶酪',
    phone: '13000000000',
    desc: '默认描述',
    other: 1
  }
]
```

图 4-55　TypeORM 插入、查询数据

4.3.6　使用 Repository 操作 CRUD

除使用实体管理器（EntityManager）来管理实体外，我们还可以使用存储库（Repository）来管理每个实体。下面通过代码演示如何进行数据库的增删改查操作。

首先是使用 Repository 进行新增和查询操作，代码如下：

```
import { AppDataSource } from "./data-source"
import { User } from "./entity/User"

async function bootstrap() {
    // 使用存储库方式
    await AppDataSource.initialize()
    const user = new User()
    user.name = "mouse2"
    user.nickname = "奶酪 2"
    user.age = 25
    user.phone = '13000000000'
    user.desc = '默认描述 2'
    user.other = 1.0
```

```
    const userRepository = AppDataSource.getRepository(User)
    // 新增
    await userRepository.save(user)
    console.log('新增用户成功');
    // 查询
    const saveUsers = await userRepository.find()
    console.log('查询用户: ', saveUsers);
}
bootstrap()
```

运行"npm run start"命令，控制台打印结果如图 4-56 所示。

图 4-56 新增并查询数据

使用 VS Code 的 MySQL 可视化插件也可以查询到这条数据，如图 4-57 所示。

图 4-57 可视化数据结果

接着尝试更新 id 为 1 的记录中的 name 和 nickname 字段。在 Repository 中，可以这样操作：

```
const userUpdate = await userRepository.findOneBy({
    id: 1,
})
userUpdate.name = 'mouse3'
userUpdate.nickname = '奶酪 3'
await userRepository.save(userUpdate)
console.log('更新成功');
```

再次运行"npm run start"命令，运行结果如图 4-58 所示，可以看到成功更新了 name 和 nickname 字段。

图 4-58　更新数据

最后，在 Repository 中，删除 id 为 1 的记录，代码如下：

```
// 删除
const userDelete = await userRepository.findOneBy({
    id: 1,
})
await userRepository.delete(userDelete)
console.log('删除成功');
```

成功删除数据，结果如图 4-59 所示。

图 4-59　删除数据

至此，我们完成了在 Repository 中进行数据库的增删改查操作，它就像操作对象一样简单。

4.3.7　使用 QueryBuilder 操作 CRUD

QueryBuilder 是 TypeOrm 中最强大的功能之一，它提供了更加底层的数据库查询操作，灵活度高，支持组合更复杂的 SQL 查询语句，如多条件查询或多表查询等，在本书的项目实战中会频繁使用它。

接下来介绍创建 QueryBuilder 的 3 种方式。

第一种是使用 DataSource 来创建，例如：

```
const user = await dataSource
    .createQueryBuilder()
    .select("user")
    .from(User, "user")
    .where("user.id = :id", { id: 1 })
    .getOne();
```

这种方式提供了较大的灵活性，适合跨实体的复杂查询操作和数据库级别的操作。

第二种是使用 Repository 来创建，例如：

```
const user = await dataSource
    .getRepository(User)
    .createQueryBuilder("user")
```

```
    .where("user.id = :id", { id: 1 })
    .getOne();
```

通过 Repository 创建的 QueryBuilder 比较常用，它直接与实体相关联，提供了比直接使用 Repository 方法更加灵活的数据库查询能力。

第三种是使用 Manager 来创建，例如：

```
const user = await dataSource.manager
    .createQueryBuilder(User, "user")
    .where("user.id = :id", { id: 1 })
    .getOne();
```

这种方式并不常用，因此这里不做过多介绍。

接下来通过代码演示 QueryBuilder 的基本用法，首先使用 DataSource 方式新增一个用户，代码如下：

```
await AppDataSource.initialize()
const user = new User()
user.name = "mouse3"
user.nickname = "奶酪 3"
user.age = 25
user.phone = '13000000000'
user.desc = '默认描述 3'
user.other = 1.0

const queryBuilder = AppDataSource.createQueryBuilder()
// 新增
await queryBuilder.insert().into(User).values(user).execute()
console.log('新增用户成功');
```

以上代码使用 insert 和 into 方法进行插入操作，语法接近原始 SQL 语句。在 values 方法中可以传入对象或数组，如果要插入多条记录，则选择传入一个数组。最后调用 execute 方法执行语句。

运行“npm run start”命令后，结果如图 4-60 所示，可以看到成功插入了一条数据。

```
query: SELECT VERSION() AS `version`
query: SELECT * FROM `INFORMATION_SCHEMA`.`COLUMNS` WHERE `TABLE_SCHEMA` = 'typeorm_mysql' AND `TABLE_NAME` = 'ty
query: COMMIT
query: INSERT INTO `user`(`id`, `name`, `age`, `nickname`, `phone`, `desc`, `other`) VALUES (DEFAULT, ?, ?, ?, ?,
ouse3",25,"奶酪3","13000000000","默认描述3",1]
新增用户成功
```

图 4-60　新增数据

接下来查询插入的数据，相关代码如下：

```
// 查询
const saveUsers = await queryBuilder.select('u').from(User, 'u').getMany()
console.log('查询用户：', saveUsers);
```

同样地，查询操作使用类似于“select from”的 SQL 语句，u 是 User 实体的别名，getMany 方法可以把多条记录查询出来，查询结果的数据字段名与实体表中定义的字段名相同，如图 4-61 所示。

```
query: SELECT `u`.`id` AS `u_id`, `u`.`name` AS `u_name`, `u`.`age` AS `u_age`, `u`.`nickname` AS `u_nickname`, `u`.`phone`
`u`.`desc` AS `u_desc`, `u`.`other` AS `u_other` FROM `user` `u`
查询用户: [
  User {
    id: 1,
    name: 'update name',
    age: 25,
    nickname: '奶酪1',
    phone: '13000000000',
    desc: '默认描述1',
    other: 1.25
  },
  User {
    id: 2,
    name: 'mouse1',
    age: 25,
    nickname: '奶酪1',
    phone: '13000000000',
    desc: '默认描述1',
    other: 1.25
```

图 4-61　getMany 查询结果

TypeORM 还提供了 getRawMany 方法，允许我们执行更底层的数据库查询并获取原始数据，查询结果如图 4-62 所示。

```
query: SELECT `u`.`id` AS `u_id`, `u`.`name` AS `u_name`, `u`.`age` AS `u_age`, `u`.`nickname` AS `u_nickname`, `u`.`phon
`u`.`desc` AS `u_desc`, `u`.`other` AS `u_other` FROM `user` `u`
查询用户: [
  {
    u_id: 1,
    u_name: 'update name',
    u_age: 25,
    u_nickname: '奶酪1',
    u_phone: '13000000000',
    u_desc: '默认描述1',
    u_other: 1.25
  },
  {
    u_id: 2,
    u_name: 'mouse1',
    u_age: 25,
    u_nickname: '奶酪1',
    u_phone: '13000000000',
    u_desc: '默认描述1',
    u_other: 1.25
  },
```

图 4-62　getRawMany 查询结果

使用此方法时，返回的数据通常包含数据库查询的实际结果，其中字段名可能会与自定义的表别名拼接在一起。

接下来修改用户信息，先查询出指定 id 为 4 的用户，修改信息后，再进行更新操作，代码如下：

```
// 修改
const userUpdate = await queryBuilder.select('u').from(User, 'u').where('u.id = :id',
{ id: 4 }).getOne()
  userUpdate.name = 'mouse4'
  userUpdate.nickname = '奶酪 4'
  queryBuilder.update(User).set(userUpdate).where('u.id = :id', { id: 4 }).execute()
  console.log('更新成功');
```

在上述代码中，首先使用 where 子句添加查询条件，并通过 getOne 方法获取满足条件的单个记录。如果找到了用户，修改其 name 和 nickname 属性，然后使用 update 和 set 方法来更新数据库中的记录，如图 4-63 所示。注意，set 方法需要一个对象，包含要更新的字段和新值。

```
query: SELECT `u`.`id` AS `u_id`, `u`.`name` AS `u_name`, `u`.`age` AS `u_age`, `u`.`nickname` AS `u_nickname`, `u`.`phone`
`u`.`desc` AS `u_desc`, `u`.`other` AS `u_other` FROM `user` `u` WHERE `u`.`id` = ? -- PARAMETERS: [4]
更新成功
query: UPDATE `user` SET `id` = ?, `name` = ?, `age` = ?, `nickname` = ?, `phone` = ?, `desc` = ?, `other` = ? WHERE `id` =
s: [4,"mouse4",25,"奶酪 4","13000000000","默认描述3",1,4]
```

图 4-63　修改用户信息

最后删除用户，把刚刚测试的 id 为 4 的用户删除，代码如下：

```
// 删除
await queryBuilder.delete().from(User).where('id = :id', { id: 4 }).execute()
console.log('删除成功')
```

运行结果如图 4-64 所示。

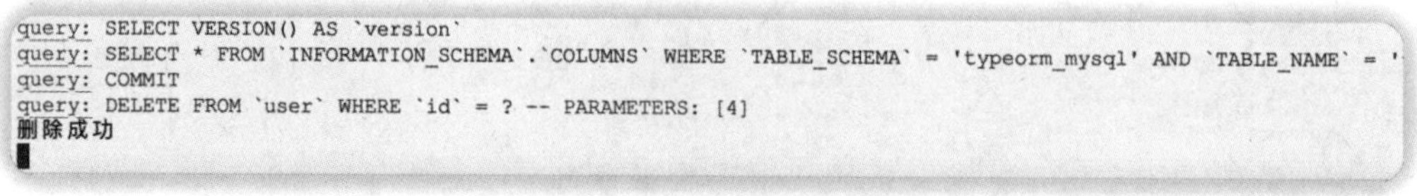

```
query: SELECT VERSION() AS `version`
query: SELECT * FROM `INFORMATION_SCHEMA`.`COLUMNS` WHERE `TABLE_SCHEMA` = 'typeorm_mysql' AND `TABLE_NAME` =
query: COMMIT
query: DELETE FROM `user` WHERE `id` = ? -- PARAMETERS: [4]
删除成功
```

图 4-64　删除用户

至此，我们完成了用 TypeORM 来创建数据模型和实体，并演示了如何连接数据库服务。最后，通过使用 TypeORM 提供的实体管理器（EntityManager）、存储库（Repository）及查询构建器（QueryBuilder）三种方式对数据库进行 CRUD 操作。这些操作不仅帮助我们进一步理解 ORM 框架的使用，也大大简化了操作数据库的过程。

4.4　使用 TypeORM 处理多表关系

MySQL 中表之间存在多种关系，如一对一、一对多/多对一以及多对多关系。本节将基于第 4.2 节中提到的实体关系示例，演示如何在 TypeORM 中通过表来描述这些关系。

4.4.1　一对一关系

本小节介绍数据库中表和表的一对一关系，以用户表和身份证表为例。在 src/entity 文件夹下创建一个 IdCard 实体，代码如下：

```
import {
  Entity,
  Column,
  PrimaryGeneratedColumn,
  OneToOne,
  JoinColumn,
} from "typeorm";
import { User } from "./User";

@Entity()
export class IdCard {
  @PrimaryGeneratedColumn()
  id: number;

  @Column()
  name: string;

  @Column()
```

```
  address: string;

  @Column()
  birthday: Date;

  @Column()
  email: string;

  @OneToOne(() => User)
  @JoinColumn()
  user: User;
}
```

在上述代码中,我们使用@OneToOne 装饰器来建立与 User 实体之间的一对一关系,同时用@JoinColumn 装饰器来定义 user 列,以维护一个外键。TypeORM 会自动生成外键的 id。

注意:请不要忘记把 IdCard 实体添加到 DataSource 配置选项 entities 中,并建议配置为./**/entity/*.ts。

接下来,在 index.ts 中添加以下的一段逻辑,分别创建用户表和身份证表,并插入数据,并确保把它们关联起来。示例代码如下:

```
// 一对一关系
// 创建一个用户
await AppDataSource.initialize()
const user = new User()
user.name = "mouse1"
user.nickname = "奶酪 1"
user.age = 25
user.phone = '13000000000'
user.desc = '默认描述 1'
user.other = 1.25
// 创建一个身份证信息
const idCard = new IdCard()
idCard.name = "my name is mouse"
idCard.address = "广东省广州市"
idCard.birthday = new Date()
idCard.email = 'jmin95@163.com'
// 关联两个实体
idCard.user = user

// 获取实体的存储库
const userRepository = AppDataSource.getRepository(User)
const idCardRepository = AppDataSource.getRepository(IdCard)
// 首先保存用户
await userRepository.save(user)

// 用户已保存。现在我们需要保存用户的身份证信息
await idCardRepository.save(idCard)

// 完成
```

```
console.log("数据已保存，并且在数据库中创建了用户与身份证之间的关联关系")
```

在这里，我们分别创建了一个名为 user 的用户和一个身份证信息，随后将它们关联起来，再调用存储库的 save 方法来保存这些信息。

运行 "npm run start" 命令，我们可以看到成功创建了两个表，如图 4-65 所示。

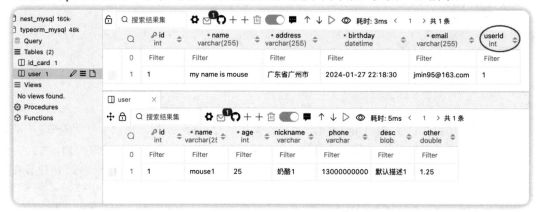

图 4-65　创建一对一表

由图 4-65 可见，id_card 表中已生成了 userId 字段（外键），它关联的是 user 表中的 id 字段。在前面的学习中我们已了解 JOIN 查询，接下来执行下面的 SQL 语句来查询数据：

```
SELECT * FROM user JOIN id_card ON user.id = id_card.`userId`;
```

执行结果如图 4-66 所示。

图 4-66　SQL 命令一对一查询

当然，我们也可以用 Repository 提供的 find 方法来查询。由于两个实体的关系是单向的，即 IdCard 实体持有 User 实体的外键列，使得它有权访问 User，但反过来 User 不知道 IdCard 的存在。因此，在这种情况下应该使用 idCardRepository 来查询，并且设置关联（relations），在 index.ts 中添加如下代码：

```
const idCardRes = await idCardRepository.find({
    relations: {
        user: true
    }
})
console.log(idCardRes);
```

再次运行 "npm run start" 命令，成功查询出两条关联了 user 字段的记录，如图 4-67 所示。

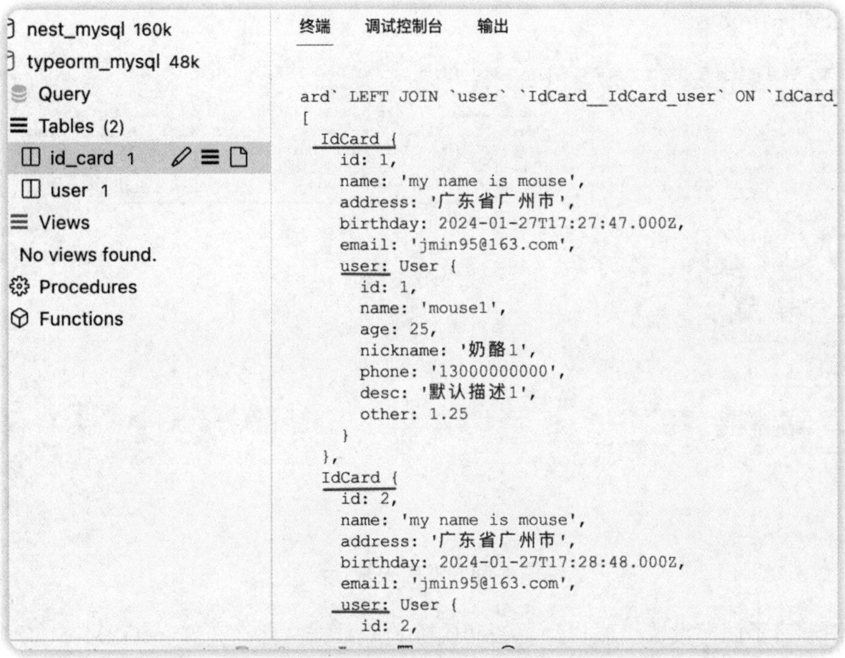

图 4-67　Repository 一对一查询

为了让 User 实体能够访问 IdCard 实体，我们需要修改实体类以建立双向关联，代码如下：

```
@Entity()
export class IdCard {
    /* 其他列 */

    @OneToOne(() => User, (user) => user.card)
    @JoinColumn()
    user: User
}


@Entity()
export class User {
    /* 其他列 */

    @OneToOne(() => IdCard, (card) => card.user)
    card: IdCard
}
```

此时，用 userRepository 加载关联表的数据，代码如下：

```
// 双向关联
const userRes = await userRepository.find({
    relations: {
        card: true
    }
})
console.log(userRes);
```

运行结果如图 4-68 所示。

```
query: COMMIT
数据已保存，并且在数据库中创建了用户与身份证之间的关联关系
query: SELECT `User`.`id` AS `User_id`, `User`.`name` AS `User_name`, `User`.`age` AS `User_age`, `User`.`nicknam
e` AS `User_nickname`, `User`.`phone` AS `User_phone`, `User`.`desc` AS `User_desc`, `User`.`other` AS `User_othe
r`, `User__User_card`.`id` AS `User__User_card_id`, `User__User_card`.`name` AS `User__User_card_name`, `User__Us
er_card`.`address` AS `User__User_card_address`, `User__User_card`.`birthday` AS `User__User_card_birthday`, `Use
r__User_card`.`email` AS `User__User_card_email`, `User__User_card`.`userId` AS `User__User_card_userId` FROM `us
er` `User` LEFT JOIN `id_card` `User__User_card` ON `User__User_card`.`userId`=`User`.`id`
[
  User {
    id: 1,
    name: 'mouse1',
    age: 25,
    nickname: '奶酪1',
    phone: '13000000000',
    desc: '默认描述1',
    other: 1.25,
    card: IdCard {
      id: 1,
      name: 'my name is mouse',
      address: '广东省广州市',
      birthday: 2024-01-28T07:02:38.000Z,
      email: 'jmin95@163.com'
    }
  }
]
```

图 4-68 双向关系查询

前面我们学习了如何设置级联自动更新和删除关联数据，那么在 TypeORM 中如何实现呢？可以在 User 实体中定义级联选项为 cascade: true，代码如下：

```
@Entity()
export class User {
    /* 其他列 */

    @OneToOne(() => IdCard, (card) => card.user, {
        cascade: true
    })
    card: IdCard;
}
```

此时保存或删除 user 对象会自动更新 idCard 对象。我们来看下面这段代码：

```
// 级联保存、更新数据
// 创建一个用户
await AppDataSource.initialize()
const user = new User()
user.name = "mouse1"
user.nickname = "奶酪1"
user.age = 25
user.phone = '13000000000'
user.desc = '默认描述1'
user.other = 1.25
// 创建一个身份证信息
const idCard = new IdCard()
idCard.name = "my name is mouse"
idCard.address = "广东省广州市"
idCard.birthday = new Date()
idCard.email = 'jmin95@163.com'
// 注意这里需要通过 user 来关联 idCard
```

```
user.card = idCard

// 获取实体的存储库
const userRepository = AppDataSource.getRepository(User)
// 保存用户
await userRepository.save(user
```

在这里，我们分别创建了一个用户和一个身份证信息，并使用 user 实体中的 card 字段关联 idCard 对象。接着通过 user 实体保存 user 实例对象。删除之前的表后，重新运行"npm run start"命令，运行结果如图 4-69 所示。

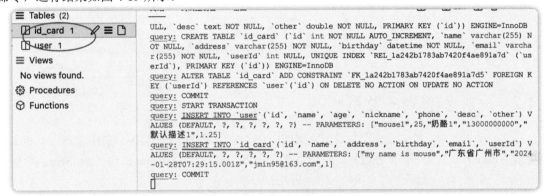

图 4-69 级联创建

从图 4-69 中可以看出，TypeORM 成功地执行了 SQL 语句并创建了两个表。有了数据后，我们接下来演示如何更新数据，代码如下：

```
// 查询 id 为 1 的数据
const loadedUser = await userRepository.findOne({
    where: {
        id: 1
    },
    relations: {
        card: true
    }
})
console.log(loadedUser);
// 更新字段
loadedUser.name = 'update name'
loadedUser.card.name = 'new idCard name'

await userRepository.save(loadedUser)
```

运行结果如图 4-70 所示，id_card 表和 user 表中的 name 字段已经被成功更新。

图 4-70　级联更新

4.4.2　一对多/多对一关系

在 TypeORM 中，要表示公司与员工之间的一对多关系（即一个公司可以有多个员工）和多对一关系（即每个员工属于一个公司），我们需要创建两个实体类：Company 和 Employee。示例代码如下：

```
import {
  Entity,
  Column,
  PrimaryGeneratedColumn,
  OneToMany,
  JoinColumn,
} from "typeorm"
import { Employee } from "./Employee"

@Entity()
export class Company {
  @PrimaryGeneratedColumn()
  id: number

  @Column()
  name: string

  @OneToMany(() => Employee, (employee) => employee.company, {
    // 设置级联自动更新关联表
    cascade: true
  })
  employees: Employee[]
}
```

在 Company 类中，我们使用@OneToMany 装饰器来定义与员工表的关联，并设置级联（cascade）选项以自动更新相关联的 Employee 记录。这意味着，当保存 Company 对象时，相关联的 Employee 记录也会自动更新。此外，我们在 Company 类中定义了一个 employees 属性，用于存储与该公司相关联的员工列表。

@OneToMany 装饰器不能单独使用，它必须与@ManyToOne 装饰器配对使用，以确保关系的双向性。现在让我们来看看 Employee 类是如何定义这种关系的：

```
import {
  Entity,
  Column,
  PrimaryGeneratedColumn,
  JoinColumn,
  ManyToOne,
} from "typeorm"
import { Company } from "./Company"

@Entity()
export class Employee {
  @PrimaryGeneratedColumn()
  id: number

  @Column()
  name: string

  @ManyToOne(() => Company, (company) => company.employees)
  @JoinColumn()
  company: Company
}
```

在上述代码中，我们同样关联了与之对应的 Company 类，并使用@JoinColumn 装饰器定义了外键列。在一对一/多对一关系中，外键列始终出现在"多"的一方。

双方实体定义完成后，接下来插入数据，在 index.ts 中添加如下代码：

```
// 一对多/多对一关系
// 创建一个公司
await AppDataSource.initialize()
const company = new Company()
company.name = "甜筒公司"
// 创建员工 1
const employee1 = new Employee()
employee1.name = "员工 1 号"
// 创建员工 2
const employee2 = new Employee()
employee2.name = "员工 2 号"
// 注意，这里需要通过 Company 来关联 Employee
company.employees = [employee1, employee2]

// 获取实体的存储库
const companyRepository = AppDataSource.getRepository(Company)
// 保存公司
await companyRepository.save(company)
```

执行"npm run start"命令，运行结果如图 4-71 所示，id 为 1 的公司下有两条员工记录，并通过 companyId 进行关联。

图 4-71　创建一对多/多对一数据

4.4.3　多对多关系

接下来，我们将创建订单（Order）和商品（Product）之间的多对多关系。在多对多关系中，一个订单可以包含多个商品，同时一个商品也可以出现在多个订单中。首先，我们需要定义 Order 实体类，它将代表数据库中的订单表。示例代码如下：

```
import {
  Entity,
  PrimaryGeneratedColumn,
  Column,
  ManyToMany,
  JoinTable,
} from "typeorm"
import { Product } from "./Product"

@Entity()
export class Order {
  @PrimaryGeneratedColumn()
  id: number

  @Column()
  name: string

  @ManyToMany(() => Product, (product) => product.orders, {
    cascade: true
  })
  products: Product[]
}
```

在上述代码中，多对多关系通过@ManyToMany 装饰器来关联，设置级联 cascade 选项会在保存订单时自动保存商品信息。而 Product 实体类的定义如下：

```
import {
  Entity,
  PrimaryGeneratedColumn,
  Column,
  ManyToMany,
```

```
    JoinTable,
} from "typeorm"
import { Order } from "./Order"

@Entity()
export class Product {
  @PrimaryGeneratedColumn()
  id: number

  @Column()
  name: string

  @ManyToMany(() => Order, (order) => order.products)
  @JoinTable()
  orders: Order[]
}
```

在这里用@JoinTable()来生成连接表（中间表），TypeORM 会创建一个中间表来管理多对多关系。我们在 index.ts 中添加如下代码，用于生成多对多关系数据。

```
// 多对多关系
// 创建订单 1
await AppDataSource.initialize()
const order = new Order()
order.name = "订单xxx"
// 创建商品 1
const product1 = new Product()
product1.name = "商品 1 号"
// 创建商品 2
const product2 = new Product()
product2.name = "商品 2 号"
// 注意这里订单 1 绑定多个商品
order.products = [product1, product2]

// 获取实体的存储库
const orderRepository = AppDataSource.getRepository(Order)
// 保存订单
await orderRepository.save(order)
```

上述代码创建了一个订单，并同时绑定了两个商品。由于使用了级联，因此在保存订单时会自动创建订单表。运行结果如图 4-72 所示。

图 4-72　创建多对多数据

由图 4-72 可见，ORM 自动创建了一个名为 product_orders_order 的中间表，生成 productId 和 orderId 列作为复合主键。同时，orderId=1 中绑定了两个商品。

除此之外，我们还可以自定义中间表的名称和列名。示例代码如下：

```
@ManyToMany(() => Order, (order) => order.products)
@JoinTable({
  name: 'order_products',
  joinColumn: {
    name: "order",
    referencedColumnName: "id"
  },
  inverseJoinColumn: {
    name: "product",
    referencedColumnName: "id"
  }
})
orders: Order[]
```

删除旧表后重新运行，结果如图 4-73 所示。

图 4-73　自定义中间表信息

至此，我们已经学习了 TypeORM 中的几种基本关系：一对一、一对多/多对一以及多对多。掌握@OneToOne、@OneToMany/@ManyToOne 和@ManyToMany 这几个装饰器的用法至关重要。理解不同表之间的关系有助于我们更好地抽象和表示现实生活中的各种复杂关系。

4.5　在 Nest 中使用 TypeORM 操作 MySQL

在掌握了 MySQL 和 TypeORM 的基础知识之后，本节将学习如何在 Nest 框架中集成 TypeORM 库来操作 MySQL 数据库。我们将从创建一个基础的 Nest 项目开始，并介绍如何在 Nest 中配置数据库连接。此外，我们还将学习 TypeORM 提供的三种操作数据库的方法：使用 EntityManager、Repository 和 QueryBuilder。

4.5.1　项目准备

运行 "nest n nest-typeorm-mysql -p pnpm" 命令，创建名为 nest-typeorm-mysql 的项目，指定包管理为 pnpm，效果如图 4-74 所示。

```
$ nest n nest-typeorm-mysql -p pnpm
⚡  We will scaffold your app in a few seconds..

CREATE nest-typeorm-mysql/.eslintrc.js (663 bytes)
CREATE nest-typeorm-mysql/.prettierrc (51 bytes)
CREATE nest-typeorm-mysql/README.md (3347 bytes)
CREATE nest-typeorm-mysql/nest-cli.json (171 bytes)
CREATE nest-typeorm-mysql/package.json (1959 bytes)
CREATE nest-typeorm-mysql/tsconfig.build.json (97 bytes)
CREATE nest-typeorm-mysql/tsconfig.json (546 bytes)
CREATE nest-typeorm-mysql/src/app.controller.spec.ts (617 bytes)
CREATE nest-typeorm-mysql/src/app.controller.ts (274 bytes)
CREATE nest-typeorm-mysql/src/app.module.ts (249 bytes)
CREATE nest-typeorm-mysql/src/app.service.ts (142 bytes)
CREATE nest-typeorm-mysql/src/main.ts (208 bytes)
CREATE nest-typeorm-mysql/test/app.e2e-spec.ts (630 bytes)
CREATE nest-typeorm-mysql/test/jest-e2e.json (183 bytes)

✔ Installation in progress... 🐢

🚀  Successfully created project nest-typeorm-mysql
```

图 4-74　创建项目

接着运行 "nest g resource user" 命令，生成 user 模块，里面包含 REST 风格的 CURD 接口和 entities 实体类，我们稍后会用到它。

接下来，安装 typeorm、mysql2 和 @nestjs/typeorm 包，运行以下命令：

```
pnpm add typeorm mysql2 @nestjs/typeorm -S
```

其中，typeorm 和 mysql2 这两个包我们之前已经使用过了，而 @nestjs/typeorm 是 Nest 对 TypeORM 的集成。它进一步封装了 TypeORM API 的用法，支持模块动态导入和依赖注入等功能。

TypeORM 提供了初始化数据库连接的方式，在入口文件 app.modules.ts 中引入：

```ts
import { Module } from '@nestjs/common';
import { AppController } from './app.controller';
import { AppService } from './app.service';
import { UserModule } from './user/user.module';
import { TypeOrmModule} from '@nestjs/typeorm';
import { User } from './user/entities/user.entity';

@Module({
  imports: [
    // 初始化 MySQL 连接
    TypeOrmModule.forRoot({
      type: 'mysql',
      host: 'localhost',
      port: 3306,
      username: 'root',
      password: 'jminjmin',
      database: 'nest_typeorm',
      entities: [User],
      synchronize: true
    }),
    UserModule
  ],
  controllers: [AppController],
  providers: [AppService],
```

```
})
export class AppModule {}
```

上述代码通过 forRoot 方法在根模块（AppModule）中注册了 TypeORM 模块。这样做可以使得 TypeORM 服务在应用程序的各个模块中都可以共享使用。需要注意的是，在开发过程中，可以在 TypeORM 配置中设置 synchronize: true，这样程序启动时会自动加载或更新数据库结构以匹配实体定义。然而，在生产环境中，我们应该禁用自动同步（synchronize: false），并采用手动迁移或其他安全的方式来管理数据库结构的变更。

接下来，在 user.entity.ts 中完善 User 实体，新增以下几列字段：

```
import { Entity, PrimaryGeneratedColumn, Column } from "typeorm";

@Entity()
export class User{

    @PrimaryGeneratedColumn()
    id: number;

    @Column()
    name: string;

    @Column()
    sex: string;

    @Column()
    createTime: Date;

    @Column()
    updateTime: Date;

}
```

运行"pnpm run start:dev"命令启动服务。TypeORM 会根据数据库配置，在内部调用 initialize 方法连接数据库，并根据实体映射自动创建 user 表，如图 4-75 所示。

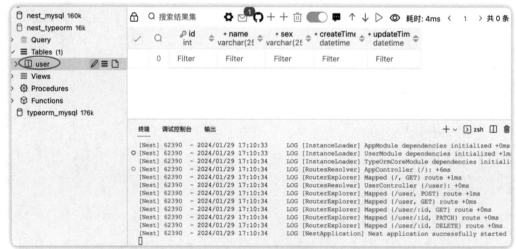

图 4-75 启动服务

4.5.2 使用 EntityManager 操作实体

一旦数据库表建立完成，我们就可以利用 EntityManager 对实体执行增删改查的操作。在 Nest 中，这些数据操作逻辑通常位于服务层（Service）。为了使 UserService 类具备操作数据库的能力，我们需要对其进行相应的修改，代码如下：

```typescript
import { Injectable } from '@nestjs/common';
import { CreateUserDto } from './dto/create-user.dto';
import { UpdateUserDto } from './dto/update-user.dto';
import { InjectEntityManager } from '@nestjs/typeorm';
import { EntityManager } from 'typeorm';
import { User } from './entities/user.entity';

@Injectable()
export class UserService {
  @InjectEntityManager()
  private manage: EntityManager
  async create(createUserDto: CreateUserDto) {
    this.manage.save(User, createUserDto)
  }

  async findAll() {
    return await this.manage.find(User);
  }

  async findOne(id: number) {
    return this.manage.findBy(User, {id})
  }

  async update(id: number, updateUserDto: UpdateUserDto) {
    return await this.manage.save(User, {
      id,
      ...updateUserDto,
    })
  }

  async remove(id: number) {
    return await this.manage.delete(User, {id})
  }
}
```

在上述代码中，我们通过@InjectEntityManager 装饰器注入了 EntityManager（实体管理器），并调用它的 API 进行增删改查操作。每次操作需要指定要处理的实体类，本例中是 User。

首先，我们创建一个用户。为了模拟前端发送请求，在根目录下创建.http 文件（使用 VS Code 的 REST Client 插件），配置请求接口如下：

```
@createdAt = {{$datetime iso8601}}

// 创建用户
POST http://localhost:3300/user
Content-Type: application/json

{
  "name": "mouse",
  "sex": "男",
```

```
"createTime": "{{createdAt}}",
"updateTime": "{{createdAt}}"
}

###

// 查询用户
GET http://localhost:3300/user
```

在这里通过 DTO 来接收 POST 请求数据，同样在 create-user.dto.ts 文件中添加以下字段：

```
export class CreateUserDto {
  name: string;
  sex: string;
  createTime: Date;
  updateTime: Date;
}
```

单击 Send Request，在数据库中可以看到新增了一条用户记录，如图 4-76 所示。

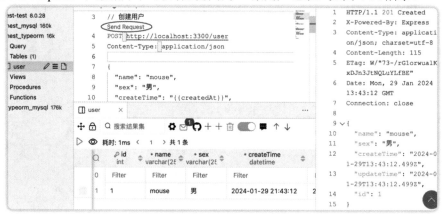

图 4-76　新增用户记录

同样，单击 Send Request，如图 4-77 所示，在右侧可以看到刚刚新建的用户数据。

图 4-77　查询用户数据

4.5.3　使用 Repository 操作实体

除了使用 EntityManager 操作实体外，我们还可以使用 TypeORM 提供的 Repository 对象来进行增删改查操作。Repository 对象拥有与 EntityManager 同样的功能，区别在于使用 Repository 时不

需要每次操作都传入实体对象。以查询为例,使用 EntityManager 的代码如下:

```
async findAll() {
  return await this.manage.find(User);
}
```

而使用 Repository 时,可以直接获取指定的实体存储库并进行操作,具体代码如下:

```
async findAll() {
  return await this.userRepository.find();
}
```

下面通过 Repository 演示更新和删除操作,我们把 user.service.ts 的代码修改如下:

```
import { Injectable } from '@nestjs/common';
import { CreateUserDto } from './dto/create-user.dto';
import { UpdateUserDto } from './dto/update-user.dto';
import { InjectRepository } from '@nestjs/typeorm';
import { Repository } from 'typeorm';
import { User } from './entities/user.entity';

@Injectable()
export class UserService {
  @InjectRepository(User) private userRepository: Repository<User>
  async create(createUserDto: CreateUserDto) {
    createUserDto.createTime = createUserDto.updateTime = new Date()
    return await this.userRepository.save(createUserDto);
  }

  async findAll() {
    return await this.userRepository.find();
  }

  async findOne(id: number) {
    return this.userRepository.findBy({
      id
    });
  }

  async update(id: number, updateUserDto: UpdateUserDto) {
    updateUserDto.updateTime = new Date()
    return await this.userRepository.update(id, updateUserDto);
  }

  async remove(id: number) {
    return await this.userRepository.delete(id);
  }
}
```

在上述代码中,通过@InjectRepository 装饰器实现依赖注入,以获取指定实体类的 Repository。然后,可以在方法中使用 userRepository 实例提供的方法进行增删改查操作。由于 Nest 的依赖注入作用域限定在模块级别,因此还需要在 Module 中通过 forFeature 方法导入并注册这些存储库。代码用法如下:

```
import { Module } from '@nestjs/common';
import { UserService } from './user.service';
import { UserController } from './user.controller';
```

```
import { TypeOrmModule } from '@nestjs/typeorm';
import { User } from './entities/user.entity';

@Module({
  imports: [
    TypeOrmModule.forFeature([User])
  ],
  controllers: [UserController],
  providers: [UserService]
})
export class UserModule {}
```

接下来修改和删除 id 为 1 的用户记录，在.http 文件中添加请求配置：

```
###

// 更新用户
PATCH http://localhost:3300/user/1
Content-Type: application/json

{
  "name": "Minnie",
  "sex": "女",
  "updateTime": "{{createdAt}}"
}

###

// 删除用户
DELETE http://localhost:3300/user/1
```

单击 Send Request，如图 4-78 所示，从右侧的响应结果可以看出已成功更新了一组数据。下方的用户记录中的 name、sex 和 updateTime 已更新。

图 4-78　更新用户数据

最后单击 Send Request，发送删除 id 为 1 的请求，此时数据库记录已经被删除，如图 4-79 所示。

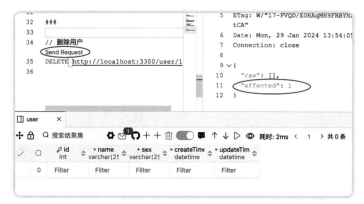

图 4-79　删除用户数据

至此，我们已经掌握了在 TypeORM 中使用 EntityManager 和 Repository 进行数据的增删改查操作。

4.5.4　使用 QueryBuilder 操作实体

QueryBuilder 是 TypeOrm 中最强大的功能之一，它提供了更加底层的数据库查询操作，具有高灵活度高，支持组合更复杂的 SQL 查询语句，如多条件查询或多表查询等。在本书的项目实战中，我们会频繁使用它。

依然以用户信息模块为例，新增用户信息。示例代码如下：

```
async create(createUserDto: CreateUserDto) {
  createUserDto.createTime = createUserDto.updateTime = new Date()
  return await this.dataSource
    .createQueryBuilder()
    .insert()
    .into(User)
    .values(createUserDto)
    .execute()
}
```

上述代码使用 TypeOrm 提供的 dataSource 对象创建查询器，将用户信息插入 user 表中，测试效果如图 4-80 所示。

图 4-80　新增数据

重点来看查询操作，如果需要查询用户名为 mouse 或者性别为"男"的记录，使用 Repository 时可以添加 where 条件。先来看下面的代码：

```
async findAll(name: string, sex: string) {
  return this.userRepository.find({
    where: {
      name,
      sex
    }
  });
}
```

事实上，这样不能查询出满足上面条件的记录，因为 where 中的条件是通过 And 连接的，而不是 Or。如果要实现上述功能，则需要使用 QueryBuilder，代码如下：

```
async findAll(name: string, sex: string) {
  return await this.dataSource
    .createQueryBuilder(User, 'u')
    .where('u.name = :name OR u.sex = :sex', { name, sex })
    .getMany()
}
```

从测试结果可以看出，所有符合条件的记录都被成功查询出来了，如图 4-81 所示。

```
0 ∨ {
1     "id": 2,
2     "name": "mouse",
3     "sex": "男",
4     "createTime": "2024-04-28T08:21:28.000Z",
5     "updateTime": "2024-04-28T08:21:28.000Z"
6   },
7 ∨ {
8     "id": 3,
9     "name": "mouse",
20    "sex": "女",
21    "createTime": "2024-04-28T09:28:14.000Z",
22    "updateTime": "2024-04-28T09:28:14.000Z"
23  },
24 ∨ {
25    "id": 4,
26    "name": "mouse2",
27    "sex": "男",
28    "createTime": "2024-04-28T09:33:45.000Z",
```

图 4-81 查询记录

QueryBuilder 的查询功能远不止如此，它在多表查询的场景中被频繁使用，这部分将在项目实战中体现。

更新和删除比较简单，使用 Repository 可以满足大部分场景的需求。示例代码如下：

```
async update(id: number, updateUserDto: UpdateUserDto) {
  updateUserDto.updateTime = new Date()
  return await this.dataSource
    .createQueryBuilder()
    .update(User)
    .set(updateUserDto)
    .where("id = :id", { id })
    .execute()
}
```

```
async remove(id: number) {
  return await this.dataSource
    .createQueryBuilder()
    .delete()
    .from(User)
    .where("id = :id", { id })
    .execute()
}
```

在上述代码中，使用 update 和 set 方法来更新具有特定 id 的用户信息，效果如图 4-82 所示。

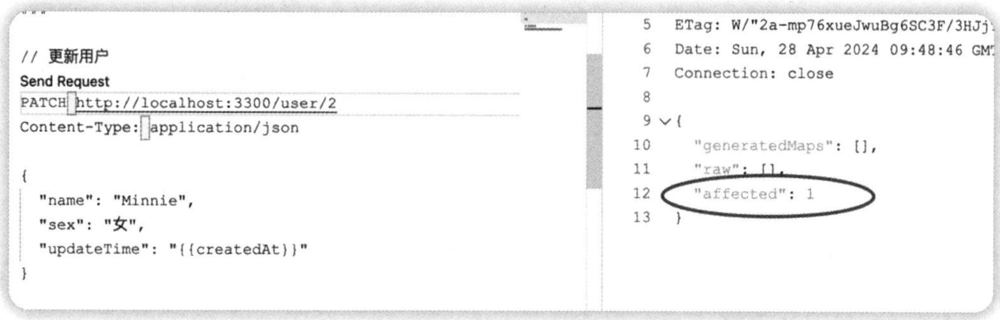

图 4-82　更新用户信息

删除用户通过 delete 方法实现，效果如图 4-83 所示。

图 4-83　删除用户信息

此时，user 表中少了 id 为 2 的记录，如图 4-84 所示。

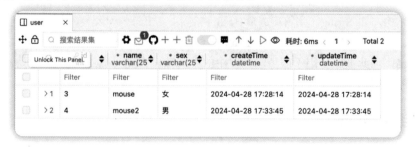

图 4-84　用户列表

以上就是在 Nest 框架中运用 TypeORM 库进行数据库操作的用法，介绍了 3 种不同的操作实体的方法，包括 EntityManager、Repository 和 QueryBuilder。读者可根据具体的查询需求和场景复杂度来选择最合适的方法进行数据库操作。

第 5 章

性能优化之数据缓存

在第 4 章中，我们学习了 MySQL 数据持久化技术，该技术通过将数据存储在硬盘上来实现数据的持久化。然而，后端服务往往要求能够快速响应，而 MySQL 由于其存储机制和查询效率等因素，在需要快速响应的情况下可能表现不尽如人意。因此，本章我们将介绍另一种存储技术——Redis。

我们将从 Redis 的基本使用开始学习，包括通过命令行和可视化工具两种方式来操作 Redis。随后，我们会学习如何在 NestJS 应用程序中引入 Redis，以提升数据处理的效率。随着课程内容的不断深入，本书的项目实战部分还将展示如何在真实场景中应用 Redis，帮助读者更好地理解和掌握这门技术。

5.1 快速上手 Redis

Redis 是一个开源的内存数据结构存储系统，采用键-值（Key-Value）对的形式存储数据。它支持多种数据结构，如 String（字符串）、Hash（哈希）、List（列表）、Set（集合）、Sorted Set（有序集合）、Geospatial（地理信息）、Bitmap（位图）和 JSON 等。Redis 将数据存储在内存中，以实现快速查询，因此被广泛用于服务端开发中的中间件。在提升应用性能和处理高并发场景方面，起着至关重要的作用。

5.1.1 安装和运行

Redis 支持两种操作方式：可视化界面和命令行界面。本节将首先指导如何在本地机器上安装 Redis，并分别介绍这两种使用方法。

对于 macOS 操作系统的用户，可以使用软件包管理器 Homebrew 来安装 Redis。Windows 用户可以直接从 Redis 官网下载安装包并解压使用，或者参考第 7 章内容了解如何使用 Docker 安装 Redis 镜像。安装过程和运行结果如图 5-1 所示。

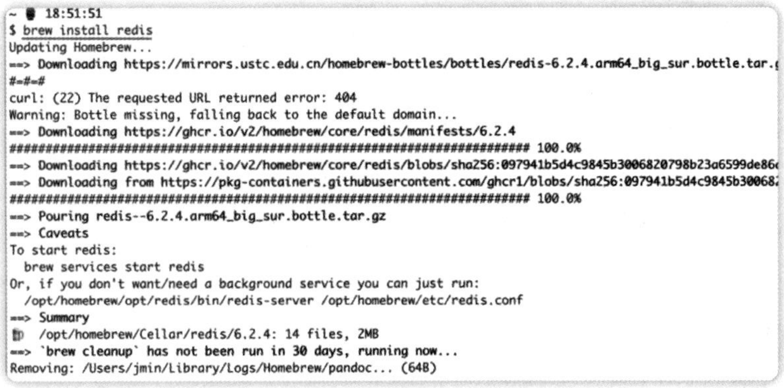

```
~ 18:51:51
$ brew install redis
Updating Homebrew...
==> Downloading https://mirrors.ustc.edu.cn/homebrew-bottles/bottles/redis-6.2.4.arm64_big_sur.bottle.tar.
#=#=#
curl: (22) The requested URL returned error: 404
Warning: Bottle missing, falling back to the default domain...
==> Downloading https://ghcr.io/v2/homebrew/core/redis/manifests/6.2.4
######################################################################## 100.0%
==> Downloading https://ghcr.io/v2/homebrew/core/redis/blobs/sha256:097941b5d4c9845b3006820798b23a6599de86
==> Downloading from https://pkg-containers.githubusercontent.com/ghcr1/blobs/sha256:097941b5d4c9845b30068
######################################################################## 100.0%
==> Pouring redis--6.2.4.arm64_big_sur.bottle.tar.gz
==> Caveats
To start redis:
  brew services start redis
Or, if you don't want/need a background service you can just run:
  /opt/homebrew/opt/redis/bin/redis-server /opt/homebrew/etc/redis.conf
==> Summary
  /opt/homebrew/Cellar/redis/6.2.4: 14 files, 2MB
==> `brew cleanup` has not been run in 30 days, running now...
Removing: /Users/jmin/Library/Logs/Homebrew/pandoc... (64B)
```

图 5-1　安装过程和运行结果

安装成功后，运行“brew services start redis”命令启动 Redis，如图 5-2 所示。

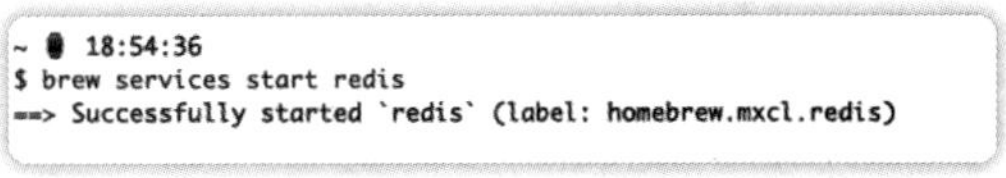

```
~ 18:54:36
$ brew services start redis
==> Successfully started `redis` (label: homebrew.mxcl.redis)
```

图 5-2　启动 Redis

Redis 内置了 redis-cli 命令行工具，运行 redis-cli 命令即可进入交互模式。接下来，将介绍如何使用这个工具来操作 Redis 支持的几种数据结构。

5.1.2　Redis 的常用命令

Redis 提供了一系列操作数据的命令，接下来我们将逐一演示并说明这些命令的使用方法。

1. 字符串操作

字符串操作命令如下：

- SET key value：设置键-值对。
- GET key：获取键对应的值。
- MGET：获取多个键的值。
- DEL key：删除键-值对。
- INCR key：将键的值增加 1。
- INCRBY key increment：将键的值增加指定的增量。

存储字符串是很常见的操作，可以在 CLI 中使用 SET、GET、MGET、DEL 命令操作字符串，如图 5-3 所示。

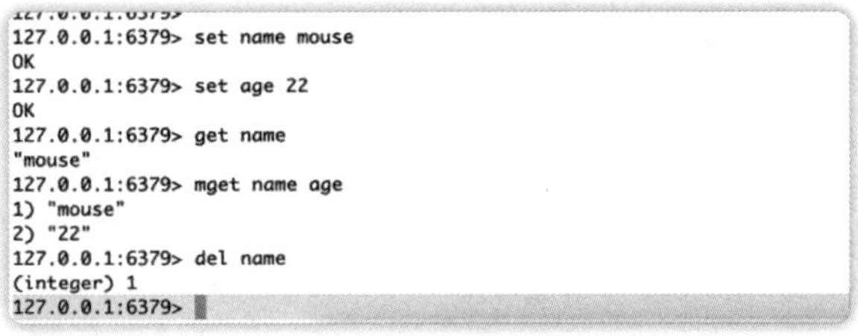

```
127.0.0.1:6379> set name mouse
OK
127.0.0.1:6379> set age 22
OK
127.0.0.1:6379> get name
"mouse"
127.0.0.1:6379> mget name age
1) "mouse"
2) "22"
127.0.0.1:6379> del name
(integer) 1
127.0.0.1:6379>
```

图 5-3　操作字符串

而 INCR 和 INCRBY 是用于计数的命令（计数器），我们常见的点赞数和浏览量就是通过它们来实现的。在 CLI 中执行这两个命令的结果如图 5-4 所示。

```
127.0.0.1:6379> incr age
(integer) 23
127.0.0.1:6379> incr age
(integer) 24
127.0.0.1:6379> incrby age 10
(integer) 34
127.0.0.1:6379>
```

图 5-4　计数器

2. 列表操作

列表操作命令如下：

- LPUSH key value：将值推入列表左侧。
- RPUSH key value：将值推入列表右侧。
- LPOP key value：从列表左侧删除值。
- RPOP key value：从列表右侧删除值。
- LLEN key：获取列表长度。
- LRANGE key start stop：获取列表指定范围的值。

在 Redis 中，列表通常应用于消息队列、任务队列、时间线等场景，它的操作类似于 JavaScript 中的数组操作，读者应该很容易理解。示例如图 5-5 所示。

```
127.0.0.1:6379> lpush list 1
(integer) 1
127.0.0.1:6379> lpush list 2
(integer) 2
127.0.0.1:6379> lpush list 3
(integer) 3
127.0.0.1:6379> rpush list 4
(integer) 4
127.0.0.1:6379> rpush list 5 6
(integer) 6
127.0.0.1:6379> llen list
(integer) 6
127.0.0.1:6379> lrange list 0 -1
1) "3"
2) "2"
3) "1"
4) "4"
5) "5"
6) "6"
127.0.0.1:6379> lpop list
"3"
127.0.0.1:6379> rpop list
"6"
127.0.0.1:6379>
```

图 5-5　操作列表

值得注意的是，lrange 用于获取列表数据，其中 0 表示索引的开始位置，而-1 表示结尾，即查询全部列表数据。

3. 集合操作

集合操作命令如下：

- SADD key member：向集合添加成员。
- SREM key member：移除集合中的成员。
- SMEMBERS key：获取集合的所有成员。

- SINTER key1 key2：获取多个集合的交集。

在 Redis 中，集合由一组无序但唯一的成员组成。使用集合可以对数据执行交集、并集、差集等操作。集合常用于标签系统，如进行文章标签和商品标签管理，从而轻松地实现标签的组合和筛选功能。

集合操作如图 5-6 所示。

图 5-6　操作集合

显然，使用 Redis-CLI 工具不够直观和方便，我们可以改用官方的 RedisInsight 可视化工具。下载并安装完成后，打开该工具，我们可以看到它已经自动连接上本地的 Redis 服务，如图 5-7 所示。

图 5-7　RedisInsight 界面

单击这个数据库，展示前面创建的各种类型的 key 和 value，如图 5-8 所示。

图 5-8　可视化 Redis 数据

接下来，通过这个 GUI 工具演示其他数据结构。

4. 有序集合操作

有序集合操作命令如下：

- ZADD key score member：向有序集合添加成员及其分数。
- ZRANGE key start stop：按分数范围获取有序集合的成员。

与集合不同，Sorted Set（有序集合，也被称为 ZSet）是由一组按照分数排序并且唯一的数据组成的，通常应用在游戏的排行榜中。在 GUI 工具中添加一个 ZSet 数据结构的 key，如图 5-9 所示。

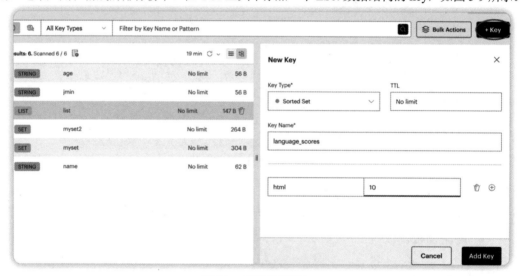

图 5-9　可视化新增 key

接下来，单击 ⊕ 按钮添加有序集合成员，并为它设置分数，如图 5-10 所示。

图 5-10　可视化新增成员

当然，如果我们需要执行一段 Redis 命令，依然可以使用命令行工具，如图 5-11 所示。

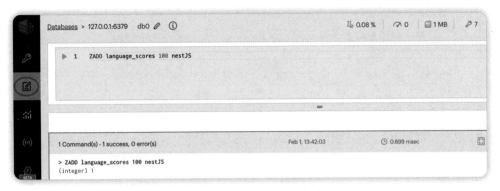

图 5-11　在 GUI 中执行命令

5. 哈希操作

哈希操作命令如下：

- HSET key field value：设置哈希字段的值。
- HGET key field：获取哈希字段的值。
- HGETALL key：获取哈希的所有字段和值。

哈希结构适用于存储复杂对象结构的数据，例如用户信息、订单信息等，每个 key 代表不同的对象，field value 代表对象的属性和值。以存储 userInfo 为例，执行如下命令设置字段：

```
HSET userInfo name mouse
HSET userInfo age 22
HSET userInfo userId 123321
```

生成的数据结构如图 5-12 所示。

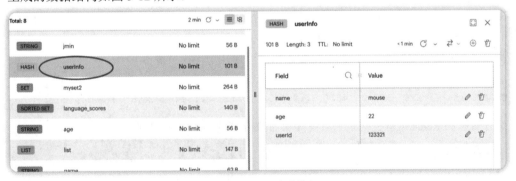

图 5-12　存储哈希数据

执行 HGET userInfo name 和 HGETALL userInfo 命令查询数据，如图 5-13 所示。

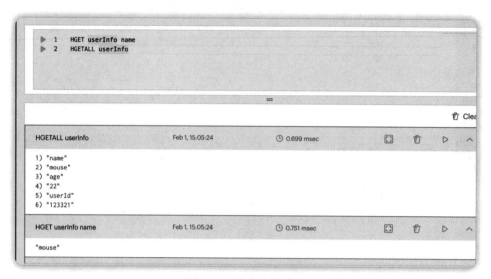

图 5-13　查询哈希数据

6. 地理空间操作

地理空间操作命令如下：

- GEOADD key longitude latitude member：根据经纬度添加坐标成员。
- GEOPOS key member [member…]：获取一个或多个成员的地理位置坐标。
- GEOSEARCH key <FROMMEMBER | FROMLONLAT > <BYRADIUS | BYBOX> <…>：根据不同条件获取成员坐标。
- GEODIST key member1 member2 [unit]：计算坐标成员之间的距离。

Redis 可以用于存储地理空间的坐标，并支持在给定半径和范围边界内进行搜索。这种功能常应用于共享汽车、自行车、充电宝等场景中。

执行以下命令给 share_cars（共享汽车）字段添加几个坐标位置：

```
GEOADD share_cars -122.27652 37.805186 car1
GEOADD share_cars -122.2674626 37.8062344 car2 -122.2469854 37.8104049 car3
```

运行后看到 Redis 地理空间是通过有序集合来存储的，如图 5-14 所示。

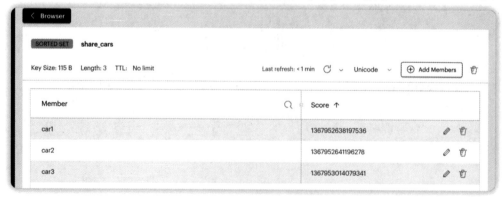

图 5-14　新增地理坐标

接下来获取 car1 与 car2 的坐标信息，查询结果如图 5-15 所示。

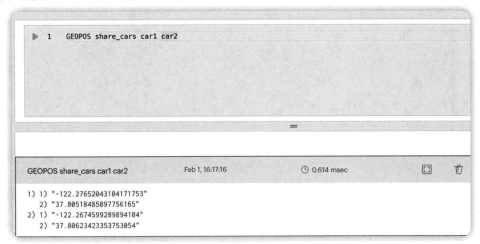

图 5-15　查询指定成员的坐标

接下来增加一些条件，给定一个经纬度坐标，查询这个坐标 5km 范围内的汽车，结果按照由近到远的顺序输出，如图 5-16 所示。

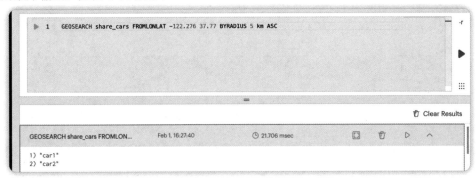

图 5-16　根据经纬度、给定范围查询数据

至此，读者应该已经理解了如何实现查找附近共享充电宝和自行车的功能原理。

7. 位图操作

位图操作命令如下：

- SETBIT key offset value：给指定偏移量的位设置 0 或 1。
- GETBIT key offset：获取指定偏移量的位。
- BITCOUNT key：获取指定位为 1 的总计数。

在 Redis 中，位图的应用也很广泛，如记录用户的活跃状态、在线状态、访问频率等。我们以统计用户 1001 在 30 天内的访问频率为例，假设他在第 10、20、25 天时有访问记录，代码如下：

```
SETBIT user:1001:visit 10 1
SETBIT user:1001:visit 20 1
SETBIT user:1001:visit 25 1
```

此时，要判断该用户在第 15 天时是否有访问记录，如图 5-17 所示，我们可以使用 GETBIT 进行查询。

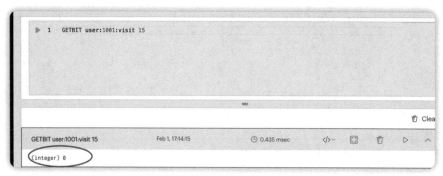

图 5-17　查询用户是否有访问记录

接下来统计用户在 30 天内的访问频率，如图 5-18 所示，可以使用 BITCOUNT 进行查询。

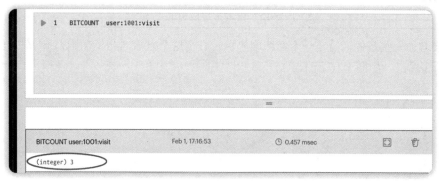

图 5-18　查询用户的访问频率

至此，Redis 中常用的数据结构与命令介绍完毕。结合实际生活中的应用场景，相信读者可以快速掌握并有效地运用 Redis。

5.2　在 Nest 中使用 Redis 缓存

在 5.1 节中，我们学习了 Redis 中各种数据结构的使用方法，并探讨了它们的多种应用场景，包括排行榜、附近的共享设备、计数器和用户访问频率等。实际上，Redis 的用途远不止这些。它还可以广泛应用于缓存实现、消息队列、单点登录和分布式锁等多种场景。Redis 的灵活性和高性能等特点使其成为一个强大而多用途的工具，为我们提供了广泛的解决方案，以满足多样化的应用需求。

接下来，我们将在 Nest 应用程序中实现 Redis 的数据缓存功能，以便更深入地了解 Redis 的实际应用。

5.2.1　项目准备

本小节实现这样一个需求：通过 Redis 缓存用户的购物车信息。当用户查询购物车信息时，首先从 Redis 中查询，如果缓存为空，再去 MySQL 中查询。当用户在购物车中增加商品数量时，需要将更新保存到 MySQL 中，并同步更新到 Redis，以确保缓存数据的一致性。

先创建一个名为 nest-redis 的项目，运行"nest n nest-redis -p pnpm"命令，指定包管理器为 pnpm。项目创建成功后，如图 5-19 所示。

```
$ nest new nest-redis -p pnpm
⚡  We will scaffold your app in a few seconds..
CREATE nest-redis/.eslintrc.js (663 bytes)
CREATE nest-redis/.prettierrc (51 bytes)
CREATE nest-redis/README.md (3347 bytes)
CREATE nest-redis/nest-cli.json (171 bytes)
CREATE nest-redis/package.json (1951 bytes)
CREATE nest-redis/tsconfig.build.json (97 bytes)
CREATE nest-redis/tsconfig.json (546 bytes)
CREATE nest-redis/src/app.controller.spec.ts (617 bytes)
CREATE nest-redis/src/app.controller.ts (274 bytes)
CREATE nest-redis/src/app.module.ts (249 bytes)
CREATE nest-redis/src/app.service.ts (142 bytes)
CREATE nest-redis/src/main.ts (208 bytes)
CREATE nest-redis/test/app.e2e-spec.ts (630 bytes)
CREATE nest-redis/test/jest-e2e.json (183 bytes)

✔ Installation in progress... 🐢

🚀  Successfully created project nest-redis
```

图 5-19　创建项目

接下来，运行"pnpm install typeorm mysql2 redis @nestjs/typeorm -S"命令，安装我们需要的依赖包。

同时，运行"nest g resource shopping-cart --no-spec"命令，生成一个名为 shopping-cart 的购物车模块，这里不生成单元测试文件，运行结果如图 5-20 所示。

```
$ nest g resource shopping-cart --no-spec
? What transport layer do you use? REST API
? Would you like to generate CRUD entry points? Yes
CREATE src/shopping-cart/shopping-cart.controller.ts (1055 bytes)
CREATE src/shopping-cart/shopping-cart.module.ts (298 bytes)
CREATE src/shopping-cart/shopping-cart.service.ts (721 bytes)
CREATE src/shopping-cart/dto/create-shopping-cart.dto.ts (38 bytes)
CREATE src/shopping-cart/dto/update-shopping-cart.dto.ts (202 bytes)
CREATE src/shopping-cart/entities/shopping-cart.entity.ts (29 bytes)
UPDATE src/app.module.ts (417 bytes)
```

图 5-20　生成购物车模块

在 VS Code 中运行"pnpm start:dev"命令启动项目，如图 5-21 所示。

```
[21:24:08] Starting compilation in watch mode...

[21:24:09] Found 0 errors. Watching for file changes.

[Nest] 29761  - 2024/02/01 21:24:09     LOG [NestFactory] Starting Nest application...
[Nest] 29761  - 2024/02/01 21:24:09     LOG [InstanceLoader] AppModule dependencies initialized +5ms
[Nest] 29761  - 2024/02/01 21:24:09     LOG [RoutesResolver] AppController {/}: +8ms
[Nest] 29761  - 2024/02/01 21:24:09     LOG [RouterExplorer] Mapped {/, GET} route +1ms
[Nest] 29761  - 2024/02/01 21:24:09     LOG [NestApplication] Nest application successfully started +0ms
```

图 5-21　启动项目

5.2.2　Redis 初始化

Redis 通常会在多个模块中使用。为了更好地管理它，我们新建一个 redis.module.ts 文件，专门用来配置和导出 Redis 模块，其他模块可以通过依赖注入的方式使用它。根据使用频率和需求，用户甚至可以把 Redis 定义为全局模块。RedisModule 的代码如下：

```
import { Module } from '@nestjs/common';
import { createClient } from 'redis';

const createRedisClient = async() => {
  return await createClient({
    socket: {
      host: 'localhost',
      port: 6379,
    }
  }).connect();
};

@Module({
  providers: [
    {
      provide: 'NEST_REDIS',
      useFactory: createRedisClient,
    },
  ],
  exports: ['NEST_REDIS'],
})
export class RedisModule {}
```

在代码实现中，createClient 方法负责根据提供的 Redis 配置信息（如主机地址、端口号等）来注册 Redis 客户端，并通过 connect 方法与 Redis 服务建立连接。在@Module 装饰器中，我们使用 providers 属性定义一个依赖项，其 token 为 'NEST_REDIS'。useFactory 属性用于定义一个工厂方法，该方法将创建并返回 Redis 客户端实例。最后，通过 exports 属性将该客户端导出，使其成为可在其他模块中使用的共享对象。

接下来，在 shopping-cart.service.ts 中通过@Inject 装饰器来导入并使用 Redis 客户端，核心代码如下：

```
@Injectable()
export class ShoppingCartService {
  @Inject('NEST_REDIS')
  private redisClient: RedisClientType

  async create(createShoppingCartDto: CreateShoppingCartDto) {
    await this.redisClient.set('xxx', JSON.stringify(createShoppingCartDto));
  }
}
```

注入 Redis 依赖后，我们可以在服务方法中通过 this.redisClient 调用它上面的所有方法。

5.2.3　建表并构建缓存

完善购物车服务逻辑之前，在 app.module.ts 中初始化 MySQL 连接，代码如下：

```typescript
import { Module } from '@nestjs/common';
import { AppController } from './app.controller';
import { AppService } from './app.service';
import { ShoppingCartModule } from './shopping-cart/shopping-cart.module';
import { TypeOrmModule } from '@nestjs/typeorm';

@Module({
  imports: [
    // 初始化 MySQL 连接
    TypeOrmModule.forRoot({
      type: 'mysql',
      host: 'localhost',
      port: 3306,
      username: 'root',
      password: 'jminjmin',
      database: 'nest_redis',
      entities: [__dirname + '/**/*.entity{.ts,.js}'],
      autoLoadEntities: true,
      synchronize: true
    }),
    ShoppingCartModule
  ],
  controllers: [AppController],
  providers: [AppService],
})
export class AppModule {}
```

这里我们用 TypeORM 的存储库模式来操作实体，完善 shopping-cart.entity.ts 文件，代码如下：

```typescript
import { Column, Entity, PrimaryGeneratedColumn } from "typeorm"
@Entity()
export class ShoppingCart {
  @PrimaryGeneratedColumn()
  id: number

  @Column()
  userId: number

  @Column({type: 'json'})
  cartData: Record<string, number>

}
```

定义 create、findOne 和 update 方法用于添加、查询、更新购物车信息，并保持 Redis 与 MySQL 数据的一致性。即在更新 MySQL 数据时，Redis 缓存必须同时更新。示例代码如下：

```typescript
import { Inject, Injectable } from '@nestjs/common';
import { CreateShoppingCartDto } from './dto/create-shopping-cart.dto';
import { UpdateShoppingCartDto } from './dto/update-shopping-cart.dto';
import { RedisClientType } from 'redis';
import { InjectRepository } from '@nestjs/typeorm';
```

```typescript
import { ShoppingCart } from './entities/shopping-cart.entity';
import { Repository } from 'typeorm';

@Injectable()
export class ShoppingCartService {
  @Inject('NEST_REDIS')
  private redisClient: RedisClientType
  @InjectRepository(ShoppingCart)
  private shoppingCartRepository: Repository<ShoppingCart>

  async create(createShoppingCartDto: CreateShoppingCartDto) {
    // 保存到 db 中
    await this.shoppingCartRepository.save(createShoppingCartDto);
    // 更新 Redis 缓存
    await this.redisClient.set(`cart:${createShoppingCartDto.userId}`, JSON.stringify(createShoppingCartDto));
    return {
      msg: '添加成功',
      success: true
    }
  }

  async findOne(id: number): Promise<ShoppingCart> {
    // 先从 Redis 中查询缓存，没有再查 db
    const data = await this.redisClient.get(`cart:${id}`)
    const cartEntity = data ? JSON.parse(data) : null
    if (cartEntity) return cartEntity
    return await this.shoppingCartRepository.findOne({
      where: {
        userId: id
      }
    });
  }

  async update(updateShoppingCartDto: UpdateShoppingCartDto) {
    const { userId, cartData: { count = 1 } } = updateShoppingCartDto
    // 查询数据
    const cartEntity = await this.findOne(userId)
    const cart = cartEntity ? cartEntity.cartData : {}
    let quality = (cart.count || 0) + count
    // 更新 count
    cart.count = quality

    // 更新 db 数据
    await this.shoppingCartRepository.update({userId}, cartEntity)
    // 更新 Redis 缓存
    await this.redisClient.set(`cart:${userId}`, JSON.stringify(cartEntity))
    return {
      msg: '更新成功',
      success: true
```

```
    }
  }
}
```

首先来看代码中用于添加购物车的 create 方法，在 createShoppingCartDto 中定义以下接口字段：

```
export class CreateShoppingCartDto {
  userId: number

  cartData: Record<string, number>
}
```

userId 将作为 Redis 中存储的键（key），而 cartData 表示购物车信息。在 create 方法中，我们首先使用 TypeORM 存储库将数据保存到 MySQL 数据库中，然后将数据设置（set）到 Redis 缓存中。userId 作为缓存键的一部分，用于保存购物车信息，这样每个用户都可以通过其 userId 获取到对应的缓存数据。

findOne 查询方法的工作原理很直观。当 NestJS 接收到请求时，它会首先尝试根据 userId 从 Redis 缓存中获取数据。如果缓存命中，则将缓存中的数据返回给前端；如果没有命中，则从数据库中获取查询结果。在处理那些用户数据更新不频繁但读取频繁的场景时，使用缓存可以显著减轻数据库的负担。

至于 update 方法，它首先会查询出指定的记录进行修改，然后再次将更新后的记录保存到数据库中，并刷新 Redis 缓存，以此完成整个更新操作。

5.2.4 运行代码

一切准备就绪后，下一步是发送请求以测试应用程序的效果。为此，我们可以使用 VS Code 的 REST Client 插件。首先，在项目的根目录下创建一个以.http 为扩展名的文件，然后在该文件中配置所需的接口。具体配置代码如下：

```
// 添加购物车
POST http://localhost:3000/shopping-cart
Content-Type: application/json

{
  "userId": 111,
  "cartData": {
    "count": 1
  }
}

###
// 更新购物车数量
PATCH http://localhost:3000/shopping-cart
Content-Type: application/json

{
  "userId": 111,
```

```
    "cartData": {
      "count": 1
    }
}
```

重新运行"pnpm start:dev"命令以启动服务。服务启动后，TypeORM 会自动创建所需的表结构，单击 Send Request 添加几条数据，如图 5-22 所示。

图 5-22　创建数据

在 VS Code 中查看数据记录，如图 5-23 所示。

图 5-23　查看数据记录

同时，Redis 缓存中也存储了相应的数据记录。要查看这些缓存数据，可以使用 RedisInsight 工具，并利用 cart*作为筛选条件来检索特定的购物车数据，如图 5-24 所示。

图 5-24　查看 Redis 缓存数据

接下来，更新数据库中的 count 字段，将 userId 为 111 的用户的购物车商品数添加为 10，单击 Send Request，结果如图 5-25 所示。

第 5 章　性能优化之数据缓存

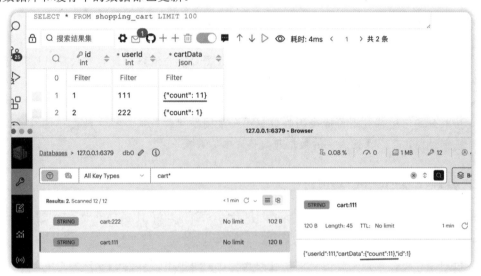

图 5-25　更新购物车商品数量

根据我们的设计方案，程序首先会在缓存中查找 userId 为 111 的用户数据，并返回这些数据。一旦数据被成功修改，程序接下来会将更新后的数据同步到数据库和缓存中。在完成这些步骤后，你可以查询数据库和缓存中的记录，以验证数据是否已经更新。图 5-26 展示了查询结果，从中可以看到数据库和缓存中的数据都已更新。

图 5-26　更新数据库及缓存

5.2.5　设置缓存有效期

在实际业务中，Redis 通常会设置缓存过期时间，以避免数据不一致或缓存长时间未访问（更新）导致内存空间浪费等问题。我们为缓存设置了 30 秒的过期时间，代码如下：

```
// 更新 Redis 缓存
await this.redisClient.set(
  `cart:${createShoppingCartDto.userId}`,
  JSON.stringify(createShoppingCartDto),
  { EX: 30 }
);
```

对于 userId 为 111 的记录，执行完更新操作后，如图 5-27 所示，其有效期（TTL）已经从 30 秒开始倒计时。

图 5-27　设置 Redis 缓存有效期

有读者可能会问：为什么要设置缓存有效期呢？笔者给出了以下几个理由：

- 释放内存空间：如果缓存长时间不被访问或更新，这部分缓存可能会持续占用大量的空间，从而不被释放，这在一定程度上会导致 Redis 频繁扩容。设置过期时间可以自动释放内存供其他缓存使用。
- 保证数据的实时性：当缓存对应的业务逻辑发生变更时，失效的缓存一直在内存中可能会导致业务逻辑错误，即我们常说的"脏数据"，这会影响系统的稳定性。设置自动过期可以保证缓存在一定时间内是有效的，有效避免这种问题。
- 保证数据的安全性：过期时间是一种容错机制，缓存长时间存活在内存中，如果遇到内存泄露或者恶意软件攻击，缓存中的隐私数据可能会泄露。设置有效期可以定时清理内存中的隐私数据，减少数据泄露风险。
- 保证数据的一致性：在并发场景或缓存服务异常时，最新缓存并未更新到内存中，此时获取到的旧缓存数据可能会因为数据不一致问题导致系统异常。设置一定的有效期让旧缓存过期，可以保证缓存的数据与数据库中的数据一致。

5.2.6　选择合理的有效期

设置 Redis 缓存的有效期并没有统一的最佳时长，这完全取决于具体的业务场景需求。通常，可以根据以下三种策略来选择：

- 短期缓存：在数据实时性要求高且频繁变动的情况下，可以设置较短的缓存时间，如几分钟或者几小时，以确保缓存数据及时与数据库同步。这常见于新闻资讯推送、热点头条及天气预报等。
- 中期缓存：对于一些变动不频繁，但要求具有一定实时性的数据，可以设置较长的缓存有效期，如几小时或几天，以尽可能减轻数据库的访问压力。这常见于电商购物车数据、用户登录数据等。
- 长期缓存：对于相对稳定且变动少的数据，可以设置较长的有效期，如几天或几周。常见于静态资源缓存、地理位置信息更新等。

通过在 Nest 中整合 Redis 缓存，能够有效减少对 MySQL 数据库的访问。同时，合理设置有效期，不仅可以提升系统性能，还能有效降低数据库的负担。

第 6 章

身份验证与授权

本章将深入探讨身份验证和授权这两个安全概念。在应用程序中，身份验证是验证用户身份的过程，而授权是确定用户是否有权限执行特定操作的过程。本章首先介绍在身份验证和授权过程中涉及的常见概念，例如 Cookie、Session、Token、JWT（JSON Web Token）和 SSO（Single Sign-On，单点登录）等，以便读者能够清晰地理解它们之间的区别和适用场景。

接下来，我们将在 Nest 中演示如何使用 JWT 和 Passport.js 进行身份验证，并使用主流的 RBAC（Role-Based Access Control，基于角色的访问控制）模型实现权限控制，最后基于 Redis 实现单点登录。

通过本章的学习，相信读者将更加深入地了解身份验证和授权的原理和实现方式。

6.1 Cookie、Session、Token、JWT、SSO 详解

身份验证和授权是设计安全、可靠系统的核心功能之一。在开始对身份验证和权限控制进行编码之前，有必要先回答以下几个问题：

- 什么是身份验证？
- 什么是授权？
- 实现身份验证或授权的方式有哪几种？
- Cookie、Session、Token、JWT、SSO、OAuth 分别是什么，它们之间有什么区别？

接下来，我们将一一解答以上问题。

6.1.1 什么是身份验证

在互联网安全领域，身份验证的通俗解释是确认当前用户确实是其自称的那个人。这个过程类似于手机上的指纹解锁功能：只有当你的指纹与手机中存储的指纹数据匹配时，手机才能被成功解锁。

常见的身份验证方式有以下几种：

- 账号密码登录验证
- 手机短信验证
- 邮箱验证码验证

- 人脸识别验证
- 指纹登录验证

6.1.2 什么是授权

用户授予第三方应用访问某些资源的权限，称为授权。例如，当你登录微信或小程序时，它会询问你是否允许授权访问昵称、微信名称、手机号等个人信息；当你首次安装 App 时，App 会询问你是否允许授权访问相机、相册、麦克风、通讯录或地理位置等个人信息。

6.1.3 什么是凭证

实现身份验证和授权的前提是需要一种媒介来标识访问者的身份。在实际生活中，身份证或护照是最为常见的证明公民身份的凭证，通过它我们可以乘坐各种交通工具或办理各种银行业务。而在互联网中，我们需要注册或登录账号才能进行下单购物、评论点赞等操作。当用户登录某个网站时，服务器会返回一个 SessionID（会话 ID）或 Token（令牌）到用户浏览器中，用于标识当前用户的身份。再次请求访问网站时带上这个凭证，服务器才能识别出该用户的身份，并授予相应的操作权限。

6.1.4 什么是 Cookie

HTTP 协议是一种无状态协议，这意味着服务端在每次请求中无法识别请求是由哪个用户发起的。为了解决这个问题，需要一种机制来存储用户的身份信息，并在用户进行后续请求时提供给服务器，以便识别是否为同一用户。这就是 Cookie 的用武之地。

Cookie 存储在客户端浏览器中，由服务器发送并以文本文件形式存在，其大小通常限制在 4KB 以内。当用户再次向同一服务器发起请求时，Cookie 会被自动包含在请求头中并发送至服务器。需要注意的是，Cookie 不支持跨域访问，即每个 Cookie 都与特定域名绑定，无法被其他域名下的页面访问，但在相同域名下的页面之间可以共享使用。有关 Cookie 工作流程如图 6-1 所示。

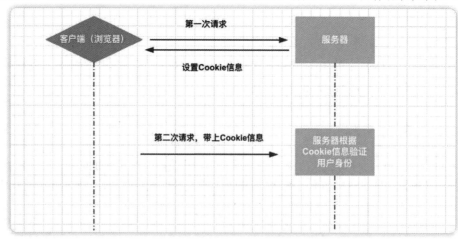

图 6-1　Cookie 的工作流程

由于 Cookie 存储在客户端，用户可以轻易地查看和修改这些信息，这可能导致用户信息泄露，带来安全隐患。为了解决这一问题，通常会采用结合使用 Session 和 Cookie 的方法。在这种方法中，用户的敏感信息存储在服务端的 Session 中，而 Cookie 仅存储 Session 的标识符。这样，即使客户端的 Cookie 被访问，攻击者也无法直接获取到敏感的用户信息。

6.1.5 什么是 Session

Session 与 Cookie 之间有着奇妙的因缘，Session 是另一种维持服务端与客户端会话状态的机制，通常基于 Cookie 实现。Session 存储在服务端中，而 SessionID 存储在客户端的 Cookie 字段中。

用户首次向服务器发起请求时，服务器根据用户信息生成对应的 Session，并将该 Session 的唯一标识 SessionID 返回到浏览器中，并保存在 Cookie 内。当用户再次向服务器发起请求时，浏览器会自动判断 Cookie 中是否存在与当前域名匹配的信息，并将其发送到服务端。服务端根据 Cookie 中的 SessionID 来查找是否存在对应的 Session 信息。如果没有找到，说明用户尚未登录或者登录失败；否则，执行用户登录后的操作。Session 工作流程如图 6-2 所示。

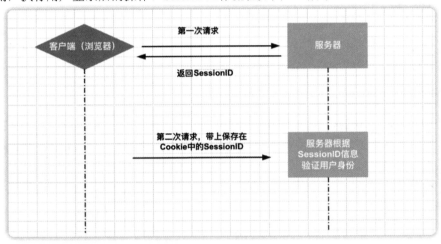

图 6-2 Session 工作流程

6.1.6 Session 与 Cookie 的区别

Session 与 Cookie 在 Web 程序中的不同用途和实现方式决定了它们之间存在以下区别：

- 存储位置：Cookie 通常以文本形式存储在客户端浏览器中，而 Session 数据则存储在服务端，可能是数据库或内存中，可以以任意数据类型存在。
- 安全性：由于 Cookie 存储在客户端，因此可以被用户查看和修改，存在安全风险。相比之下，Session 数据存储在服务端，用户无法直接访问或修改，因此相对更安全。不过，Session 仍然可能面临 CSRF（Cross-Site Request Forgery，跨站请求伪造）攻击的风险，这种攻击经常通过钓鱼网站进行。
- 生命周期：Cookie 可以设置一个较长的有效期，实现如默认登录等功能。而 Session 的生命周期通常较短，会在客户端关闭或 Session 超时后失效（默认情况下）。
- 存储大小：单个 Cookie 的数据存储大小通常限制在 4KB 以内，而 Session 所能存储的数据量远大于 Cookie，没有 4KB 的限制。

结合使用 Session 和 Cookie 的方案相较于仅使用 Cookie 具有明显优势，但它并非没有缺陷。除了容易受到 CSRF（跨站请求伪造）攻击之外，Session 由于存储在服务端，还面临着一些挑战。特别是在多服务器环境中，当出现高并发时，用户的请求可能会被负载均衡器分配到不同的服务器上。如果用户的 Session 信息仅保存在某一台服务器上，那么其他服务器将无法访问该 Session 信息，从而导致会话管理问题。

为了解决这一问题，通常需要采用特殊的方案在多台服务器之间同步 Session 信息。一种常见的做法是使用像 Redis 这样的分布式缓存系统来存储 Session 数据，以确保所有服务器都能访问到最新的用户会话状态。

除了 Session 和 Cookie 的结合使用，我们还可以考虑使用 Token（如 JWT）来实现服务端的无状态化。无状态化意味着用户的会话信息被编码在 Token 中，每次请求时由客户端发送，服务端通过解析 Token 来验证用户身份，无须在服务端存储 Session 信息。

6.1.7　什么是 Token

Token 是访问受限资源时需要携带的一种凭证，通常是一个加密字符串，常见的 Token 有以下几种。

1. 访问令牌

访问令牌（Access Token）是一种用于授权用户访问特定资源或执行特定操作的凭证。用户登录成功后，服务器会颁发一个访问令牌，用户在访问受保护的资源时需要携带该令牌进行验证。访问令牌的工作流程如图 6-3 所示。

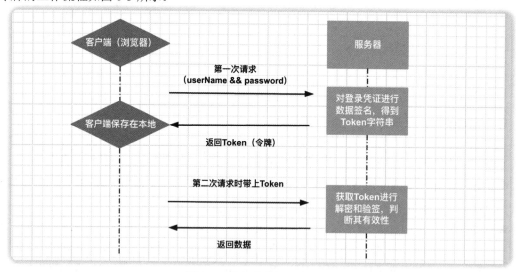

图 6-3　访问令牌的工作流程

图 6-3 中的流程详细描述如下：

（1）客户端发送登录请求：客户端发送包含用户名和密码的登录请求。

（2）服务端验证用户名和密码：服务端接收到登录请求后，对用户名和密码进行验证。

（3）服务端签发 Token：账号和密码验证成功后，服务端会生成一个包含用户身份信息的访问令牌，并将其发送给客户端。

（4）客户端存储 Token：客户端收到 Token 后，通常会将其存储在 Cookie、localStorage 或 sessionStorage 等地方，以便在后续的请求中使用。

（5）客户端请求接口：客户端在发送请求时，会将 Token 作为身份凭证附加在请求的头部（通常是 Authorization 头部）或者请求参数中。

（6）服务端验证 Token：服务端在接收到客户端的请求后，会从请求中提取 Token 并进行验证。如果 Token 有效且包含足够的权限，服务端会处理请求并返回相应的数据。

2. 刷新令牌

刷新令牌（Refresh Token）专门用于刷新访问令牌的 Token，它的有效期通常比访问令牌长，以保证用户登录状态的持久性。如果没有刷新令牌，用户每次访问时都需要输入账号和密码以重新验证，这显然很麻烦。因此，客户端通常会维护一个刷新令牌，定时刷新访问令牌，从而提升网站的用户体验。刷新令牌的工作流程如图 6-4 所示。

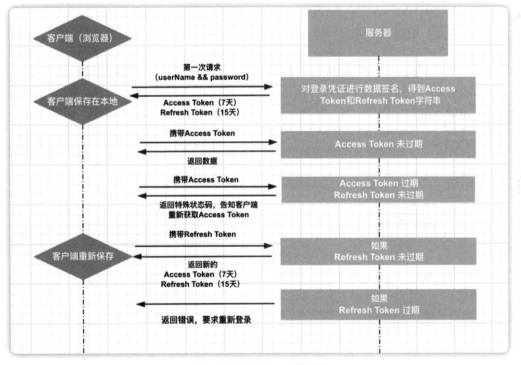

图 6-4　双 Token 工作流程

图 6-4 中的流程详细描述如下：

（1）客户端发送登录请求：客户端发送包含用户名和密码的登录请求。

（2）服务端验证用户名和密码：服务端接收到登录请求后，对用户名和密码进行验证。

（3）服务端签发访问令牌和刷新令牌：如果账号和密码验证成功，服务端会生成一个包含用户身份信息的访问令牌和一个用于刷新访问令牌的刷新令牌，并将它们一起发送给客户端。

（4）客户端存储 Token：客户端收到 Token 后，通常会将访问令牌存储在 Cookie、localStorage 或 sessionStorage 等地方，刷新令牌则存储在安全的地方，比如 HTTP Only 的 Cookie 中，以免被 XSS（Cross-Site Scripting，跨站脚本）攻击。

（5）客户端请求接口：客户端在发送请求时，会将访问令牌作为身份凭证附加在请求的头部（通常是 Authorization 头部）或者请求参数中。

（6）服务端验证访问令牌：服务端在接收到客户端的请求后，从请求中提取访问令牌并进行验证。如果访问令牌有效，服务端处理请求并返回相应的数据。如果访问令牌过期或无效，则返回指定状态提示或错误，告知客户端用刷新令牌获取新的访问令牌。

（7）客户端用刷新令牌获取新的访问令牌：客户端发送一个特殊的请求，携带刷新令牌向服务端请求新的访问令牌。

（8）服务端验证刷新令牌：服务端接收到客户端的请求后，验证刷新令牌的有效性。如果刷新令牌有效且未过期，服务端会生成一个新的访问令牌和刷新令牌，并一同发送给客户端。

（9）客户端更新 Token：客户端收到新的 Token 后，更新旧的 Token，并使用新的 Token 进行后续请求。

（10）重新登录：如果刷新令牌失效或过期，则返回错误信息，要求用户重新登录。

6.1.8 什么是 JWT

JWT 是一种特殊的 Token，遵循基于 JSON 对象结构的开放标准（RFC 7519），用于安全地在网络应用中传递认证信息。JWT 允许在用户和服务端之间传递经认证的用户信息，而无需暴露用户的认证凭据。

用户登录成功后，服务端会认证用户的身份，并把一个 JWT 返回给客户端。此后，客户端在请求受保护资源时需要将 JWT 作为认证信息包含在请求中。服务端会验证 JWT 的有效性，以确认用户是否有权访问所请求的资源。JWT 的工作流程如图 6-5 所示。

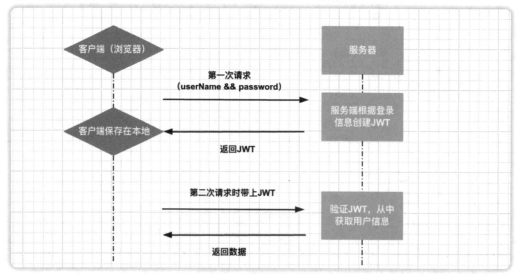

图 6-5　JWT 的工作流程

图 6-5 中的流程详细描述如下：

（1）客户端发送登录请求：客户端发送包含用户名和密码的登录请求。

（2）服务端验证用户名和密码：服务端接收到登录请求后，对用户名和密码进行验证。

（3）服务端签发 JWT：账号和密码验证成功后，服务端会创建一个包含用户身份信息的 JWT，并将其发送给客户端。

（4）客户端存储 JWT：客户端收到 Token 后，通常会将其存储在 localStorage 或 Cookie 等地方，以便在后续的请求中使用。

（5）客户端请求接口：客户端在发送请求时，会在请求头部的 Authorization 字段中使用 Bearer 模式添加 JWT，如 Authorization: Bearer <token>。

（6）服务端验证 JWT：服务端在接收到客户端的请求后，会从请求中提取 JWT 并进行验证。如果 JWT 有效且包含足够的权限，那么服务端会处理请求并返回相应的数据。

细心的读者可能会发现，JWT 与传统的 Token 认证方式在工作流程上有相似之处：在首次认证响应时，服务器都会将认证信息返回给客户端，客户端随后将其存储起来；当客户端再次请求服务器资源时，都需要携带这些认证信息以便进行验证。尽管工作流程相似，但它们之间还是存在一些关键的区别，下一小节将详细介绍。

6.1.9　JWT 与 Token 的区别

JWT 与传统 Token 在身份验证和资源访问控制中都扮演着令牌的角色，它们有一些共同点。

- 访问资源的令牌：无论是 JWT 还是传统 Token，它们都被用作访问资源的令牌。
- 记录用户认证信息：它们都用于记录用户的认证信息，以便服务端进行验证。
- 服务端无状态化：两者都支持服务端的无状态化，即服务端不需要存储会话信息。
- 跨域资源共享：JWT 和 Token 都能够支持跨域资源共享（Cross-Origin Resource Sharing，CORS）。

然而，JWT 与传统 Token 之间也存在一些关键的区别。

- 鉴权机制：传统 Token 鉴权通常需要服务端解析客户端发送的 Token，并根据该 Token 去数据库查询用户信息，以此验证 Token 的有效性。
- 多样性：Token 可以采取多种形式，例如随机字符串、OAuth 令牌、Session Token 等。而 JWT 是 Token 的一种具体实现形式。
- 信息存储与加密：JWT 在发行时会将用户信息加密并存储在 Token 自身中，客户端携带 JWT 进行请求，服务端通过密钥进行解密校验，无须再次查询数据库。

6.1.10　什么是 SSO

SSO（Single Sign On，单点登录）是在多个应用系统中，只需要登录一次，就可以访问其他相互信任的应用系统。

在企业中，多个应用系统可能在同一域名下，子系统通过二级域名来区分。当然，多个应用系统也可能分布在不同域名下，下面分别来说明这两种场景中单点登录的工作流程。

1. 同域名下的单点登录

企业通常拥有一个主域名，并利用二级域名来区分其不同的子系统。例如，如果企业的顶级域名是 xxx.com，那么业务系统 A 和 B 可能分别位于 a.xxx.com 和 b.xxx.com。为了实现单点登录，企业需要一个集中的登录系统，比如位于 sso.xxx.com。

在单点登录机制下，用户只需在 sso.xxx.com 进行一次登录，随后便能无缝访问系统 A 和 B，无须重复登录。这种登录认证机制允许用户在多个相关联的系统间自由切换，同时保持会话的连续性。具体的工作流程如图 6-6 所示。

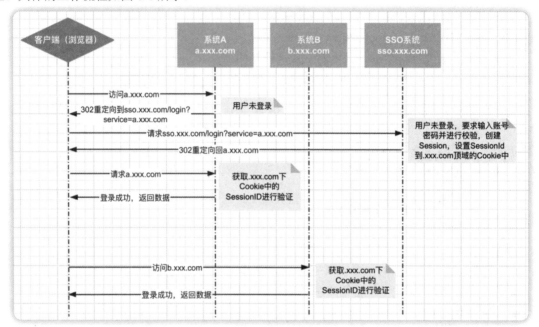

图 6-6 单点登录的工作流程

图 6-6 中的流程详细描述如下：

（1）客户端访问系统 A：用户尝试访问系统 A，由于在顶级域名 xxx.com 下未检测到用户的 Cookie 信息，系统触发重定向流程。

（2）重定向至 SSO 系统：用户被重定向到 SSO 登录页面（sso.xxx.com），并附上系统 A 的标识或重定向 URL。用户此时需输入账号和密码进行登录。登录成功后，系统在顶级域名下设置 Cookie 信息。

（3）重定向回系统 A：SSO 系统根据用户输入的重定向 URL，将用户带回系统 A，并携带 SessionID 的 Cookie。系统 A 通过 Cookie 中的 SessionID 验证用户身份，确认无误后允许用户登录并访问数据。

（4）客户端访问系统 B：用户访问系统 B 时，由于顶级域名 xxx.com 下的 Cookie 中已存在 SessionID，系统 B 通过该 SessionID 查询并验证用户的 Session 信息。身份验证通过后，用户即可成功登录并获取数据。

（5）退出登录：用户在任一系统中登出时，系统需清除顶级域名下的 Cookie 信息，以确保用户在所有相关系统中都已登出。

需要注意的是，SSO、系统 A 和 B 虽然是不同的应用，但它们的 Session 存储在自己的服务器上，并不默认共享。当用户登录 SSO 系统后，若要访问系统 A，系统 A 需要能够识别和验证 SSO 系统设置的 Cookie 中的 SessionID。为解决这一问题，可以采用如 Redis 这样的工具来实现 Session 的共享。此外，也可以使用 Token 机制来维护用户的认证信息，Token 由客户端存储，从而使服务端达到无状态化。

理解了同域名下的 SSO 原理后，我们可以进一步探讨不同域名下的 SSO 实现方式。

2. 不同域名下的单点登录

在相同域名下实现单点登录时，可以利用顶级域名 Cookie 的特性来共享会话信息。然而，不同域名的网站由于浏览器的同源策略限制，通常无法直接共享 Cookie。要解决这一问题，我们可以采用中央认证服务（Central Authentication Service，CAS）来集中管理身份认证。跨域 SSO 的工作流程如图 6-7 所示。

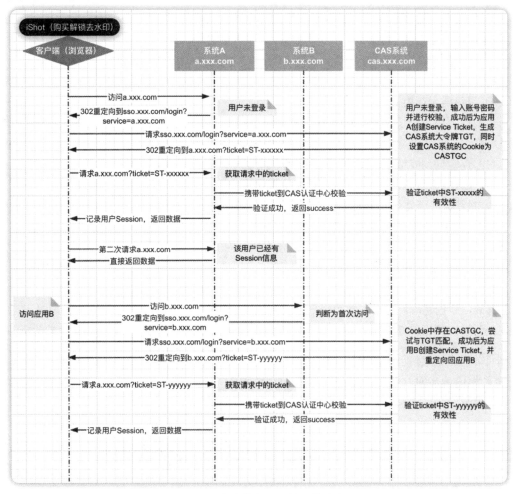

图 6-7　跨域单点登录的工作流程

图 6-7 中的流程详细描述如下：

（1）用户访问应用 A：用户在浏览器中输入应用 A 的网址并尝试访问。

（2）检查用户登录状态：应用 A 检查用户是否已登录。如果用户未登录，则跳转到统一 CAS 认证中心进行登录。

（3）重定向到 CAS 的登录页面：应用 A 将用户重定向到 CAS 的登录页面，并传递应用 A 的标识信息（例如应用 A 的标识符或回调 URL）。

（4）用户登录：用户在 CAS 系统的登录页面输入其凭证（例如用户名和密码），然后进行身份验证。

（5）生成令牌：验证成功后，CAS 系统将生成一个 TGT（Ticket Granted Ticket，俗称大令牌或票根），该令牌可以用于签发 ST（Service Ticket，小令牌）。为了隔离不同应用的验证过程，系统会根据 TGT 再生成 ST，并将该小令牌绑定到重定向 URL 中，随后返回给应用 A。同时，CAS 系统会设置一个名为 CASTGC 的 Cookie，该 Cookie 与 TGT 的关系类似于 Web 会话（Session）与会话 ID（SessionID）之间的关系。通过 CASTGC，系统可以定位到相应的 TGT。

（6）携带令牌返回应用 A：CAS 系统用户重定向到应用 A 的回调 URL。

（7）应用 A 验证令牌：应用 A 获取 URL 后面携带的 Ticket，向 CAS 服务器发起 HTTP 请求，以验证 Ticket 的有效性。如果令牌有效，则创建用户会话，并将用户标记为已登录状态。

（8）再次访问应用 A 的资源：用户现在被视为已登录应用 A，并且可以访问其受保护的资源。

（9）用户访问应用 B：用户在同一浏览器会话中尝试访问应用 B。

（10）检查用户登录状态：应用 B 会在守卫或拦截器中（以 Nest 为例）检查用户是否已登录，发现用户首次访问，于是发起重定向，去 CAS 系统登录。

（11）重定向到 CAS 系统：应用 B 将用户重定向到 CAS 系统，并传递应用 B 的回调 URL。

（12）CAS 系统验证会话：CAS 系统检查用户的会话状态，发现之前已经访问过一次了，因此 Cookie 中会携带一个 CASTGC。

（13）生成新令牌：CAS 系统生成一个新令牌，并将其与用户的身份相关信息关联起来。

（14）返回令牌：CAS 系统通过 TGC 查找到对应的 TGT，于是用 TGT 签发一个 ST，回调 URL 携带上 ST，并将用户重定向到应用 B。

（15）应用 B 验证令牌：应用 B 获取到 URL 后面的 Ticket 后，验证 Ticket 令牌的有效性。如果令牌有效，则创建用户会话，并将用户标记为已登录状态。

（16）访问应用 B 的资源：用户现在被视为已登录应用 B，并且可以访问其受保护的资源。

到目前为止，我们已经学习了身份验证中常见的概念，如 Cookie、Session、Token、JWT 以及 SSO 等，相信读者对这些知识已经有了更加系统和深入的理解。下一节将通过案例演示如何在 Nest 中使用身份验证。

6.2 基于 Passport 和 JWT 实现身份验证

本节将介绍如何在 Nest 中结合 Passport（通行证）和 JWT 进行身份验证。首先，详细介绍 Passport 的基本概念，包括其作用、使用方式及其与 JWT 的关系。然后，创建一个 Nest 项目，安装并集成 Passport 与 JWT，用 Passport 提供的本地策略来实现账号和密码的登录验证，并结合

JWT 策略实现接口的验证功能。

6.2.1 基本概念

Passport 是最流行的 Node.js 身份验证中间件，广泛应用于许多生产应用中。它支持多种身份验证策略，包括本地认证（用户名和密码）、OAuth 策略、OAuth2 策略、OpenID Connect 策略、JWT 策略等，可以满足不同场景下的身份认证需求。

Passport 基于策略模式扩展不同的验证方式，它不直接提供 JWT 的具体实现，而是扩展了一个名为 passport-jwt 的策略包，用于在 Node.js 应用中实现基于 JWT 的无状态身份验证机制。JWT 被用作验证令牌（Token），在客户端与服务端之间传递用户身份信息，而 Passport 负责验证和解析 JWT 中的 Payload，将用户信息添加到请求对象中，以便在后续的路由处理程序中通过请求对象来获取用户信息，进行访问控制和权限判断。

需要注意的是，在 Nest 中依旧可以单独用 JWT 进行身份认证，使用 Passport 是因为它提供了进一步的抽象，以简化身份认证流程。此外，Passport 与 Nest 框架的分层理念相契合，这使得在开发大型应用时，身份认证的维护和扩展变得更加容易。

6.2.2 项目准备

为了更好地说明 Passport 与 JWT 的用法，我们将创建一个 Nest 项目来实践应用它。首先，运行“nest new nest-passport-jwt -p pnpm”命令，创建完成后的结果如图 6-8 所示。

```
$ nest new nest-passport-jwt -p pnpm
⚡  We will scaffold your app in a few seconds..

CREATE nest-passport-jwt/.eslintrc.js (663 bytes)
CREATE nest-passport-jwt/.prettierrc (51 bytes)
CREATE nest-passport-jwt/README.md (3347 bytes)
CREATE nest-passport-jwt/nest-cli.json (171 bytes)
CREATE nest-passport-jwt/package.json (1958 bytes)
CREATE nest-passport-jwt/tsconfig.build.json (97 bytes)
CREATE nest-passport-jwt/tsconfig.json (546 bytes)
CREATE nest-passport-jwt/src/app.controller.spec.ts (617 bytes)
CREATE nest-passport-jwt/src/app.controller.ts (274 bytes)
CREATE nest-passport-jwt/src/app.module.ts (249 bytes)
CREATE nest-passport-jwt/src/app.service.ts (142 bytes)
CREATE nest-passport-jwt/src/main.ts (208 bytes)
CREATE nest-passport-jwt/test/app.e2e-spec.ts (630 bytes)
CREATE nest-passport-jwt/test/jest-e2e.json (183 bytes)

✔ Installation in progress... ☕

🚀  Successfully created project nest-passport-jwt
```

图 6-8　创建项目

要实现用户登录验证，需要使用 Passport 提供的本地策略包（passport-local），它实现了用户名和密码的验证机制。同时，还需安装 Passport 的其他依赖包。执行如下命令：

```
pnpm add --save @nestjs/passport passport passport-local
pnpm add --save-dev @types/passport-local
```

然后创建两个功能模块：auth 和 user。运行如下命令：

```
nest g mo auth
nest g s auth --no-spec
nest g mo user
nest g s user --no-spec
```

6.2.3 用本地策略实现用户登录

创建成功后，auth 模块用于实现验证用户身份的相关逻辑，而 user 模块用于用户的增删改查操作。为了方便说明，本节将使用模拟数据来代替从数据库中查询用户。下面是 user.service.ts 的代码：

```
import { Injectable } from '@nestjs/common';

export type User = {
  id: number;
  userName: string;
  password: string;
};

@Injectable()
export class UserService {
  private users: User[] = [
    { id: 1, userName: 'user1', password: 'user111' },
    { id: 2, userName: 'user2', password: 'user222' },
    { id: 3, userName: 'user3', password: 'user333' },
  ];

  async findOneByUserName(userName: string): Promise<User | undefined> {
    return this.users.find((item) => item.userName === userName);
  }
}
```

模拟数据库中有三条数据，分别为 user1、user2 和 user3 三个用户，并且定义了 findOneByUserName 方法，用于根据用户名查找记录。

在查询到用户记录后，在 auth.service.ts 中定义 validateUser 方法，用于验证请求中用户传递的密码是否匹配。如果匹配，则返回除密码之外的其他字段，否则返回 null。示例代码如下：

```
import { Injectable } from '@nestjs/common';
import { UserService } from '../user/user.service';

@Injectable()
export class AuthService {
  constructor(private usersService: UserService) {}

  async validateUser(userName: string, passord: string): Promise<any> {
    const user = await this.usersService.findOneByUserName(userName);
    if (user && user.password === password) {
      const { password, ...result } = user;
      return result;
    }
    return null;
  }
}
```

有了验证方法，现在我们可以实现本地身份验证策略了。在 auth 文件夹下创建一个名为 local.strategy.ts 的文件，定义 LocalStrategy 类并继承 PassportStrategy。示例代码如下：

```
import { Strategy } from 'passport-local';
import { PassportStrategy } from '@nestjs/passport';
import { Injectable, UnauthorizedException } from '@nestjs/common';
import { AuthService } from './auth.service';

@Injectable()
export class LocalStrategy extends PassportStrategy(Strategy) {
  constructor(private authService: AuthService) {
    super({
      usernameField: 'userName',
    });
  }

  async validate(userName: string, password: string): Promise<any> {
    const user = await this.authService.validateUser(userName, password);
    if (!user) {
      throw new UnauthorizedException();
    }
    return user;
  }
}
```

无论是定义本地策略还是即将实现的 JWT 策略，都需要继承 PassportStrategy 类，并实现其 validate 方法。默认情况下，本地策略需要从请求中获取名为 username 和 password 的两个属性。然而，在上述示例代码中，笔者采用了驼峰命名法，将字段名改为 userName。因此，需要在调用 super 方法时传递一个选项对象，例如{ usernameField: 'userName' }，以重写默认的字段名。这种做法在邮箱密码登录场景下特别有用。

此外，通过调用先前定义的 validateUser 方法来验证用户身份。如果用户不存在，则抛出 UnauthorizedException 异常。

然后，在 auth.module.ts 文件中引入本地策略，并导入 PassportModule 模块。相关代码如下：

```
import { Module } from '@nestjs/common';
import { AuthService } from './auth.service';
import { LocalStrategy } from './local.strategy';
import { UserModule } from 'src/user/user.module';
import { PassportModule } from '@nestjs/passport';

@Module({
  imports: [UserModule, PassportModule],
  providers: [AuthService, LocalStrategy]
})
export class AuthModule {}
```

最后，在 app.controller.ts 中实现登录方法（login），使用 Nest 中的守卫（Guard）进行切面判断，并指定路由程序的认证策略为 local，代码如下：

```
import { Controller, Post, UseGuards } from '@nestjs/common';
import { Req } from '@nestjs/common/decorators';
```

```typescript
import { AuthGuard } from '@nestjs/passport';
import { AppService } from './app.service';

@Controller()
export class AppController {
  constructor(private readonly appService: AppService) {}

  @UseGuards(AuthGuard('local'))
  @Post('auth/login')
  async login(@Req() req) {
    return req.user;
  }
}
```

接下来，对实现的身份验证功能进行测试。首先，运行"pnpm start:dev"命令以启动开发服务。如果服务启动成功，将继续进行测试。在之前的章节中，我们使用了 .http 文件来发送请求进行测试。现在，使用 macOS 系统自带的 curl 工具来进行测试（对于 Windows 系统用户，需要先下载并安装 curl 工具），执行如下命令：

```
curl -X POST http://localhost:3000/auth/login -d '{"userName": "user1", "password": "user111"}' -H "Content-Type: application/json"
```

运行结果如图 6-9 所示。

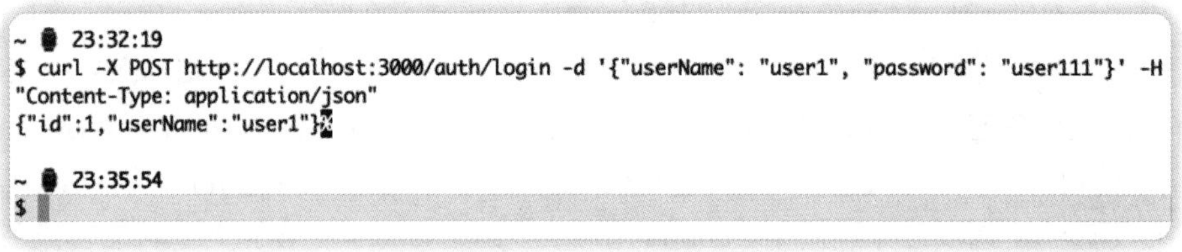

```
~  23:32:19
$ curl -X POST http://localhost:3000/auth/login -d '{"userName": "user1", "password": "user111"}' -H
"Content-Type: application/json"
{"id":1,"userName":"user1"}
~  23:35:54
$
```

图 6-9　用户登录认证成功

此时输入错误的密码 userxxx，再次运行命令，运行结果如图 6-10 所示，返回 401 授权认证失败提示。

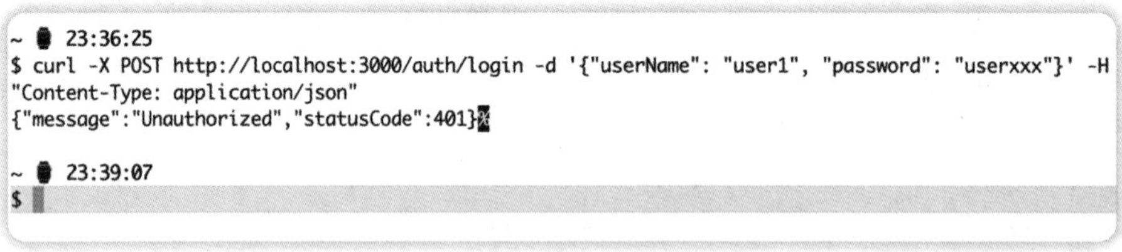

```
~  23:36:25
$ curl -X POST http://localhost:3000/auth/login -d '{"userName": "user1", "password": "userxxx"}' -H
"Content-Type: application/json"
{"message":"Unauthorized","statusCode":401}
~  23:39:07
$
```

图 6-10　用户登录认证失败

我们已经实现了用户的身份验证功能，但在实际应用中，通常只在登录时使用用户名和密码进行验证。对于其他接口的请求，我们使用令牌（Token）来进行身份验证，这时就需要引入 JWT 验证机制。

6.2.4　用 JWT 策略实现接口校验

要想在 Passport 的基础上实现 JWT 验证，我们还需要安装其对应的依赖，运行如下命令：

```
pnpm add --save @nestjs/jwt passport-jwt
pnpm add --save-dev @types/passport-jwt
```

其中，passport-jwt 是实现 JWT 策略的 Passport 包，而 Nest 集成了 JWT 相关操作，方便实现依赖注入。

现在，我们在 AuthModule 中引入 JwtModule，代码如下：

```
import { Module } from '@nestjs/common';
import { AuthService } from './auth.service';
import { LocalStrategy } from './local.strategy';
import { UserModule } from 'src/user/user.module';
import { PassportModule } from '@nestjs/passport';
import { JwtModule } from '@nestjs/jwt';
import { JwtStrategy } from './jwt.strategy';

@Module({
  imports: [
    UserModule,
    PassportModule,
    // 新增 JWT 模块
    JwtModule.register({
      // 这里应该读取配置中的 secret
      secret: 'jwt-secret',
      signOptions: { expiresIn: '7d' },
    }),
  ],
  providers: [AuthService, LocalStrategy, JwtStrategy],
  exports: [AuthService],
})
export class AuthModule {}
```

有了 JWT 模块，还需要有对应的 JWT 策略，在 auth 文件夹下创建一个名为 jwt-strategy.ts 的文件，同样定义 JwtStrategy 类并继承 PassportStrategy。示例代码如下：

```
import { ExtractJwt, Strategy } from 'passport-jwt';
import { PassportStrategy } from '@nestjs/passport';
import { Injectable } from '@nestjs/common';

@Injectable()
export class JwtStrategy extends PassportStrategy(Strategy) {
  constructor() {
    super({
      // 表示从 header 中的 Authorization 的 Bearer 表头中获取 token 值
      jwtFromRequest: ExtractJwt.fromAuthHeaderAsBearerToken(),
      // 不忽视 token 过期的情况，过期会返回 401
      ignoreExpiration: false,
      // 这里应该读取配置中的 secret
      secretOrKey: 'jwt-secret',
    });
  }

  async validate(payload: any) {
    return { id: payload.sub, username: payload.userName };
  }
```

}
```

在上述代码中，jwtFromRequest 表示从 Request 对象中获取 Token 的方式，包含 header、body、URLparameter 等，secretOrKey 是 JWT 密钥，ignoreExpiration 设置为 false 会在 Token 过期时返回 401 Unauthorized 提示。最后实现 validate 方法并返回一个 user 对象。

策略实现完成后，接下来在 AppController 中定义一个名为 getUserInfo 的路由方法。这个方法明显是受保护的资源，需要进行身份验证才能访问。同时，我们需要修改 login 方法，以便在用户登录成功后返回一个 access_token 给前端进行保存。示例代码如下：

```
import { Controller, Post, UseGuards } from '@nestjs/common';
import { Get, Req } from '@nestjs/common/decorators';
import { AuthGuard } from '@nestjs/passport';
import { AuthService } from './auth/auth.service';

@Controller()
export class AppController {
  constructor(private readonly authService: AuthService) {}

  @UseGuards(AuthGuard('local'))
  @Post('auth/login')
  async login(@Req() req) {
    // 调用 auth 中 login 方法返回 access_token
    return this.authService.login(req.user);
  }

  // 新增获取用户信息的方法
  @UseGuards(AuthGuard('jwt'))
  @Get('getUserInfo')
  getUserInfo(@Req() req) {
    return req.user;
  }
}
```

可见，我们使用了 AuthGuard('jwt')来指定该路由方法应用 JWT 认证策略，并且调用了 AuthService 中的方法来返回 access_token。AuthService 中 login 方法的实现代码如下：

```
import { Injectable } from '@nestjs/common';
import { UserService, User } from '../user/user.service';
import { JwtService } from '@nestjs/jwt';

@Injectable()
export class AuthService {
  constructor(
    private usersService: UserService,
    private jwtService: JwtService,
  ) {}

  async validateUser(userName: string, password: string): Promise<any> {
    const user = await this.usersService.findOneByUserName(userName);
    if (user && user.password === password) {
      const { password, ...result } = user;
      return result;
```
```

```
    }
    return null;
  }
  // 新增 login 方法，生成 token
  async login(user: User) {
    const payload = { username: user.userName, sub: user.id };
    return {
      access_token: this.jwtService.sign(payload),
    };
  }
}
```

我们最终调用 JwtService 的 sign 方法生成了一个 token，其中定义 username 和 sub 是为了与 JWT 标准保持一致。

至此，我们已经完成了所有的认证编码过程。接下来测试效果，运行以下命令尝试获取用户信息：

```
curl -X GET http://localhost:3000/getUserInfo
```

显然，在未登录情况下，如果没有携带 access_token，服务器将返回 401 错误，如图 6-11 所示。

```
~  7:45:39
$ curl -X GET http://localhost:3000/getUserInfo
{"message":"Unauthorized","statusCode":401}

~  7:45:40
$
```

图 6-11　获取用户信息失败

接下来执行用户登录操作，获取 access_token，运行如下命令：

```
curl -X POST http://localhost:3000/auth/login -d '{"userName": "user1", "password": "user111"}' -H "Content-Type: application/json"
```

返回了一串格式为 xxx.xxx.xxx 的 JWT 加密字符串，如图 6-12 所示。

```
~  7:55:26
$ curl -X POST http://localhost:3000/auth/login -d '{"userName": "user1", "password": "user111"}' -H
"Content-Type: application/json"
{"access_token":"eyJhbGciOiJIUzI1NiIsInR5cCI6IkpXVCJ9.eyJ1c2VybmFtZSI6InVzZXIxIiwic3ViIjoxLCJpYXQiOjE
3MDgzODY5MzIsImV4cCI6MTcwODk5MTczMn0.JPJIAYxKD3fhcz9SmQhc5Gz8vgq4VS9bTVtZQK6f7TY"}
```

图 6-12　获取 access_token 成功

我们在请求头中添加这串 access_token，运行如下命令：

```
curl -X GET http://localhost:3000/getUserInfo -H "Authorization: Bearer
eyJhbGciOiJIUzI1NiIsInR5cCI6IkpXVCJ9.eyJ1c2VybmFtZSI6InVzZXIxIiwic3ViIjoxLCJpYXQiOjE3MDgz
ODY5MzIsImV4cCI6MTcwODk5MTczMn0.JPJIAYxKD3fhcz9SmQhc5Gz8vgq4VS9bTVtZQK6f7TY"
```

可以看到权限验证成功，返回了正确的数据，如图 6-13 所示。

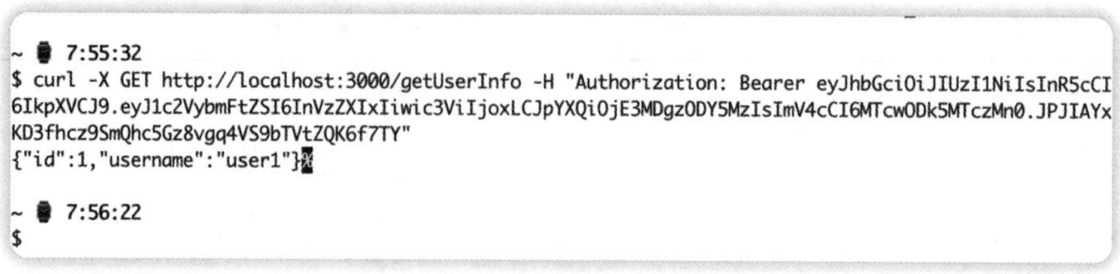

图 6-13 获取用户信息成功

至此，我们基本完成了 JWT 身份验证的核心过程。但前面的部分代码仍不够优雅，下一小节跟随笔者来优化它们。

6.2.5 代码优化

在 AppController 中，在路由方法上定义认证策略时，我们使用 AuthGuard('local') 或 AuthGuard('jwt')来指定使用哪个策略，事实上这无形之中引入了魔术字符串，并不好维护，可以把它抽离成单独的类来管理。在 auth 文件夹下新建 jwt-auth.guard.ts 文件，代码如下：

```
import { Injectable } from '@nestjs/common';
import { AuthGuard } from '@nestjs/passport';

@Injectable()
export class JwtAuthGuard extends AuthGuard('jwt') {}
```

同样，修改本地策略，在同级目录下新建 local-auth.guard.ts 文件，代码如下：

```
import { Injectable } from '@nestjs/common';
import { AuthGuard } from '@nestjs/passport';

@Injectable()
export class LocalAuthGuard extends AuthGuard('local') {}
```

接下来，在 AppController 中引入它们，代码如下：

```
import { LocalAuthGuard } from './auth/local-auth.guard';
import { JwtAuthGuard } from './auth/jwt-auth.guard';

@Controller()
export class AppController {
  ...
  @UseGuards(LocalAuthGuard)
  ...

  @UseGuards(JwtAuthGuard)
  ...
}
```

除此之外，细心的读者可能会留意到，在配置 JWT 密钥时，我们把 jwt-secret 字符串硬编码到代码中。然而，在生产环境中并不建议这样做，而是应该使用密钥库、环境变量或配置服务来管理此密钥（类似管理数据库账号和密码）。为此，我们可以在项目根目录下创建一个.env 文件来维护它：

```
JWT_SECRET=my-jwt-secret
JWT_EXPIRE_TIME=7d
```

为了获取配置中的信息，还需要安装@nestjs/config 包。在 AppModule 中引入 ConfigModule，并将配置模块注入全局，代码如下：

```
import { Module } from '@nestjs/common';
import { AppController } from './app.controller';
import { AppService } from './app.service';
import { AuthModule } from './auth/auth.module';
import { UserModule } from './user/user.module';
import { JwtAuthGuard } from './auth/jwt-auth.guard';
import { ConfigModule } from '@nestjs/config';

@Module({
  imports: [
    AuthModule,
    UserModule,
    // 将配置模块注入全局
    ConfigModule.forRoot({
      isGlobal: true,
    }),
  ],
  controllers: [AppController],
  providers: [
    AppService,
    JwtAuthGuard
  ],
})
export class AppModule {}
```

此时，为了更好地获取配置信息，我们需要把 JWT 模块改为使用工厂函数方式进行注册，并注入 ConfigService 获取配置信息。示例代码如下：

```
import { Module } from '@nestjs/common';
import { AuthService } from './auth.service';
import { LocalStrategy } from './local.strategy';
import { UserModule } from 'src/user/user.module';
import { PassportModule } from '@nestjs/passport';
import { JwtModule } from '@nestjs/jwt';
import { JwtStrategy } from './jwt.strategy';
import { ConfigModule, ConfigService } from '@nestjs/config';

@Module({
  imports: [
    UserModule,
    PassportModule,
    // 新增 JWT 模块
    JwtModule.registerAsync({
      useFactory: async (configService: ConfigService) => (
        {
          // 读取配置中的 secret
          secret: configService.get<string>('JWT_SECRET'),
          signOptions: {
            expiresIn: configService.get<string>('JWT_EXPIRE_TIME')
```

```
      },
    }
  ),
    // 将 ConfigService 注入工厂函数中
    inject: [ConfigService],
  }),
],
providers: [AuthService, LocalStrategy, JwtStrategy],
exports: [AuthService]
})
export class AuthModule {}
```

另外，我们还需要在 JWT 策略中修改密钥获取方式，将构造器代码修改如下：

```
import { ExtractJwt, Strategy } from 'passport-jwt';
import { PassportStrategy } from '@nestjs/passport';
import { Injectable } from '@nestjs/common';
// 新增依赖
import { ConfigService } from '@nestjs/config';

@Injectable()
export class JwtStrategy extends PassportStrategy(Strategy) {
  // 注入配置服务
  constructor(configService: ConfigService) {
    super({
      // 表示从 header 中的 Authorization 的 Bearer 表头中获取 token 值
      jwtFromRequest: ExtractJwt.fromAuthHeaderAsBearerToken(),
      // 不忽视 token 过期的情况，过期会返回 401
      ignoreExpiration: false,
      // 读取配置中的 secret
      secretOrKey: configService.get<string>('JWT_SECRET'),
    });
  }

  async validate(payload: any) {
    return { id: payload.sub, username: payload.username };
  }
}
```

至此，我们已经完成了所有的优化工作。最后来测试一下，登录并成功获取用户信息，效果如图 6-14 所示。

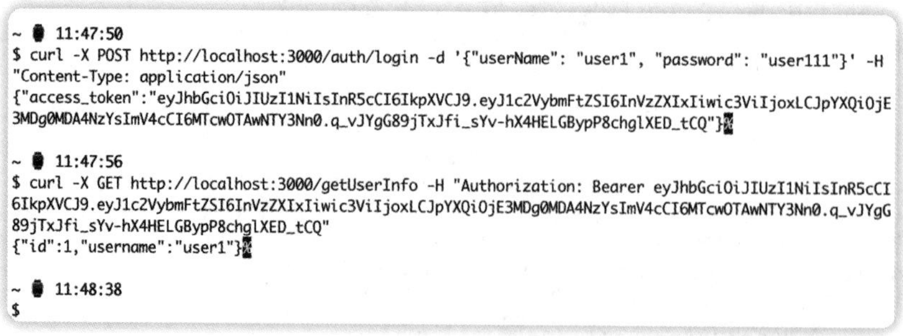

图 6-14 运行成功

综上所述，我们详细介绍了如何结合 Nest 中的 Passport 模块和 JWT 实现身份验证。通过

Passport 提供的策略和中间件，我们能够轻松实现多种身份验证策略，而 JWT 提供了一种安全可靠的身份认证方式，使得在无状态的分布式环境下进行身份验证更加便捷和高效。

6.3 基于 RBAC 实现权限控制

在实际应用中，基于角色的访问控制（Role Based Access Control，RBAC）是当前应用最广泛的角色控制模型。RBAC 通过将用户分配到不同的角色，并为每个角色分配不同的权限来管理用户对系统的访问权。

本节将介绍在 Nest 中如何使用 RBAC 模型，在接口级别实现权限控制，以确保每个用户只能访问其具备权限的接口，从而实现一个简单可扩展的权限系统。

6.3.1 基本概念

在 RBAC 模型中，通常包含以下几个核心概念。

- 用户（User）：系统中的一个操作实体，是系统的使用者，被赋予一个或多个角色来获得对应的权限。
- 角色（Role）：角色是一组权限的集合，用于定义用户在系统中的身份或地位。用户可以被分配一个或多个角色，不同的角色有不同的权限，例如店长拥有删除数据的权限，而普通员工没有。
- 权限（Permission）：权限是指用户在系统中执行某个操作的能力，如删除、编辑、查看等。

它们之间的关系概括如图 6-15 所示。

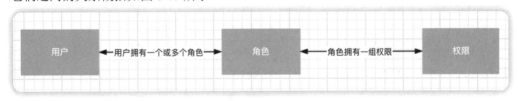

图 6-15　RBAC 的基本关系

可见，用户与角色之间是一对多或多对多的关系，角色与权限之间是多对多的关系。

有读者可能会有疑惑，为什么不直接把权限分配给用户呢？

实际上，是可以直接把权限分配给用户的，只是这样做会少一层关系，这在一定程度上削弱了系统的扩展性、维护性、安全性等，因此适用于用户量少、角色类型少的平台。

通过角色来管理权限的分配，这种设计有以下 3 个优点。

（1）扩展性：通过角色来分配权限，在大型复杂的系统中，可以扩展角色层次，如部门经理、技术总监、组长以及组员，不同角色之间存在包含关系。

（2）安全性：有了角色管理，系统超级管理员只需要为角色分配一组权限集，这意味着无须操作庞大的用户为其分配权限，减少了操作失误的风险，同时也能够提升操作效率。

（3）维护性：当用户的角色权限发生变化时，只需更新角色与权限之间的关系，而无须为每个用户单独修改权限设置，从而降低维护成本。

现在我们对 RBAC 有了基本的认识，接下来在 Nest 中演示如何使用它实现权限控制。

6.3.2 数据表设计

以常用的 RBAC 管理系统（Admin）为例，角色与权限相关联，用户通过拥有合适的角色获得相应的权限。我们通过图 6-16 来加深理解。

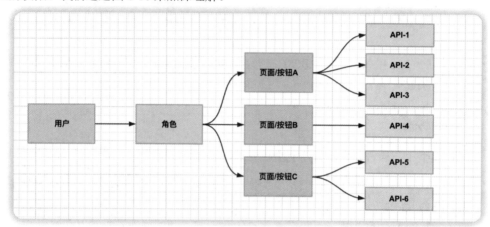

图 6-16　RBAC 实际关系图

可见，角色会得到一组页面或按钮的权限集，每个页面或按钮背后可能包含一个或多个接口，前端通过控制页面 UI 来限制用户是否有权限调用相关 API，后端同样需要根据用户角色判断是否有指定接口的权限，这样才能成功调用 API。安全是双方的事情。

理清关系之后，接下来设计数据库表结构，如图 6-17 所示。

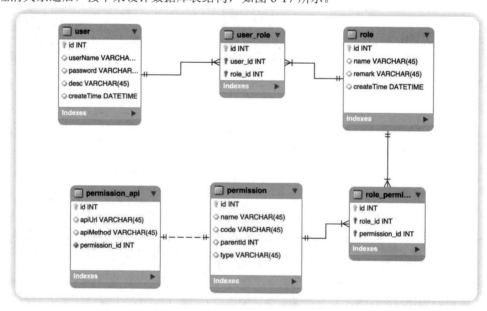

图 6-17　RBAC 关系表

分别创建 user 用户表、role 角色表、permission 权限表、permission_api 权限表。另外，user_role 表用于关联 user 表与 role 表，使用 role_permission 表来关联 role 表与 permission 表。为了契合实际开发，本节将不使用外键，而是采用业务逻辑来管理表与表之间的关联。

6.3.3 项目准备

前面已经实现了登录和接口身份验证。为了方便起见，本小节基于 nest-passport-jwt 项目实现 RBAC 权限控制，我们复制项目并将其重命名为 nest-rbac。

有了 user 模块，我们还需要创建 role 和 permission 模块。运行如下命令：

```
nest g resource permission --no-spec
nest g resource role --no-spec
```

接下来安装依赖，需要用到 typeorm 和 mysql2 @nestjs/typeorm path-to-regexp 包。运行"pnpm add path-to-regexp typeorm mysql2 @nestjs/typeorm -S"命令安装所需依赖。

安装完成后，需要初始化数据库连接。在 AppModule 中导入 TypeOrmModule，核心代码如下：

```
imports: [
  // 初始化 MySQL 连接
  TypeOrmModule.forRootAsync({
    inject: [ConfigService],
    useFactory: (config: ConfigService) => ({
      type: 'mysql',
      host: config.get<string>('MYSQL_HOST'),
      port: config.get<number>('MYSQL_PORT'),
      username: config.get<string>('MYSQL_USER'),
      password: config.get<string>('MYSQL_PASSWORD'),
      database: config.get<string>('MYSQL_DATABASE'),
      entities: [__dirname + '/**/*.entity{.ts,.js}'],
      autoLoadEntities: true,
      // 在生产环境中禁止开启，应该使用数据迁移
      synchronize: true
    })
  }),
  AuthModule,
  UserModule,
  // 将配置模块注入全局
  ConfigModule.forRoot({
    isGlobal: true,
  }),
  PermissionModule,
  RoleModule,
],
```

在上述代码中，通过 ConfigService 获取数据库配置，这些配置已经抽离到 .env 环境变量中统一管理。代码如下：

```
# mysql 配置
MYSQL_HOST=localhost
MYSQL_PORT=3306
MYSQL_USER=root
MYSQL_PASSWORD=xxxxx
```

```
MYSQL_DATABASE=nest_rbac
```

6.3.4 创建实体

配置完毕后,接下来按照图 6-16 的结构来设计表结构。每个模块下都有一个 entities 文件夹(user 模块需要手动创建),需要完善对应的 entity 实体。启动服务时,系统会自动创建数据库表。

在 user/entities 中分别创建 user.entity.ts 和 user-role.entity.ts 文件。user-role 实体用于关联用户与角色,是示例代码如下:

```
// user.entity.ts
import { Entity, PrimaryGeneratedColumn, Column } from "typeorm";

@Entity()
export class User{

    @PrimaryGeneratedColumn()
    id: number;

    @Column()
    userName: string;

    @Column()
    password: string;

    @Column()
    desc: string;

    @Column()
    createTime: Date;

}

// user-role.entity.ts
import { Entity, PrimaryGeneratedColumn, Column } from "typeorm";

@Entity()
export class UserRole{

    @PrimaryGeneratedColumn()
    id: number;

    @Column()
    userId: number;

    @Column()
    roleId: number;

}
```

同样,在 role/entities 下的 role.entity.ts 和 role-permission 文件中分别定义角色、关联角色与权限实体。示例代码如下:

```
// role.entity.ts
```

```
import { Entity, PrimaryGeneratedColumn, Column } from "typeorm";

@Entity()
export class Role{

    @PrimaryGeneratedColumn()
    id: number;

    @Column()
    name: string;

    @Column()
    remark: string;

    @Column()
    createTime: Date;

}
// role-permission.entity.ts
import { Entity, PrimaryGeneratedColumn, Column } from "typeorm";

@Entity()
export class RolePermission{

    @PrimaryGeneratedColumn()
    id: number;

    @Column()
    role_id: number;

    @Column()
    permission_id: number;

}
```

最后，在 permission/entities 下完善权限相关的实体定义。这包括 permission.entity.ts 和 permission-api.entity.ts 两个文件。示例代码如下：

```
// permission.entity.ts
import { Entity, PrimaryGeneratedColumn, Column } from "typeorm";

@Entity()
export class Permission{

    @PrimaryGeneratedColumn()
    id: number;

    @Column()
    name: string;

    @Column()
    code: string;

    @Column()
    parentId: number;
```

```
    @Column()
    type: string;

}
// permission-api.entity.ts
import { Entity, PrimaryGeneratedColumn, Column } from "typeorm";

@Entity()
export class PermissionApi{

    @PrimaryGeneratedColumn()
    id: number;

    @Column()
    apiUrl: string;

    @Column()
    apiMethod: string;

    @Column()
    permission_id: number;

}
```

6.3.5 启动服务

我们已经设计并创建好了数据库表。运行"pnpm start:dev"命令启动服务后，可以在 VS Code 的可视化数据库工具中看到成功创建的数据表，如图 6-18 所示。

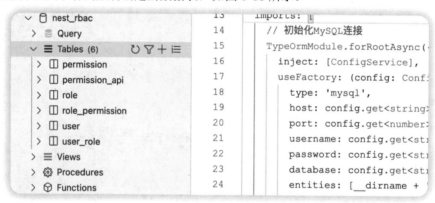

图 6-18　成功启动服务

6.3.6 实现角色守卫控制

为了实现不同用户角色管理不同接口权限的功能，我们首先考虑通过守卫（Guard）来实现这一需求。守卫的工作流程如下：

（1）判断当前请求的接口是否需要进行 Token 验证（例如，登录和注册接口通常不需要 Token 验证）。

（2）判断该接口是否需要进行权限验证（一些向外部开放的接口可能不需要进行权限验证）。

（3）检查当前用户是否具有访问当前接口的权限。

在明确了这一逻辑之后，我们可以在 auth 文件夹下创建一个新的守卫文件 role-auth.guard.ts。示例代码如下：

```
import { CanActivate, Inject, ExecutionContext, Injectable, ForbiddenException } from '@nestjs/common'
import { Reflector } from '@nestjs/core'
import { pathToRegexp } from 'path-to-regexp'
import { ALLOW_NO_PERMISSION } from '../decorators/permission.decorator'
import { PermissionService } from 'src/permission/permission.service'
import { ALLOW_NO_TOKEN } from 'src/decorators/noToken.decorator'

@Injectable()
export class RoleAuthGuard implements CanActivate {
  constructor(
    private readonly reflector: Reflector,
    @Inject(PermissionService)
    private readonly permissionService: PermissionService,
  ) {}

  async canActivate(ctx: ExecutionContext): Promise<boolean> {
    // 若函数请求头配置@AllowNoToken()装饰器，则无须验证 token 权限
    const allowNoToken = this.reflector.getAllAndOverride<boolean>(ALLOW_NO_TOKEN, [ctx.getHandler(), ctx.getClass()])
    if (allowNoToken) return true

    // 若函数请求头配置@AllowNoPermission()装饰器，则无须验证权限
    const allowNoPerm = this.reflector.getAllAndOverride<boolean>(ALLOW_NO_PERMISSION, [ctx.getHandler(), ctx.getClass()])
    if (allowNoPerm) return true

    const req = ctx.switchToHttp().getRequest()
    const user = req.user
    // 若没有携带 token，则直接返回 false
    if (!user) return false

    // 获取该用户所拥有的接口权限
    const userApis = await this.permissionService.findUserApis(user.id)
    console.log('当前用户拥有的 URL 权限集：',userApis);

    const index = userApis.findIndex((route) => {
      // 请求方法类型相同
      if (req.method.toUpperCase() === route.method.toUpperCase()) {
        // 比较当前请求 url 是否存在于用户接口权限集中
        const reqUrl = req.url.split('?')[0]
        console.log('当前请求 URL: ', reqUrl);

        return !!pathToRegexp(route.path).exec(reqUrl)
      }
      return false
    })
```

```typescript
      if (index === -1) throw new ForbiddenException('您无权限访问该接口')
      return true
    }
  }
```

同时,在 PermissionService 中,我们需要实现一个接口,用于查询用户所拥有的权限集,代码如下:

```typescript
import { Injectable } from '@nestjs/common';
import { CreatePermissionDto } from './dto/create-permission.dto';
import { UpdatePermissionDto } from './dto/update-permission.dto';
import { DataSource } from 'typeorm'
import { RouteDto } from './dto/route.dto';

@Injectable()
export class PermissionService {
  constructor(
    private dataSource: DataSource
  ) {}
  // 查找用户接口权限集
  async findUserApis(userId: string): Promise<RouteDto[]> {
    const permsResult = await this.dataSource
      .createQueryBuilder()
      .select(['pa.apiUrl', 'pa.apiMethod'])
      .from('user_role', 'ur')
      .leftJoin('role_permission', 'rp', 'ur.roleId = rp.role_id')
      .leftJoin('permission_api', 'pa', 'rp.permission_id = pa.permission_id')
      .where('ur.userId = :userId', { userId })
      .groupBy('pa.apiUrl')
      .addGroupBy('pa.apiMethod')
      .getRawMany()
    const perms = permsResult.map(item => ({path: item.pa_apiUrl, method: item.pa_apiMethod}));
    return perms
  }
  create(createPermissionDto: CreatePermissionDto) {
    return 'This action adds a new permission';
  }

  findAll() {
    return `This action returns all permission`;
  }

  findOne(id: number) {
    return `This action returns a #${id} permission`;
  }

  update(id: number, updatePermissionDto: UpdatePermissionDto) {
    return `This action updates a #${id} permission`;
  }

  remove(id: number) {
    return `This action removes a #${id} permission`;
  }
}
```

由于角色判断是通用逻辑，因此我们应该把它注册为全局守卫，而无须在每个路由方法中绑定，核心代码如下：

```
providers: [
  AppService,
  {
    provide: APP_GUARD,
    useClass: JwtAuthGuard
  },
  {
    provide: APP_GUARD,
    useClass: RoleAuthGuard,
  },
],
```

别忘记，在前面的 RoleAuthGuard 中调用了 PermissionService 服务。接下来，在 PermissionModule 中将其导出：

```
@Module({
  controllers: [PermissionController],
  providers: [PermissionService],
  exports: [PermissionService]
})
```

此外，细心的读者可能会发现，在上面的代码中，我们需要实现 @AllowNoToken() 和 @AllowNoPermission() 装饰器。这些装饰器用于对特定接口进行放行，允许它们绕过 Token 验证和权限验证。

为了组织这些装饰器，我们可以在 src 目录下新增一个 decorators 文件夹，并在其中创建 token.decorator.ts 和 permission.decorator.ts 文件。实现代码如下：

```
// token.decorator.ts
import { SetMetadata } from '@nestjs/common'

/**
 * 接口允许 Token 访问
 */
export const ALLOW_NO_TOKEN = 'allowNoToken'

export const AllowNoToken = () => SetMetadata(ALLOW_NO_TOKEN, true)
// permission.decorator.ts
import { SetMetadata } from '@nestjs/common'

/**
 * 接口允许无权限访问
 */
export const ALLOW_NO_PERMISSION = 'allowNoPerm'

export const AllowNoPermission = () => SetMetadata(ALLOW_NO_PERMISSION, true)
```

然后，在 AppController 的路由程序中使用它们，代码如下：

```
@Post('auth/login')
// 登录接口无须 Token 验证
@AllowNoToken()
```

```
@UseGuards(LocalAuthGuard)
async login(@Req() req) {
  // 调用 auth 中的 login 方法返回 access_token
  return this.authService.login(req.user);
}
// 开放 API 无须角色验证
@Get('xxx')
@AllowNoPermission()
OpenXxxApi() {
  return []
}
```

至此，我们通过守卫实现了不同场景下的接口权限控制。

6.3.7 生成测试数据

开始测试之前，先在数据库各表中插入一些数据，以便进行测试。执行如下 SQL 语句：

```sql
-- Active: 1705225958726@@127.0.0.1@3306@nest_rbac
-- 新增权限数据
INSERT INTO `permission` VALUES (1, '新增员工', 'user_add', 1, 2);
INSERT INTO `permission` VALUES (2, '删除员工', 'user_delete', 1, 2);
INSERT INTO `permission` VALUES (3, '编辑员工', 'user_edit', 1, 2);
INSERT INTO `permission` VALUES (4, '员工列表', 'user_list', 1, 1);
INSERT INTO `permission` VALUES (5, '角色管理', 'role_list', 2, 1);
INSERT INTO `permission` VALUES (6, '编辑角色', 'role_edit', 2, 2);
INSERT INTO `permission` VALUES (7, '删除角色', 'role_delete', 2, 2);

-- 新增权限 API
INSERT INTO `permission_api` VALUES (1, '/user/add', 'POST', 1);
INSERT INTO `permission_api` VALUES (2, '/user/delete', 'GET', 2);
INSERT INTO `permission_api` VALUES (3, '/user/edit', 'POST', 3);
INSERT INTO `permission_api` VALUES (4, '/user/list', 'GET', 4);
INSERT INTO `permission_api` VALUES (5, '/role/list', 'GET', 5);
INSERT INTO `permission_api` VALUES (6, '/role/edit', 'GET', 6);
INSERT INTO `permission_api` VALUES (7, '/role/delete', 'GET', 7);

-- 新增角色

INSERT INTO `role` VALUES (1, '管理员', '系统管理员，拥有全部权限', '2024-02-23');
INSERT INTO `role` VALUES (2, '普通用户', '普通用户，拥有部分权限', '2024-02-23');


-- 新增用户

INSERT INTO `user` VALUES (1, 'admin', 'admin123', '我是管理员', '2024-02-23');
INSERT INTO `user` VALUES (2, 'user1', 'user111', '我是user1', '2024-02-23');
INSERT INTO `user` VALUES (3, 'user2', 'user222', '我是user2', '2024-02-23');

-- 新增用户角色关联数据：1 个管理员和 2 个用户

INSERT INTO `user_role` VALUES (1, 1, 1);
INSERT INTO `user_role` VALUES (2, 2, 2);
INSERT INTO `user_role` VALUES (3, 3, 2);
```

```sql
-- 新增角色权限关联数据
---管理员拥有全部权限，用户拥有读写权限，没有删除权限
-- 管理员【1,2,3,4,5,6,7】
-- 用户 1【1,3,4】
-- 用户 2【1,5,6】
INSERT INTO `role_permission` VALUES (1, 1, 1);
INSERT INTO `role_permission` VALUES (2, 1, 2);
INSERT INTO `role_permission` VALUES (3, 1, 3);
INSERT INTO `role_permission` VALUES (4, 1, 4);
INSERT INTO `role_permission` VALUES (5, 1, 5);
INSERT INTO `role_permission` VALUES (6, 1, 6);
INSERT INTO `role_permission` VALUES (7, 1, 7);
INSERT INTO `role_permission` VALUES (8, 2, 1);
INSERT INTO `role_permission` VALUES (9, 2, 3);
INSERT INTO `role_permission` VALUES (10, 2, 4);
INSERT INTO `role_permission` VALUES (11, 3, 1);
INSERT INTO `role_permission` VALUES (12, 3, 5);
INSERT INTO `role_permission` VALUES (13, 3, 6);
```

有了真实数据后，我们就可以替换 6.2 节中 UserService 遗留的模拟数据逻辑，改用 Repository 来查询用户数据，代码如下：

```typescript
import { Injectable } from '@nestjs/common';
import { InjectRepository } from '@nestjs/typeorm'
import { User as UserEntity } from './entities/user.entity';
import { Repository } from 'typeorm';

export type User = {
  id: number;
  userName: string;
  password: string;
};

@Injectable()
export class UserService {
  constructor(
    @InjectRepository(UserEntity)
    private readonly userRepo: Repository<UserEntity>,
  ) {}
  async findOneByUserName(userName: string): Promise<User | undefined> {
    return this.userRepo.findOne({where: { userName }});
  }
}
```

同时，在 AppController 中新增路由方法，用于模拟管理员和普通用户的操作请求，代码如下：

```typescript
// 新增获取用户列表的方法
@Get('user/list')
getUserList() {
  return [];
}

// 新增删除用户的方法
@Get('user/delete')
```

```
deleteUser() {
  return "删除成功";
}
```

6.3.8　测试效果

接下来，我们通过 curl 工具登录管理员账号获取 access_token，先后请求 user/list 接口和需要管理员权限的 user/delete 接口，如图 6-19 所示。

图 6-19　管理员请求结果

从图中可以明显看出，管理员账号访问时能够正常返回预期的结果。同样地，如果使用普通用户账号登录并发送相同的请求，其效果将如图 6-20 所示。

图 6-20　普通用户请求的结果

从图中可以明显看出，当普通用户请求/user/list 时，能够返回预期的结果。然而，当尝试请求权限管理员访问的/user/delete 接口时，系统返回了 403 错误，提示缺少权限。这表明我们的守卫成功拦截了未授权的请求。

通过这一过程，我们成功实现了基于角色的访问控制（RBAC）系统。我们通过设置守卫和自定义装饰器来区分不同场景下的接口访问控制，并根据不同角色的权限集来管理和分发系统权限。掌握了本节内容后，相信你将完全有能力设计出一套包括超级管理员、总监、财务、行政、销售、仓管等角色的复杂权限系统。

第 7 章

系统部署与扩展

系统部署是许多读者关注的重要环节,同时也是非运维(后端)人员相对较少涉足的领域。特别是在涉及前后端统一部署时,关于这一主题的网络资料通常不够充分。本章将深入探讨系统部署所涉及的容器技术——Docker。我们将学习如何在 Nest 中编写自定义的 Dockerfile 来创建镜像,并最终通过 Docker Compose 来实操部署一个完整的 Nest 应用程序。

7.1 快速上手 Docker

在后端应用的部署中,我们通常需要部署多种服务,例如 MySQL 数据库、Redis 缓存服务、Nginx 作为反向代理网关等,同时还包括我们自己开发的业务服务。如果采用微服务架构,可能还需要部署数十甚至数百个微服务。要将这些应用部署到多台服务器上时,需要对它们执行一系列相同的环境配置和依赖安装操作。这不仅工作量巨大,而且容易出错。

Docker 容器技术可以帮助我们高效地解决这些问题。本节将介绍 Docker 的安装和基本使用方法,涵盖通过命令行界面(CLI)和可视化工具两种方式操作 Docker,为后续 Nest 项目的应用提供更好的技术支持。

7.1.1 初识 Docker

在 Docker 普及之前,将后端服务如 Redis、Nginx、MySQL 以及消息队列服务部署到多台服务器时,通常使用虚拟机(Virtual Machine,VM)实现服务部署和隔离。然而,虚拟机存在一些明显的缺点,包括资源消耗大、启动速度慢以及隔离和扩展的复杂性较高。Docker 等容器技术的出现有效地解决了这些问题,因此迅速获得了业界的广泛认可和采用。

Docker 的设计理念图如图 7-1 所示。

图 7-1 Docker 设计理念

接下来，介绍 Docker 中的三个核心概念。

- 镜像（Image）：镜像就像静止等待运送的集装箱，它包含了运行某个程序所需的一切依赖项、库文件和配置信息等，被看作容器的模板。
- 容器（Container）：容器就像运送中的集装箱，是镜像运行时的实例。每个容器都拥有自己的文件系统、CPU、内存和网络资源，是一个独立的小型虚拟机，它比传统虚拟机更轻量、启动更快。通俗地说，每个容器都是一个独立隔离的空间，彼此互不影响。
- 仓库（Registry）：仓库类似于集中存储和分发集装箱的超级码头，汇聚了世界各地的集装箱，可以把集装箱分发到世界各地。Docker 中最流行的是 Docker Hub，我们可以从仓库中拉取（Pull）所需的镜像或者将镜像推送（Push）到仓库中以供他人使用。

了解以上概念后，接下来探讨 Docker 的工作原理。

Docker 的运行过程如下：首先，从远程仓库中拉取（Pull）所需的镜像文件到本地环境。然后，利用这个镜像文件，Docker 可以创建并启动一个轻量级的虚拟容器。每个容器都可以独立运行一个或多个服务，类似于我们在自己的计算机上可以同时运行 MySQL、Redis 等不同的服务。这意味着可以在任何支持 Docker 的机器上运行指定的镜像，从而确保不同环境中都能使用相同配置和依赖的服务。

对 Docker 有了基本的认识之后，下一步我们将安装并学习如何使用 Docker。

7.1.2 安装 Docker

Docker 是开源的商业产品，提供社区版和企业版，个人开发者选择社区版即可。以 Mac 系统为例（M1 芯片-Apple Chip），读者可以选择适合自身系统的 Docker Desktop 版本进行下载和安装，如图 7-2 所示。

图 7-2　下载 Docker Desktop

安装完成后，在命令行窗口输入"docker -v"或"docker –h"命令，查看是否安装成功，如图 7-3 所示。

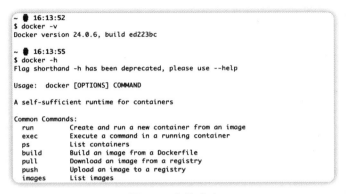

图 7-3 安装成功

接下来，打开 Docker Desktop 可视化操作界面，如图 7-4 所示。

图 7-4 Docker 可视化操作界面

其中，Images 是 Docker 在本地的所有镜像，Containers 是运行镜像之后的容器，Volumes 是持久化容器内的应用程序运行时产生的数据，存储在宿主机文件系统中。

7.1.3 Docker 的使用

为了演示 Docker 的使用方法，我们从仓库中拉取 Nginx 进行测试，在搜索框中输入 nginx，如图 7-5 所示。

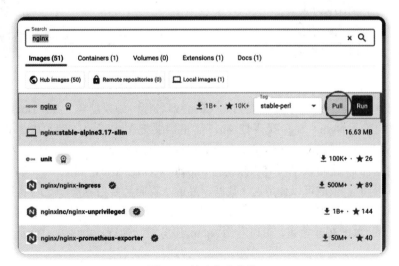

图 7-5 搜索指定镜像

单击 Pull 按钮将所需的镜像拉取到本地，成功后，你可以在 Images 中看到这个镜像，如图 7-6 所示。

图 7-6　本地镜像列表

接下来，运行这个镜像。单击 Run 按钮之后，填写容器的运行配置信息，如图 7-7 所示。

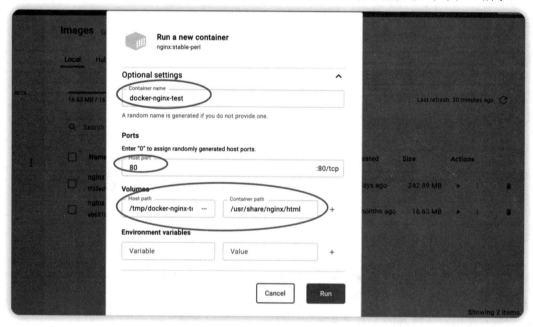

图 7-7　填写容器的运行配置信息

其中，docker-nginx-test 是自定义的容器名称。如果没有设置，Docker 会自动生成随机的名称。容器中的 Nginx 端口默认是 80，这里需要把宿主机上的端口映射为容器的端口才能进行访问。Volumes 在前文介绍过了，表示数据卷。容器运行后的数据会保存在这个数据卷中，并持久化到宿主机的某个目录下（笔者挂载到 /tmp/docker-nginx-test 目录下，读者可以自定义）。在下一次容器运行时，依旧可以使用这个数据。最后，我们需要为容器设置环境变量，这些环境变量包含容器运行过程中所需的参数。这个过程类似于在运行 Node.js 应用程序时为其设置参数。

单击 Run 按钮，可以看到 Nginx 服务成功运行，如图 7-8 所示。

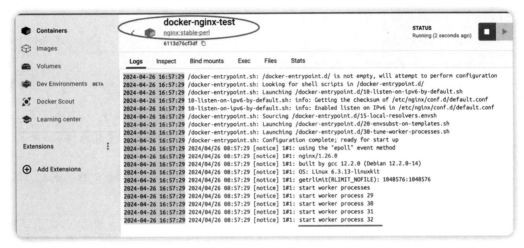

图 7-8　运行镜像

Nginx 通常用于托管静态资源，我们在/tmp/docker-nginx-test 目录下创建一个 index.html 文件，运行 echo 'hello docker' > ./index.tml 命令，随后尝试访问 http://localhost，结果如图 7-9 所示。

图 7-9　访问 index.html

此时，运行在 Docker 容器上的 Nginx 服务可以正常托管我们的静态资源。切换至 Files 标签页，在容器目录（/usr/share/nginx/html）下可以看到刚刚添加在宿主机目录（/tmp/docker-nginx-test）下的 index.html 文件，这表明文件已经成功被挂载到容器中，如图 7-10 所示。

图 7-10　容器目录文件同步更新

虽然使用可视化工具操作 Docker 非常方便，但我们仍需要掌握 Docker 的基本命令行操作，以便在没有可视化工具的服务器环境中进行 Docker 管理。

下面通过命令行方式对前面的流程进行类比说明。首先是拉取镜像的操作，这等同于执行以下命令：

```
docker pull nginx:latest
```

运行镜像前需要配置信息，执行以下命令：

```
docker run --name docker-nginx-test -p 80:80 -v /tmp/docker-nginx-
test:/usr/share/nginx/html -e key1=value1
```

其中，--name 用于设置容器名称，-p 用于指定映射端口，-v 用于设置数据卷挂载目录，-e 用于设置容器环境变量。

执行成功后会创建一个容器，我们可以执行“docker ps -a”命令获取本地创建的所有容器列表，如图 7-11 所示。

```
$ docker ps -a
CONTAINER ID   IMAGE                          COMMAND                  CREATED          STATUS                      PORTS                 NAMES
348ef47be53e   nginx:stable-perl              "/docker-entrypoint.…"   46 minutes ago   Up 19 minutes               0.0.0.0:80->80/tcp    docker-nginx-test
619cdb79e34b   nginx:stable-alpine3.17-slim   "/docker-entrypoint.…"   6 months ago     Exited (137) 50 minutes ago                       dockerTest
```

图 7-11　查看容器列表

然后执行“docker images”或“docker image ls”命令，获取本地所有的镜像列表，如图 7-12 所示。

```
$ docker images
REPOSITORY   TAG                          IMAGE ID       CREATED        SIZE
nginx        stable-perl                  ff53ed93b7c9   2 days ago     243MB
nginx        stable-alpine3.17-slim       ab6510b890a4   6 months ago   16.6MB
```

图 7-12　查看镜像列表

接着执行“docker logs docker-nginx-test”命令，查看容器的运行日志，如图 7-13 所示。

```
2024/04/26 09:10:56 [notice] 1#1: worker process 25 exited with code 0
2024/04/26 09:10:56 [notice] 1#1: worker process 26 exited with code 0
2024/04/26 09:10:56 [notice] 1#1: signal 29 (SIGIO) received
/docker-entrypoint.sh: /docker-entrypoint.d/ is not empty, will attempt to perform configuration
/docker-entrypoint.sh: Looking for shell scripts in /docker-entrypoint.d/
/docker-entrypoint.sh: Launching /docker-entrypoint.d/10-listen-on-ipv6-by-default.sh
10-listen-on-ipv6-by-default.sh: info: IPv6 listen already enabled
/docker-entrypoint.sh: Launching /docker-entrypoint.d/20-envsubst-on-templates.sh
/docker-entrypoint.sh: Launching /docker-entrypoint.d/30-tune-worker-processes.sh
/docker-entrypoint.sh: Configuration complete; ready for start up
2024/04/26 10:10:12 [notice] 1#1: using the "epoll" event method
2024/04/26 10:10:12 [notice] 1#1: nginx/1.24.0
2024/04/26 10:10:12 [notice] 1#1: built by gcc 12.2.1 20220924 (Alpine 12.2.1_git20220924-r4)
2024/04/26 10:10:12 [notice] 1#1: OS: Linux 6.3.13-linuxkit
2024/04/26 10:10:12 [notice] 1#1: getrlimit(RLIMIT_NOFILE): 1048576:1048576
2024/04/26 10:10:12 [notice] 1#1: start worker processes
2024/04/26 10:10:12 [notice] 1#1: start worker process 22
2024/04/26 10:10:12 [notice] 1#1: start worker process 23
2024/04/26 10:10:12 [notice] 1#1: start worker process 24
2024/04/26 10:10:12 [notice] 1#1: start worker process 25
```

图 7-13　查看容器日志

接着执行“docker volume”命令，管理容器的数据卷，如图 7-14 所示。

```
$ docker volume

Usage:  docker volume COMMAND

Manage volumes

Commands:
  create      Create a volume
  inspect     Display detailed information on one or more volumes
  ls          List volumes
  prune       Remove unused local volumes
  rm          Remove one or more volumes

Run 'docker volume COMMAND --help' for more information on a command.
```

图 7-14　管理数据卷

最后，对容器进行启动、停止和删除操作，分别对应的命令为"docker start docker-nginx-test""docker stop docker-nginx-test""docker rm docker-nginx-test"，如图 7-15 所示。

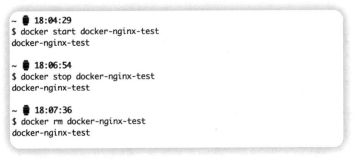

图 7-15　容器的启动、停止与删除

以上就是 Docker 中常用的命令。在后续使用 Docker 时还会涉及其他命令，在后续的章节中会进行介绍。

至此，我们对 Docker 有了一个全新的认识。本节通过"Docker Desktop"拉取远程仓库的镜像演示了 Docker 的基本使用方法。然而在实际业务中，我们需要根据自身的需求自定义镜像。接下来的 7.2 节将介绍如何构建一个自定义的 Docker 镜像。

7.2　快速上手 Dockerfile

远程仓库中拥有丰富的 Docker 镜像，如 Nginx、MySQL 和 Kafka 等。我们可以直接拉取并使用它们。除此之外，我们还需要学会自定义业务镜像，以便在开发、测试和生产各阶段使用同一份镜像，确保环境的高度一致，从而减少因环境差异而引起的运行问题。本节将介绍 Dockerfile 的基本概念和语法，同时创建一个实际的 Nest 项目，并为其构建一个自定义的 Docker 镜像。

7.2.1　Docker 的基本概念

在 Docker 中，我们所说的镜像是一个二进制文件。当镜像运行起来时，它变成一个容器实例，容器实例本身也是一个文件，称为容器文件。镜像可以从远程仓库拉取，用户也可以自定义镜像，这时就需要用到 Dockerfile。Dockerfile 是一个文本文件，用于描述如何构建一个 Docker 镜像，里面包含若干指令，这些指令按照一定的顺序排序。Docker 根据这些指令一步一步生成镜像文件。简单来说，镜像是基于 Dockerfile 中定义的指令构建生成的产物。

7.2.2 Dockerfile 的基本语法

以某个项目的 Dockerfile 为例，代码如下：

```
FROM node:18
CMD [ "mkdir", "/upload" ]
WORKDIR /servers
COPY . .
ENV TZ=Asia/guangzhou
RUN npm i --registry=https://registry.npmmirror.com -g pnpm && pnpm i && pnpm run build
EXPOSE 8080
```

上面一共有 7 行指令，这些指令的含义解析如下。

- FROM node:18：该 image 文件指定了基础镜像，基于 node:18 镜像进行构建。
- CMD ["mkdir","/upload"]：该 CMD 指令用于指定容器启动时需要执行的命令，当容器启动时，在容器根目录下创建一个名为 upload 的目录。一个 Dockerfile 中只能运行一个 CMD 指令。
- WORKDIR /servers：WORKDIR 指令用于设置容器的工作目录，这意味着后续的命令都会在/servers 目录下执行。
- COPY ..：COPY 指令用于将构建上下文中的文件或目录复制到镜像中。
- ENV TZ=Asia/Guangzhou：ENV 指的是设置环境变量，其中 TZ（Time Zone，时区）指定了运行该镜像容器时默认使用广州时区。
- RUN npm：RUN 指令用于在镜像中执行命令。这里首先使用 npm 全局安装了 pnpm 包管理器，然后使用 pnpm 安装依赖并进行打包构建，这些依赖会被一同打包到镜像中。一个 Dockerfile 中可以运行多条 RUN 指令。
- EXPOSE 8080：EXPOSE 指令声明了容器在运行时监听的网络端口，这里声明容器将监听 8080 端口。

7.2.3 Dockerfile 实践

接下来，我们创建一个名为 nest-docker 的 Nest 项目进行演示，运行"nest n nest-docker -p pnpm"命令，结果如图 7-16 所示。

```
$ nest n nest-docker -p pnpm
  We will scaffold your app in a few seconds..
CREATE nest-docker/.eslintrc.js (663 bytes)
CREATE nest-docker/.prettierrc (51 bytes)
CREATE nest-docker/README.md (3347 bytes)
CREATE nest-docker/nest-cli.json (171 bytes)
CREATE nest-docker/package.json (1952 bytes)
CREATE nest-docker/tsconfig.build.json (97 bytes)
CREATE nest-docker/tsconfig.json (546 bytes)
CREATE nest-docker/src/app.controller.spec.ts (617 bytes)
CREATE nest-docker/src/app.controller.ts (274 bytes)
CREATE nest-docker/src/app.module.ts (249 bytes)
CREATE nest-docker/src/app.service.ts (142 bytes)
CREATE nest-docker/src/main.ts (208 bytes)
CREATE nest-docker/test/app.e2e-spec.ts (630 bytes)
CREATE nest-docker/test/jest-e2e.json (183 bytes)

✓ Installation in progress... 🐿

  Successfully created project nest-docker
```

图 7-16　创建项目

在根目录下创建 .dockerignore 文件，排除不需要打包的文件或目录，示例如下：

```
.git
node_modules
dist
package-lock.json
npm-debug.log
```

在同一目录下创建 Dockerfile 文件，输入 Docker 镜像构建指令，代码如下：

```
FROM node:18
WORKDIR /servers
COPY . .
ENV TZ=Asia/Guangzhou
RUN npm set registry=https://registry.npmmirror.com
RUN npm i -g pnpm && pnpm i && pnpm run build
EXPOSE 3000
CMD node dist/main.js
```

在上面的代码中，与前面示例不同的是，将暴露端口设置为 3000，并且在容器启动时自动运行 CMD 命令 "node dist/main.js" 以启动 Nest 服务。

Dockerfile 编写完毕之后，运行 "docker build -t nest-docker:0.0.1 ." 命令构建镜像，结果如图 7-17 所示。

```
$ docker build -t nest-docker:0.0.1 .
[+] Building 42.8s (10/10) FINISHED                                        docker:desktop-linux
 => [internal] load .dockerignore                                                           0.0s
 => => transferring context: 2B                                                             0.0s
 => [internal] load build definition from Dockerfile                                        0.0s
 => => transferring dockerfile: 238B                                                        0.0s
 => [internal] load metadata for docker.io/library/node:18                                  2.1s
 => [1/5] FROM docker.io/library/node:18@sha256:98218110d09c63b72376137860d1f30a4f61ce029d7de4caf2e8c00f3  0.0s
 => [internal] load build context                                                           0.6s
 => => transferring context: 2.79MB                                                         0.5s
 => CACHED [2/5] WORKDIR /servers                                                           0.0s
 => [3/5] COPY . .                                                                          2.0s
 => [4/5] RUN npm set registry=https://registry.npmmirror.com                               0.5s
 => [5/5] RUN npm i -g pnpm && pnpm i && pnpm run build                                    36.5s
 => exporting to image                                                                      1.0s
 => => exporting layers                                                                     1.0s
 => => writing image sha256:06669be0aac7c0caac5dc8d62e88c8e22182b8c1ff95c2ffe58fad882843387b  0.0s
 => => naming to docker.io/library/nest-docker:0.0.1                                        0.0s
```

图 7-17　构建镜像

其中，-t 参数用来指定镜像文件的名称，后面可以用冒号指定标签，如果不指定，默认表示 latest。命令中的最后一个点（.）指示 Docker 构建上下文的位置，即 Dockerfile 所在的目录。在上例中，点（.）表示 Dockerfile 位于执行"docker build"命令的当前目录。

运行成功之后，执行"docker images"命令，可以看到生成了新的 docker-test 镜像文件，如图 7-18 所示。

```
$ docker images
REPOSITORY    TAG                      IMAGE ID       CREATED       SIZE
nest-docker   0.0.1                    06669be0aac7   2 hours ago   1.24GB
nginx         stable-perl              ff53ed93b7c9   3 days ago    243MB
nginx         stable-alpine3.17-slim   ab6510b890a4   6 months ago  16.6MB
```

图 7-18　成功构建镜像

图 7-18 中包含镜像名称、标签、ID 和体积等信息。接下来运行镜像文件以启动一个新的容器，执行如下命令：

```
docker container run -p 8000:3000 -it nest-docker:0.0.1
```

此时可以看到 Nest 服务被启动了。这意味着容器在启动时成功执行了 CMD 命令，如图 7-19 所示。

```
$ docker container run -p 8000:3000 -it nest-docker:0.0.1
[Nest] 8  - 04/27/2024, 3:32:53 AM     LOG [NestFactory] Starting Nest application...
[Nest] 8  - 04/27/2024, 3:32:53 AM     LOG [InstanceLoader] AppModule dependencies initialized +5ms
[Nest] 8  - 04/27/2024, 3:32:53 AM     LOG [RoutesResolver] AppController {/}: +5ms
[Nest] 8  - 04/27/2024, 3:32:53 AM     LOG [RouterExplorer] Mapped {/, GET} route +1ms
[Nest] 8  - 04/27/2024, 3:32:53 AM     LOG [NestApplication] Nest application successfully started +2ms
```

图 7-19　运行容器并自动启动服务

在浏览器中输入 http://localhost:8000/ 访问 Nest 服务，可以看到正常返回了"Hello World！"文本，说明 Nest 服务已成功部署到 Docker 上，如图 7-20 所示。

```
← → C  ⓘ http://localhost:8000

Hello World!
```

图 7-20　访问 Nest 接口服务

通过执行"docker container ls"命令可以看到正在运行中的容器，如图 7-21 所示。

```
$ docker container ls
CONTAINER ID   IMAGE               COMMAND                  CREATED         STATUS         PORTS                    NAMES
6a8c62aa8611   nest-docker:0.0.1   "docker-entrypoint.s…"   8 minutes ago   Up 8 minutes   0.0.0.0:8000->3000/tcp   eager_dubinsky
```

图 7-21　查看运行中的容器

当然，如果你想查看所有容器列表，包含终止运行的容器，可以运行"docker container ls –all"命令。

执行上述命令后，Docker 会输出容器的 ID、名称等信息。我们可以通过这个 ID 来终止容器运行，命令格式为"docker container kill [containerID]"。终止运行的容器仍然会占用磁盘空间。此时可以使用"docker container rm [containerID]"命令将其删除。或者在容器运行时加上--rm 参数，这样容器在终止运行后将自动删除其文件系统和网络设置。对应的命令为"docker container run --

rm -p 8000:3000 -it nest-docker"。

除此之外，如果不想在容器启动时自动启动 Nest 服务，第一种方式是在 Dockerfile 中删除 CMD 命令，第二种方式是在启动容器时添加参数覆盖 CMD 命令，例如：

```
docker container run --rm -p 8000:3000 -it nest-docker /bin/bash
```

执行上述命令后，会返回一个命令提示符，这表示系统当前已经处于容器内的 Bash Shell 环境，并且提示符是容器内部的 Shell 提示符，如图 7-22 所示。

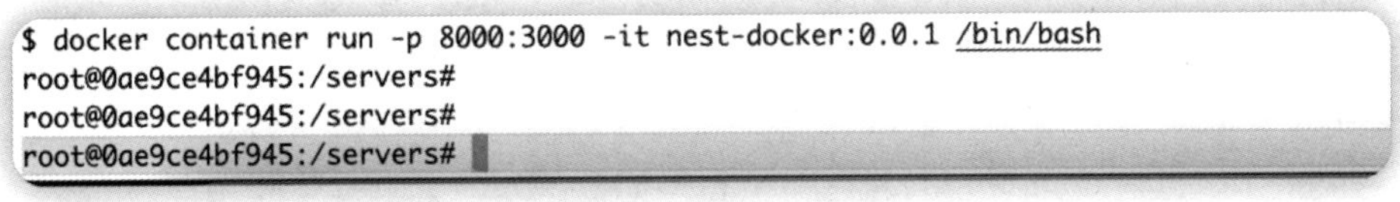

图 7-22 容器 Shell 界面

手动执行"node dist/main"命令即可启动 Nest 服务，运行结果如图 7-23 所示。

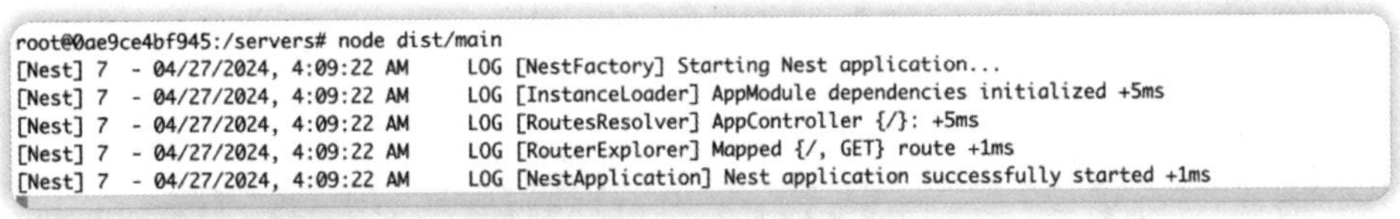

图 7-23 手动启动 Nest 服务

注意：在实际的自动化部署（CI/CD）流程中，如果希望容器一旦启动就立即提供服务而不需要手动干预，可以设置自动启动以确保容器启动后立即进入工作状态。

最后，通过图 7-18 可以看到镜像体积高达 1.24GB，导致镜像体积过大，这影响了构建性能。为了提升性能，建议采用体积更小的 alpine 版本镜像。有关 alpine 版本镜像的详细信息，可以参考图 7-24 所示的站点进行查找。

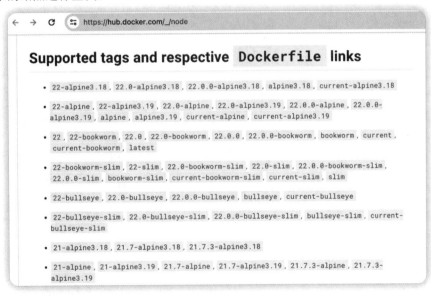

图 7-24 选择 alpine 版本

在本案例中，我们使用 18-alpine3.18 作为基础镜像。在 Dockerfile 中，我们将基础镜像修改为 FROM node:18-alpine3.18，并重新构建标签为 1.0.0 的镜像。镜像构建前后的体积大小如图 7-25 所示。

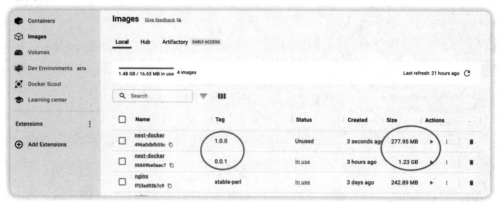

图 7-25　优化前后体积对比

可以看出，使用不同版本的基础镜像构建出的产物大小差异显著。

通过本节的学习，我们不仅掌握了 Dockerfile 的基本概念和语法，还完成了 Nest 项目的 Dockerfile 文件编写，成功构建了镜像，并部署运行了 Nest 服务。此外，选择使用 alpine 版本的基础镜像可以显著提升构建速度并减少镜像体积，从而优化整体的部署效率。

第 3 部分

扩 展 篇

第 3 部分我们将深入学习 Nest 框架生态体系中的各种测试方法，探讨如何通过单元测试、集成测试和端到端测试来提升工程质量。同时，我们也将了解如何构建高效的日志系统和完善的错误处理机制，以提升系统的稳定性。通过全面掌握这些技术，我们可以打造出更健壮、更可靠的应用程序。

第 8 章

单元测试与端到端测试

本章将深入探讨软件测试中的三个关键测试方法：单元测试、集成测试和端到端测试。首先介绍单元测试，这是一种针对代码中最小可测试单元的测试方法，通常由开发人员编写，用于验证代码的行为是否符合预期。接着扩大测试范围，进行模块的集成测试，这涉及将各个独立测试完成的模块聚合到一起，确保它们作为一个整体协同工作。最后，探讨端到端测试，这是一种模拟真实用户操作流程的测试方法，旨在验证整个软件系统在实际环境中的行为。通过学习这几种测试方法，读者将能够更好地保证软件质量，并确保系统在不同层面上的稳定性和可靠性。

8.1 重新认识单元测试

单元测试是一个经常被讨论的话题，但许多人可能对它持有一种敬而远之的态度。尽管单元测试的重要性被广泛认可，但在实际应用中，它可能没有得到充分的利用。本节将从多个角度深入探讨单元测试。首先介绍单元测试的基本概念，包括它的定义和重要性。然后探讨为什么我们应该编写单元测试，以及它能够为我们带来哪些好处。接下来讨论在实践中是应该先编写测试还是先编写代码，因为这两种方法在效果上存在显著差异。最后介绍测试驱动开发（Test-Driven Development，TDD），这是一种在编写功能代码之前先编写测试代码的编程实践方法，有助于提高代码质量和开发效率。

8.1.1 什么是单元测试

所谓单元，通常指的是代码中的最小可测试部分，例如一个函数、方法或类中的一个方法。单元测试是一种软件测试方法，其目的是独立地测试应用程序中的每个代码单元，以验证它们是否按照开发阶段的预期设计正常工作。这个过程包括为每个单元编写测试用例，执行这些测试，并验证结果是否符合预期。单元测试在软件开发生命周期中扮演着至关重要的角色，它有助于在开发早期发现和修复错误，从而提高代码质量并减少后期的维护成本。

8.1.2 为什么大部分公司没有进行单元测试

根据笔者的从业经验及任职过的公司来看,绝大多数公司对程序员编写单元测试的要求并不高。笔者认为主要有以下几点原因。

1. 没有真正感受到单元测试带来的好处

很多人可能听说单元测试很有作用,事实上,甚至一些公司的管理层也未体验过单元测试带来的价值。潜意识中可能仍然认为这会影响需求评估的耗时,并增加了单元测试的成本。时常会出现这种声音:"这个版本把需求完成已经十分困难了,更不用说加上单元测试。"

2. 潜意识认为发现 BUG 是测试人员的工作

在互联网软件公司中,测试团队通常承担着产品质量的最终把关责任,他们面临的压力并不亚于开发团队。开发人员可能没有意识到,不编写单元测试会增加测试阶段的工作量和压力,因为他们可能认为测试团队会负责发现并修复 BUG。这种做法导致开发人员在完成基本需求后匆忙进入测试阶段,而测试过程中的 BUG 修复可能引发更多的问题,即所谓的"改崩了"现象。这种情况通常源于开发人员在理解需求和进行代码设计时花费了大量时间,而没有足够的时间进行单元测试。

为了改善这一状况,公司可以鼓励开发团队在编写代码的同时进行单元测试,以提前发现和修复 BUG,减少后期测试阶段的压力。此外,通过提高开发人员对单元测试重要性的认识,可以帮助他们更高效地完成需求,并提高产品质量。

3. 投入产出比(Return on Investment,ROI)考量

在一些公司中,单元测试可能被认为不如项目交付和满足客户需求重要,因此在面临高产量需求时,可能会被认为不那么重要。一旦忽视了单元测试,重新启动并实施这项工作将变得更加具有挑战性。

8.1.3 为什么要编写单元测试

需要明确指出,编写单元测试是一项需要投入大量精力和时间的工作。许多人编写单元测试是因为领导或公司的要求,目的是达到一定的测试覆盖率,从而使相关数据看起来更加令人满意。此外,有些人认为编写单元测试的主要目的是减少 BUG 的产生,但在面对复杂功能时,他们可能会选择性地编写一些测试,以求得心理上的安慰,毕竟有质量保证(Quality Assurance,QA)团队作为最终的测试保障。然而,许多优秀的程序员深刻意识到单元测试的巨大作用,并从内心重视其编写。单元测试究竟能为我们带来哪些好处呢?以下是笔者认为的几个关键点。

1. 验证代码的正确性

通常情况下,我们编码完成后需要自己测试一遍才能交付给测试团队。最常见的方式是运行程序,简单测试主要分支场景。如果测试通过,我们可能会认为自己的代码没有问题。然而,这种方式很难覆盖所有特殊场景或边界情况。单元测试可以轻松构建各种测试场景,特别是在后续的维护中,直接运行测试用例,以确保我们的代码可以交付测试。

2. 确保重构的可行性

在实际工作中，我们往往不敢随意重构旧系统，因为不确定它的影响范围，担心改动逻辑后影响其他模块的正常工作。单元测试可以帮助我们了解代码改动后的影响，只需运行测试用例，从而让我们自信地进行代码修改。

3. 加深对业务的理解

单元测试为我们提供了设计测试用例时考虑各种边界条件和业务场景的机会。换句话说，如果我们不清楚业务的具体实现，就无法知道代码应该如何运行，应该输出什么，也就无法编写出完善的测试用例。

4. 完善研发流程

在研发闭环流程中，我们应尽可能通过单元测试让 BUG 在方案设计阶段、编码阶段和测试阶段暴露出来。因为 BUG 发现得越晚，修复成本越高。例如，当 BUG 出现在预上线环境中时，可能涉及环境、数据兼容、配置等因素，修复成本随之增大。

5. 单元测试是最好的开发文档

单元测试覆盖了接口的所有使用方法和边界条件，是最佳的示例代码。它清晰地告诉我们输入什么数据会得到什么输出。示例代码如下：

```
describe('numeral.add', () => {
  test('加法', () => {
    expect(numeral.add(1, 1)).toBe(2);
    expect(numeral.add(0.1, 0.2)).toBe(0.3);
    expect(numeral.add(0.7, 0.1)).toBe(0.8);
    expect(numeral.add(0.2, 0.4)).toBe(0.6);
    expect(numeral.add(35.41, 19.9)).toBe(55.31);
  });
});
```

在上述代码中，numeral.add 方法解决了 JavaScript 中浮点数运算的精度问题，如 0.1 加 0.2 在 JavaScript 中可能会得到 0.30000000000000004，而 numeral.add 确保了结果为 0.3，从而解决了精度丢失的问题。

8.1.4 先编写单元测试还是先编写代码

先编写代码后编写单元测试是一种常见的做法，目的是确保所编写的代码能够正常运行。而测试驱动开发（Test-Driven Development，TDD）则是一种不同的方法，它从业务需求出发，先设计测试用例再编写代码，这种方法鼓励开发者从不同的角度思考问题。

采用先编写代码后编写单元测试的方式，虽然能够验证代码的基本功能，但可能无法全面覆盖所有边界条件和场景。这种方法可能导致开发者在编写测试用例时受到原有代码逻辑的限制，形成思维定势，从而难以发现潜在的问题。我们通常将这种方法视为一种"防守"策略，即在代码编写后通过测试来防止错误。

相比之下，TDD 是一种"进攻"策略，它要求开发者在编写任何代码之前先从业务场景出发设计测试用例。这种方法与 QA 工程师的工作方式相似，他们通过各种测试手段来验证代码的正确性，目标是发现并修复潜在的错误。TDD 鼓励开发者在编写代码之前就深入思考业务需求和可能的测试场景，从而提高代码的质量和可维护性。

8.1.5 测试驱动开发

测试驱动开发是有步骤的，通常分为红、绿、重构三部分，图解流程如图 8-1 所示。

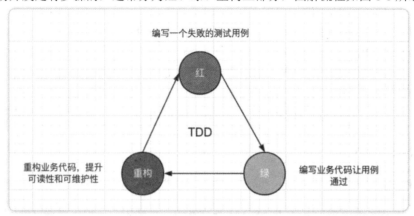

图 8-1　测试驱动开发的 TDD 原理

图 8-1 中的流程详细说明如下：

（1）编写失败的测试用例：根据业务逻辑编写一个新的测试用例，由于未编写实际代码，这个测试用例通常无法通过。

（2）编写实际代码：编写足够的代码让测试用例通过。

（3）重构实际代码：重构刚写的代码，以提高代码的质量和可维护性。如果在重构的过程中测试失败，我们应该修复问题，使测试再次通过并恢复到绿色状态。只有当测试全部通过，我们才能认为重构流程成功完成。

有些人可能不太习惯或接受先测试后编码的开发模式，认为在项目初期投入时间编写测试并不值得。这种观点可能源于对 TDD 优势的不了解。实际上，TDD 在特定场景下可以显著提高开发效率和代码质量。

TDD 特别适用于以下几种场景：

（1）编写纯函数：例如前面提到的 add 方法，它是一个纯函数，具有明确的输入/输出关系，使得编写测试用例变得相对简单。这类函数通常用于工具方法中。

（2）编写新的模块：以 Nest 框架为例，TDD 可以帮助我们实现 API 接口、数据库读写、中间件及守卫等功能的开发，确保输入/输出的一致性。

（3）修复 BUG：面对 BUG 时，TDD 允许我们通过编写测试用例来复现问题，并在测试通过后确保 BUG 被修复。

从这些场景中，我们可以得出结论：TDD 特别适合那些逻辑复杂但输入输出关系明确的开发任务。

通过重新审视单元测试，我们更深入地理解了编写单元测试的目的、方法和流程。在下一节中，我们将通过 Nest 框架的实际案例，演示如何遵循 TDD 流程编写有效的测试用例，以提升 API 接口的可靠性和稳定性。

8.2 在 Nest 中使用 Jest 编写单元测试

在深入理解单元测试的重要性之后，本节将首先介绍主流的单元测试库——Jest。然后，将应用测试驱动开发（TDD）流程，以编写一个常见的功能——登录账号密码验证的测试用例。在编写测试用例后，我们将根据测试用例的要求来实现具体的业务逻辑代码。最后，为了进一步提升代码的质量和可维护性，我们将对实现的代码进行重构。

通过本节的学习，读者将了解如何利用 Jest 和 TDD 流程来提高软件开发的质量和效率。

8.2.1 初识 Jest

1. 基本概念

Jest 是由 Facebook 开发的开源 JavaScript 测试框架，主要用于自动化测试以确保代码按预期运行。它不仅广泛应用于前端和后端的测试场景，还支持单元测试、集成测试和端对端测试等多种测试类型。

Jest 提供了包括丰富的断言库、模拟功能和详尽的测试报告在内的多种工具，帮助开发者进行有效的测试。此外，它还内置了代码覆盖率分析、快照测试和并行测试等高级功能，以全面检测代码质量并提高开发效率。

2. 常用的 API 介绍

Jest 提供了丰富的 API 供开发者使用，下面简单介绍一下常用的 API 及其作用，以便让读者有一个初步的认识。

1）describe

describe 函数用于对测试用例进行分组，以帮助我们更好地组织用例代码。我们可以用它来管理同一类型的用例。示例代码如下：

```
const myBeverage = {
  delicious: true,
  sour: false,
};

describe('my beverage', () => {
  test('is delicious', () => {
    expect(myBeverage.delicious).toBeTruthy();
  });

  test('is not sour', () => {
    expect(myBeverage.sour).toBeFalsy();
  });
```

});
```

**2）test 和 it**

test 和 it 函数的作用一致，用于编写单个测试用例，多个 test 或 it 函数可以使用 describe 进行包裹。示例代码如下：

```
test('adds 1 + 2 to equal 3', () => {
  expect(1 + 2).toBe(3);
});
```

3）expect

expect 函数用于断言，对实际的结果进行判断，即期望结果应该是什么样的。示例代码如下：

```
expect(sum(1, 2)).toBe(3);
expect(sum(0.1, 0.2)).toBe(0.3);
```

4）mock

jest.mock 函数用于模拟依赖项，例如模拟调用外部 API 或模拟模块引用。假设有一个 userManager 用户管理模块，我们需要测试其中的 getUserName 方法，但它依赖于 userService 中的 getUser 方法，用于从数据库中获取用户信息。在这种情况下，我们就可以用 mock 函数来模拟 userService.getUser 方法，以便在测试中独立验证 userManager.getUserName 函数的行为。示例代码如下：

```
// userService.js
export const userService = {
  getUser: (userId) => {
    // 这里进行数据库查询操作
    // 返回用户信息
  }
};

// userManager.js
import { userService } from './userService';
// 获取用户名称
export const getUserName = (userId) => {
  const user = userService.getUser(userId);
  return user.name;
};

// userManager.test.js
import { userService } from './userService';
import { getUserName } from './userManager';

// 用 Jest 的 mock 函数来模拟 userService.getUser 方法
jest.mock('./userService', () => ({
  userService: {
    getUser: jest.fn().mockReturnValue({ name: '小铭同学' })
  }
}));

test('getUserName returns the correct user name', () => {
  const userName = getUserName(123);
  expect(userName).toBe('小铭同学');
```
```

```
    expect(userService.getUser).toHaveBeenCalledWith(123);
});
```

5）spyOn

jest.spyOn 函数用于创建一个被监视的函数，跟踪函数的调用和返回结果。依旧使用上面的案例，我们可以监视是否调用了 userService.getUser 方法以及是否返回预期结果。示例代码如下：

```
// userService.js
export const userService = {
  getUser: (userId) => {
    // 实际的数据库查询操作
    // 返回用户信息
    return { name: '小铭同学' }
  }
};

// userManager.js
import { userService } from './userService';

export const getUserName = (userId) => {
  const user = userService.getUser(userId);
  return user.name;
};

// userManager.test.js
import { userService } from './userService';
import { getUserName } from './userManager';

test('getUserName calls userService.getUser and returns the correct user name', () => {
    // 使用 jest.spyOn 来创建一个对 userService.getUser 方法的监视器
    const spy = jest.spyOn(userService, 'getUser');

    // 调用 getUserName 方法
    const userName = getUserName(123);

    // 验证 userService.getUser 方法是否被调用
    expect(spy).toHaveBeenCalledWith(123);

    // 验证 getUserName 方法是否返回了正确的用户名称
    expect(userName).toBe('小铭同学');
});
```

6）beforeEach 和 afterEach

beforeEach 和 afterEach 函数用于在每个测试用例运行前后执行特定的操作，例如初始化数据库和清理数据库，确保每个测试用例都处于相同的初识状态。示例代码如下：

```
beforeEach(() => {
  // 在每个测试运行之前初始化数据库
  initializeDatabase();
});

afterEach(() => {
  // 在每个测试运行之后清理数据库
```

```
    clearDatabase();
});
```

7）beforeAll 和 afterAll

beforeAll 和 afterAll 函数用于在所有测试用例运行前后执行特定的操作，例如设置全局测试数据和清理全局数据。示例代码如下：

```
beforeAll(() => {
  // 在所有测试运行之前执行设置全局测试数据的操作
  initializeData();
});

afterAll(() => {
  // 在所有测试运行之后执行清理全局测试数据的操作
  clearData();
});
```

8.2.2 项目准备

现在，我们对 Jest 已经有了大概的了解，接下来在 Nest 中创建用户测试。在此之前，先明确需要实现的需求，使用 TypeOrm 实现通过账号和密码创建用户，其中密码必须经过哈希处理，并且用户信息需保存到数据库中。

运行"nest n nest-jest -p pnpm"命令创建项目，指定包管理器为 pnpm，创建成功后如图 8-2 所示。

```
$ nest n nest-jest -p pnpm
⚡ We will scaffold your app in a few seconds..

CREATE nest-jest/.eslintrc.js (663 bytes)
CREATE nest-jest/.prettierrc (51 bytes)
CREATE nest-jest/README.md (3347 bytes)
CREATE nest-jest/nest-cli.json (171 bytes)
CREATE nest-jest/package.json (1950 bytes)
CREATE nest-jest/tsconfig.build.json (97 bytes)
CREATE nest-jest/tsconfig.json (546 bytes)
CREATE nest-jest/src/app.controller.spec.ts (617 bytes)
CREATE nest-jest/src/app.controller.ts (274 bytes)
CREATE nest-jest/src/app.module.ts (249 bytes)
CREATE nest-jest/src/app.service.ts (142 bytes)
CREATE nest-jest/src/main.ts (208 bytes)
CREATE nest-jest/test/app.e2e-spec.ts (630 bytes)
CREATE nest-jest/test/jest-e2e.json (183 bytes)

✔ Installation in progress... ☕

🚀  Successfully created project nest-jest
```

图 8-2 创建项目

接着运行"nest g resource user"命令，创建 user 模块。细心的读者会发现，Nest CLI 在初始化项目和生成指定模块时，已经为用户自动创建了各种以.spec 命名的单元测试文件，并提供了部分测试用例，读者可以在此基础上快速扩展代码逻辑，如图 8-3 所示。

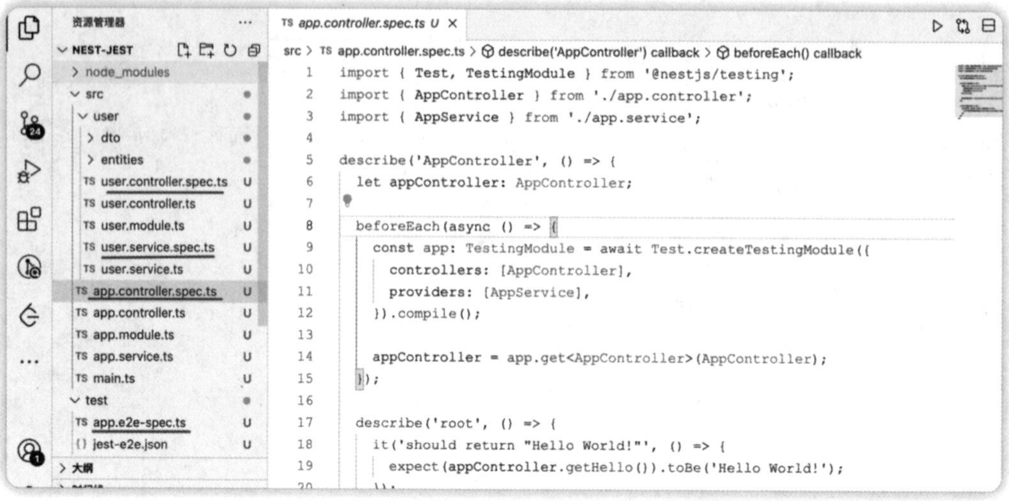

图 8-3　自动生成测试文件

8.2.3　编写测试用例

在 user.service.spec.ts 文件中新增测试用例，代码如下：

```
// 测试 createUser 函数
it('createUser 方法中必须用 hash 对用户密码进行加密', async () => {
  // 创建 DTO
  const createUserDto: CreateUserDto = {
    username: 'mouse',
    password: 'mouse123'
  }
  // 保存用户
  const saveUser: User = await service.createUser(createUserDto)
  const passFlag: boolean = await bcrypt.compare(
    createUserDto.password,
    saveUser.password
  )
  // 保存动作应该成功
  expect(saveUser).toBeDefined()
  // 保存前后账号应该相同
  expect(saveUser.username).toBe(createUserDto.username)
  // 比较输入密码与数据库中的密码的哈希值，一致则为 true
  expect(passFlag).toBeTruthy()
})
```

在上述代码中，我们创建了 CreateUserDto 类来模拟客户端请求的数据结构。然后，调用 service.createUser 方法以保存用户信息，并获取返回的已保存数据。接下来，对用户输入的密码执行哈希处理，并将其与数据库中存储的哈希密码进行比较。最后，对保存结果、用户名以及哈希密码的比较结果进行断言。

由于尚未实现业务逻辑，包括 createUser 方法的定义、bcrypt 包的安装、CreateUserDto 类的定义以及 User 实体类的定义，预期该测试用例将失败。运行"pnpm test:watch"命令以启动测试的监视模式，测试结果如图 8-4 所示。

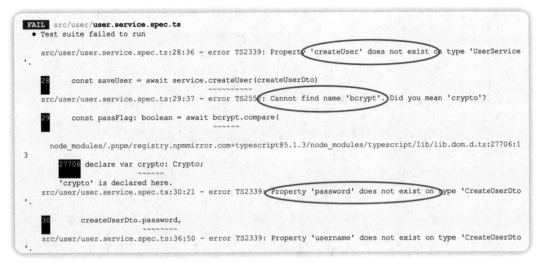

图 8-4　运行测试用例

根据测试结果的输出，我们可以看到测试用例未能通过，原因是出现了一系列的依赖项未找到或未安装的错误。这通常意味着项目中缺少必要的库或模块，需要通过安装相应的依赖来解决。

8.2.4　实现业务代码

有了初步运行结果后，接下来编写实际业务代码让测试用例通过。首先安装 TypeORM、@nestjs/typeorm 和 bcrypt 依赖，运行下面的命令：

```
pnpm add typeorm @nestjs/typeorm mysql2 bcryptjs @types/bcryptjs -S
```

安装完成后，使用 typeorm 初始化数据库，在 AppModule 中引入 TypeOrmModule，代码如下：

```
import { Module } from '@nestjs/common';
import { AppController } from './app.controller';
import { AppService } from './app.service';
import { UserModule } from './user/user.module';
import { TypeOrmModule } from '@nestjs/typeorm'

@Module({
  imports: [
    UserModule,
    // 初始化 MySQL 连接
    TypeOrmModule.forRoot({
      type: 'mysql',
      host: 'localhost',
      port: 3306,
      username: 'root',
      password: 'jminjmin',
      database: 'nest_jest',
      entities: [__dirname + '/**/*.entity{.ts,.js}'],
      autoLoadEntities: true,
      // 在生产环境中禁止开启，应该使用数据迁移
      synchronize: true
    })
```

```
  ],
  controllers: [AppController],
  providers: [AppService],
})
export class AppModule {}
```

在 VS Code 中的可视化数据库插件中创建名为 nest_jest 的数据库，如图 8-5 所示。

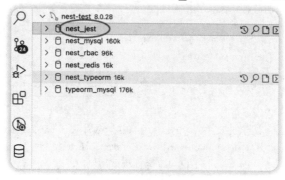

图 8-5　创建数据库

然后，在 user.entity.ts 实体文件中设计数据表结构，代码如下：

```
import { Column, Entity, PrimaryGeneratedColumn } from "typeorm";

@Entity()
export class User {
  @PrimaryGeneratedColumn()
  id: number;

  @Column()
  username: string;

  @Column()
  password: string;
}
```

同时，别忘记完善创建用户的 DTO 类，代码如下：

```
export class CreateUserDto {
  username: string;
  password: string;
}
```

最后，在 user.service.ts 中实现 createUser 函数逻辑。由于前面我们编写了测试用例，这时读者应该清楚在 createUser 方法中将对密码进行哈希加密，然后将数据保存到数据库并返回保存结果，代码如下：

```
import { Injectable } from '@nestjs/common';
import { CreateUserDto } from './dto/create-user.dto';
import { InjectRepository } from '@nestjs/typeorm';
import { User } from './entities/user.entity';
import { Repository } from 'typeorm';
import bcrypt from 'bcryptjs'
```

```
@Injectable()
export class UserService {
  @InjectRepository(User)
  private userRepository: Repository<User>

  async createUser(createUserDto: CreateUserDto) {
    createUserDto.password = await bcrypt.hash(createUserDto.password, 10)
    const saveUser = await this.userRepository.save(createUserDto)
    return saveUser;
  }
}
```

别忘记在 UserController 中定义 createUser 路由方法：

```
@Post()
create(@Body() createUserDto: CreateUserDto) {
  return this.userService.createUser(createUserDto);
}
```

运行 pnpm start:dev 命令启动服务，程序正常运行，如图 8-6 所示。

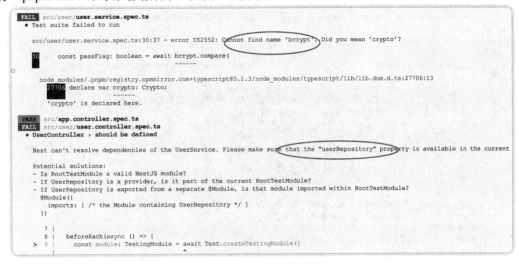

图 8-6 启动服务成功

服务正常运行只能说明当前运行的业务代码没有语法错误，但并不意味着可以通过单元测试，运行"pnpm test"命令来看看效果，如图 8-7 所示。

图 8-7 单元测试结果

在图 8-7 中，第一个错误是由于缺少 bcrypt 依赖，可以通过引入 bcrypt 依赖即可解决。需要注意的是，这里需要导入 bcrypt 的整个模块，使用"import * as bcrypt from 'bcryptjs'"导入即可。

第二个错误表示在 UserController 和 UserService 两个测试模块中都未找到 userRepository 属性，原因在于我们使用 TypeOrm 的存储库来操作数据库，而 Jest 无法自动模拟这个依赖。为了解决这个问题，我们需要通过模拟来注入外部依赖。在两个测试模块中都加入以下代码：

```
// 模拟外部存储库
class UserRepository {
  save(user: User) {
    return user;
  }
}

beforeEach(async () => {
  const module: TestingModule = await Test.createTestingModule({
    providers: [
      UserService,
      // 注入依赖
      {
        provide: getRepositoryToken(User),
        useClass: UserRepository
      }
    ],
  }).compile();

  service = module.get<UserService>(UserService);
});
```

此时，控制台中还有一个错误，提示接收到的 passFlag 值为 false，与期望值 true 冲突，如图 8-8 所示。

```
PASS  src/app.controller.spec.ts
PASS  src/user/user.controller.spec.ts
FAIL  src/user/user.service.spec.ts
 ● UserService › createUser方法中必须使用hash对用户密码进行加密

    expect(received).toBeTruthy()

    Received: false

      53 |       expect(saveUser.username).toBe(createUserDto.username)
      54 |       // 比较输入密码与数据库中的密码的哈希值，一致则为true
    > 55 |       expect(passFlag).toBeTruthy()
         |                        ^
      56 |     })
      57 |   });
      58 |

      at Object.<anonymous> (user/user.service.spec.ts:55:22)

Test Suites: 1 failed, 2 passed, 3 total
Tests:       1 failed, 3 passed, 4 total
Snapshots:   0 total
Time:        2.269 s, estimated 3 s
Ran all test suites related to changed files.
```

图 8-8　测试用例错误

显然，这表示用户输入的密码在经过哈希加密后与存储在数据库中的密码不一致，导致无法通过单元测试。通过审查代码，发现 createUserDto 对象在 createUser 方法调用后被修改为数据库返回的数据，导致前后加密不匹配。下面来优化这段代码：

```
async createUser(createUserDto: CreateUserDto) {
  const password = await bcrypt.hash(createUserDto.password, 10)
  const user: User = new User()
  user.username = createUserDto.username;
  user.password = password

  const saveUser = await this.userRepository.save(user)
  return saveUser;
}
```

通过新建一个 user 实例，我们解决了之前的问题。现在，控制台显示所有测试用例都已成功通过，如图 8-9 所示。

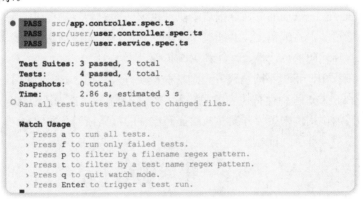

图 8-9　单元测试通过

可以看到测试用例全部通过，笔者感到非常欣慰，因为这意味着我们的代码现在更加健壮，更接近最终的目标。

8.2.5　重构代码

通过不断迭代修改业务代码和测试用例，我们最终通过了单元测试。这个过程虽然反复，但确保了代码的质量和功能的正确性。现在，我们可以开始重构（优化）业务代码。幸运的是，单元测试可以随时告诉我们程序是否存在问题，并帮助我们快速定位问题所在。

此时，可能会有读者问：我的功能已经实现了，接下来应该如何重构或优化代码呢？实际上，重构和优化并没有绝对的标准。对于简单的功能，一旦测试用例通过，我们可能就完成了 TDD 流程。然而，对于复杂的功能，即使测试用例通过，我们仍然可以审查并优化代码。

这里提供一种思路：当测试用例通过后，我们可以审查代码是否符合当前企业要求的编码规范，包括 JavaScript/TypeScript 编码规范、性能规范、接口规范、日志规范、数据库规范以及缓存规范等。这正是许多企业中 CodeReview 的规范。

例如，为了验证用户通过 DTO 发送的数据，我们可以安装 class-validator 和 class-transformer（版本 0.3.1）依赖，并为 username 和 password 添加验证器。示例代码如下：

```typescript
import { IsNotEmpty, IsString } from "class-validator";

export class CreateUserDto {
  @IsString({message: "username 必须是字符串类型"})
  @IsNotEmpty()
  username: string;

  @IsString({message: "password 必须是字符串类型"})
  @IsNotEmpty()
  password: string;
}
```

至此，我们已经完成了功能的单元测试部分。通过遵循测试驱动开发（TDD）流程，我们首先分析了需求并进行了功能设计。接下来，我们编写了第一个注定失败的测试用例，这是 TDD 流程的起点。测试失败提供了反馈，指导我们开始完善业务代码，引入所需的依赖，并实现缺失的方法。

我们运行了"pnpm test:watch"命令来启动热测试服务，它为我们提供了实时的反馈，让我们知道测试用例何时通过。一旦编码完成并通过所有测试用例，我们根据企业的编码规范进行了代码的重构和优化，以提高代码的可读性、性能和可维护性。

在整个 TDD 流程中，我们不断迭代，直到功能模块开发完成。这个过程不仅帮助我们确保代码质量，也让我们在开发过程中保持清晰的思路和高效的问题解决能力。TDD 是一种有效的开发实践，它通过先编写测试用例来指导开发，从而提高软件的可靠性和开发效率。

8.3 集成测试

集成测试是一种关键的软件测试方法，它超越了单元测试的范畴，专注于验证多个单元或组件在合并后的功能是否能够协同工作。与我们熟悉的 Vue 或 React 组件化开发类似，集成测试可以看作是将独立开发的组件组合到一起的过程。在组件化开发中，我们分别开发每个组件，然后将它们组装起来，形成完整的用户界面。类似地，集成测试将不同的模块或服务融合在一起，测试它们作为一个整体时是否能够正常交互和运行。

本节将继续上一节的代码，通过编写集成测试来进一步完善我们的应用程序。我们将采用真实数据库进行测试，以确保用户信息能够正确地保存和检索。这不仅验证了模块间的交互，也确保了我们的应用程序在实际运行环境中的可靠性。

集成测试的目的是确保应用程序的各个部分在合并后能够作为一个协调一致的整体运行，从而提高软件质量和用户满意度。通过这种测试，我们可以发现并修复那些在单元测试阶段可能被忽略的问题。

8.3.1 编写测试用例

为了方便读者阅读代码，我们对上一节中的 nest-jest 进行备份，并将其命名为 nest-jest-integration。我们将 UserModule 与 TypeOrmModule 模块结合起来，以测试新用户是否保存在数据库中。

同时，将 user.service.spec.ts 文件重新命名为 user.integration-spec.ts，以明确这是集成测试文件。修改后的代码如下：

```typescript
import { Test, TestingModule } from '@nestjs/testing';
import { UserService } from './user.service';
import { CreateUserDto } from './dto/create-user.dto';
import { User } from './entities/user.entity';
import { TypeOrmModule } from '@nestjs/typeorm';
import * as bcrypt from 'bcryptjs'
import { UserModule } from './user.module';

describe('测试 user 服务与数据库交互', () => {
  let service: UserService;

  beforeEach(async () => {
    const module: TestingModule = await Test.createTestingModule({
      imports: [
        TypeOrmModule.forRoot({
          type: 'mysql',
          host: 'localhost',
          port: 3306,
          username: 'root',
          password: 'jminjmin',
          database: 'nest_jest',
          entities: [__dirname + '/**/*.entity{.ts,.js}'],
          autoLoadEntities: true,
          // 在生产环境中禁止开启，应该使用数据迁移
          synchronize: true
        }),
        UserModule,
      ],
    }).compile();

    service = module.get<UserService>(UserService);
  });
  it('should be defined', () => {
    expect(service).toBeDefined();
  });

  // 测试 createUser 函数
  it('createUser 方法中必须使用 hash 对用户密码进行加密', async () => {
    // 创建 DTO
    const createUserDto: CreateUserDto = {
      username: 'mouse',
      password: 'mouse123'
    }
    // 保存用户
    const saveUser: User = await service.createUser(createUserDto)
    const passFlag: boolean = await bcrypt.compare(
      createUserDto.password,
      saveUser.password
    )
    // 保存动作应该成功
    expect(saveUser).toBeDefined()
    // 保存前后账号应该相同
```

```
      expect(saveUser.username).toBe(createUserDto.username)
      // 比较输入密码与数据库中的密码的哈希值比较，一致则为 true
      expect(passFlag).toBeTruthy()
    })
  });
```

上述代码创建了一个测试模块，导入 TypeOrmModule 与 UserModule 并连接了测试数据库 nest_jest，将两者组合到一起进行测试。

8.3.2 测试效果

运行"pnpm test"测试命令。当测试用例通过后，我们可以看到指定数据已成功插入到数据库中，如图 8-10 所示。

图 8-10 成功插入数据

可见，用户密码成功被哈希加密，并且用户信息已经成功存储到数据库中。这种模块组合称为集成测试，使用这种方式可以不断丰富集成测试用例，例如引入 Redis 缓存等。

然而，Nest CLI 在生成项目和模块的时候，并不会生成类似 xxx.integration-spec.ts 这样的模块测试文件。笔者认为这可能有两个原因：

- 其一是 Nest 框架的核心逻辑主要由控制器、服务和其他提供者进行处理，而模块作为管理依赖的入口，通常并不包含太多业务逻辑。
- 其二是 Nest 提供了端对端测试文件（xxx.e2e-spec.ts），用于模拟真实用户的系统行为，提供更全面的测试反馈。

关于端对端测试的更多细节，我们将在下一节进行介绍。

8.4 端到端测试

与测试单个类或模块的单元测试和集成测试不同，端对端（e2e）测试涵盖类和模块在更高聚合程度上的交互，即我们需要知道真实用户在操作应用系统时会发生什么事情，端对端测试提供了这种手段，让我们可以模拟真实用户的行为，以确保项目在真实场景下正常运行。

本节依旧使用账号和密码创建用户功能，模拟用户发送 HTTP 请求，对用户输入的信息进行合法性校验测试，并返回相应的信息提示。

8.4.1　编写测试用例

沿用 8.2 节的 nest-jest 项目，将其备份并重命名为 nest-jest-e2e。在 test 目录下，新建 user.e2e-spec.ts 文件，用于对 user 模块进行 e2e 测试。我们将使用 supertest 提供的 request 方法来模拟对 createUser 方法的请求，并期望返回 201 状态码。以下是测试用例的代码：

```typescript
import { NestApplication } from '@nestjs/core'
import { Test, TestingModule } from '@nestjs/testing'
import { TypeOrmModule } from '@nestjs/typeorm'
import { AppController } from '../src/app.controller'
import { UserModule } from '../src/user/user.module'
import * as request from 'supertest'
import { AppService } from '../src/app.service'

describe('user module e2e test, () => {
  let app: NestApplication

  beforeAll(async () => {
    const moduleFixture: TestingModule = await Test.createTestingModule({
      imports: [
        TypeOrmModule.forRoot({
          type: 'mysql',
          host: 'localhost',
          port: 3306,
          username: 'root',
          password: 'jminjmin',
          database: 'nest_jest',
          entities: [__dirname + '/**/*.entity{.ts,.js}'],
          autoLoadEntities: true,
          // 在生产环境中禁止开启，应该使用数据迁移
          synchronize: true
        }),
        UserModule,
      ],
      controllers: [AppController],
      providers: [AppService]
    }).compile()

    app = moduleFixture.createNestApplication()
    await app.init()
  })

  afterAll(async () => {
    await app.close()
  })

  describe('/user/create (POST)', () => {
    it('should return 201', async () => {
      const requestBody = {
        username: 'mouse100',
        password: '12345678',
      }
      await request(app.getHttpServer())
        .post('/user/create')
```

```
      .send(requestBody)
      .expect(201)
  })
})
```

运行"pnpm test:e2e"命令启动端对端测试，结果如图 8-11 所示。

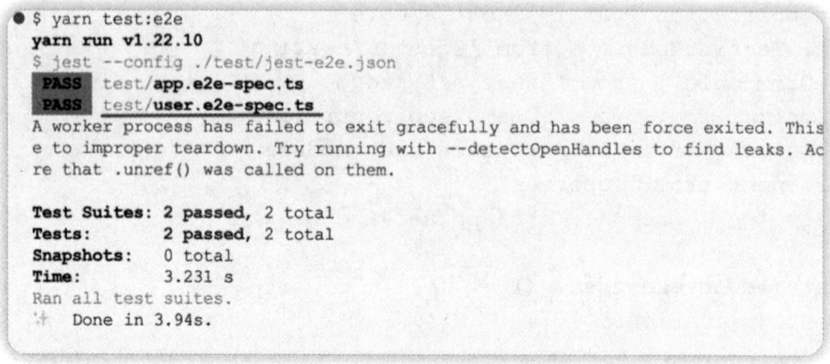

图 8-11　运行 e2e 测试

测试用例成功运行并通过测试，在数据库的可视化界面中可以看到用户信息已成功插入，如图 8-12 所示。

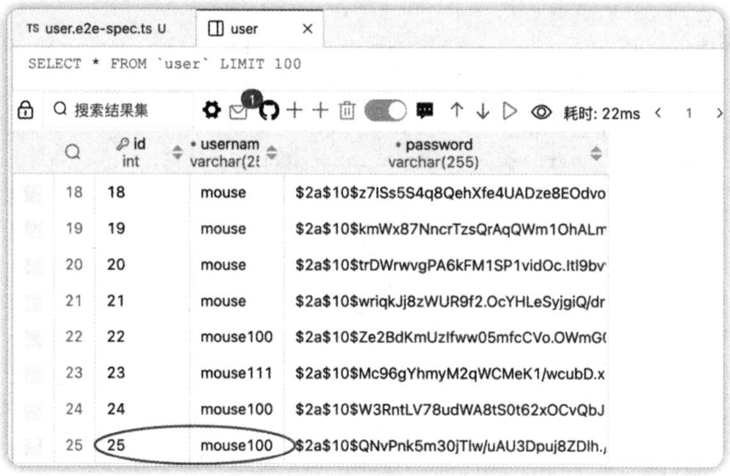

图 8-12　e2e 测试插入数据

到目前为止，我们的用户创建功能还不完善，需要添加更多的场景，比如当账号已存在时或在根据账号查找用户时账号输入错误，都应返回 400 错误。接下来我们来补充测试用例：

```
// 创建用户
describe('/user/create (POST)', () => {
  // 正常创建用户
  it('should return 201', async () => {
    const requestBody = {
      username: 'mouse200',
      password: '12345678',
    }
    await request(app.getHttpServer())
      .post('/user/create')
```

```
      .send(requestBody)
      .expect(201)
  })

  // 传入已存在的用户名，返回 400
  it('should return 400 given exist username', async() => {
    const requestBody = {
      username: 'mouse100',
      password: '12345678',
    }
    await request(app.getHttpServer())
      .post('/user/create')
      .send(requestBody)
      .expect(400)
  })
})

// 查询用户
describe('/user/findOne (GET)', () => {
  // 正常查询用户
  it('should return 200', async() => {
    await request(app.getHttpServer())
      .get('/user/findOne')
      .query({username: 'mouse111'})
      .expect(200)
  })
  // 查询不存在的用户，返回 400
  it('should return 400 given not exist username', async() => {
    await request(app.getHttpServer())
      .get('/user/findOne')
      .query({username: 'mouse000'})
      .expect(400)
  })
})
```

在原有测试的基础上，我们新增了三个测试用例，分别用于验证以下场景：用户名重复时的行为、正常条件下查询用户的功能，以及查询一个不存在的用户。这些测试用例的设计旨在全面覆盖用户管理功能的不同方面。

运行这些测试后，按预期结果应该会看到 1 个测试通过（在测试报告中通常以绿色显示），而另外 3 个测试未通过（通常以红色显示）。出现这种结果是因为我们尚未实现这三个新增测试用例所对应的业务逻辑。

观察到预期的测试结果后，接下来我们将实现这些测试用例所对应的业务逻辑，并重新运行测试，以验证实现的正确性。

8.4.2 实现业务代码

我们依然按照 TDD 的"红-绿-重构"流程来完善代码。接下来，我们将编写业务代码以通过测试（即让测试变"绿"）。在 createUser 方法中，我们添加了用户名重复的判断逻辑，代码如下：

```
async createUser(createUserDto: CreateUserDto) {
  // 判断用户是否已经存在
  let existUser: User = await this.userRepository.findOneBy({username: createUserDto.
```

```
username})
    if (existUser && existUser.username) throw new BadRequestException('用户名已存在')

    const password = await bcrypt.hash(createUserDto.password, 10)
    const user: User = new User()
    user.username = createUserDto.username;
    user.password = password

    const saveUser = await this.userRepository.save(user)
    return saveUser;
}
```

接着实现查询用户功能，在 UserController 中添加 findOne 路由方法，接收 username 作为查询参数，并调用 UserService 中对应的 findOne 方法进行查询，代码如下：

```
// user.controller.ts
@Get('/findOne')
findOne(@Query('username') username: string) {
    return this.userService.findOne(username);
}

// user.service.ts
async findOne(username: string) {
    const user: User = await this.userRepository.findOneBy({username: username});
    if (!user || user.username !== username) throw new BadRequestException('当前用户不存在')
    return user
}
```

再次运行测试，可以看到新增的所有测试用例都通过了，如图 8-13 所示。

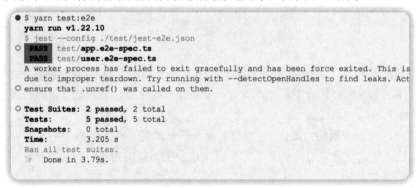

图 8-13 成功通过 e2e 测试

至此，我们已经实现了用户模块的端对端测试。在实际项目中，我们可以将测试命令添加到 CI/CD 流水线中。当检测到分支更新时，自动运行端对端测试并生成测试覆盖率，以实现持续集成。

第 9 章

日志与错误处理

本章将深入探讨后端开发中的日志记录和错误处理机制，这些是构建稳定系统的关键组成部分。本章首先强调日志记录在监控系统运行和捕获潜在问题中的重要性，并介绍记录日志的最佳实践，这将帮助用户确保系统的健康和可维护性。然后，深入探讨 Nest 框架提供的内置日志器。通过学习其基本用法，读者将了解如何有效地在应用程序中实现日志记录功能。为了进一步提升日志管理的灵活性和功能，最后将介绍 Winston，这是一个功能强大的第三方日志管理器。我们将探讨 Winston 的优势，包括其高度可配置性、支持多种日志级别和传输机制，以及如何将其集成到 Nest 应用程序中。

通过本章的学习，读者将获得必要的知识和工具，以建立一个健壮的日志记录系统，这对于维护现代后端服务的稳定性和可靠性至关重要。

9.1 如何在 Nest 中记录日志

在日常开发过程中，我们经常使用浏览器环境提供的 console 对象来输出信息，这有助于调试程序和记录日志。同样，在服务器端开发中，Nest 框架也提供了一套强大的基于文本的日志系统。它不仅可以记录程序运行中的错误、警告和调试信息，还支持定制化功能，如自定义日期格式、日志加密、压缩，以及将日志上报至外部存储系统等，满足多样化的业务需求。

本节内容不仅教读者如何使用 Nest 的日志记录功能，还将深入探讨日志记录的重要性，并介绍一系列最佳实践方法。这些方法包括但不限于日志级别的合理使用、日志的可读性与结构化以及日志的监控和告警策略。掌握这些方法后，读者可以有效地将它们应用到实际项目中，提高项目的可维护性和稳定性。

通过本节的学习，读者将更加深入地理解日志记录的价值，并掌握如何通过 Nest 框架的日志系统来优化应用程序的日志记录策略。

9.1.1 为什么要记录日志

作为开发人员，在开发环境中调试和定位问题相对容易。然而，在生产环境中，通常无法方便地附加调试器来追踪 BUG。此时，日志记录的作用就显得至关重要。

日志记录就像一部相机，能够捕捉程序运行时的各个方面的事件。如果能够正确地使用和维护，它将成为我们进行故障排除和程序诊断的强大工具。通过日志，我们可以回溯问题发生时的具体情况，从而快速定位并解决问题。

在运行 Node 应用程序时，通常需要多个服务器和组件，如数据库或 Redis 服务，它们需要相互协作。如果这些服务因某些问题而未能正常运行，没有日志记录，我们可能永远无法得知服务器失败的真正原因。这就像是公路上发生了交通事故，如果没有监控设备记录下事故的经过，那么事故的调查和责任认定将变得极为困难。

因此，建立一个全面且有效的日志记录系统对于任何应用程序的生产运行都是必不可少的。它不仅帮助我们监控应用程序的状态，还能在出现问题时提供关键的信息，帮助我们快速响应并恢复服务。

9.1.2 内置日志器 Logger

在开始探索 Nest 内置的 Logger 日志器之前，我们首先需要创建一个新的 Nest 项目。可以通过运行 "nest n nest-logger -p pnpm" 命令来完成，结果如图 9-1 所示。

图 9-1 创建项目

Nest 中默认开启了日志记录（Logger），当我们运行 "pnpm start:dev" 命令启动 Nest 应用时，Logger 会记录启动过程中发生的事情，如图 9-2 所示。

图 9-2 启动日志记录

可见，日志信息包含时间戳、日志级别和上下文等重要内容。根据这些信息，可以确认程序运行的状态。默认格式如下：

```
[AppName] [PID] [Timestamp] [LogLevel] [Context] Message [+ms]
```

其中，AppName 为应用程序名称，固定为 Nest；PID 为系统分配的进程编号；Timestamp 为输出的当前系统时间。

Logger 分为多种级别，包括'log'、'error'、'warn'、'debug' 和 'verbose'。用户可以指定任意组合来启用记录，例如只在出现错误或警告时打印日志。示例代码如下：

```
async function bootstrap() {
  const app = await NestFactory.create(AppModule, {
    logger: ['error', 'warn'],
  });
  await app.listen(3000);
}
```

Context 上下文表示当前日志的产生阶段，例如应用启动阶段或者路由程序执行阶段。

在许多情况下，我们需要使用 Logger 手动记录自定义事件或消息，Nest 会以相同的格式和行为把它记录下来。例如，在 AppService 的 getHello 方法中记录错误，代码如下：

```
import { Injectable, Logger } from '@nestjs/common';

@Injectable()
export class AppService {

  private logger = new Logger(AppService.name);

  getHello(): string {
    this.logger.error('getHello error')
    return 'Hello World!';
  }
}
```

在上述示例代码中，我们通过向 Nest 的 Logger 类传递上下文来创建了一个 logger 实例。这样做可以帮助我们在日志条目中标识它们所属的模块或功能区。接着，我们调用了 logger 实例的 error 方法来记录错误信息。

为了模拟对服务器的 GET 请求，我们执行了"curl -X GET http://localhost:3000/"命令。这个请求触发了应用程序中的 getHello 方法，控制台中记录了一条格式化的错误日志消息，如图 9-3 所示。

```
[Nest] 73613  - 2024/03/11 09:39:26    ERROR [AppService] getHello error
```

图 9-3　记录错误日志

有些读者可能会注意到，使用这种方式记录日志存在一些问题，例如：

- 每次使用 Logger 都需要先实例化它。
- 无法记录其他服务信息，如 ConfigService 环境配置。

那么，能否进行依赖注入呢？答案是可以的。此时，我们需要扩展内置日志器。

9.1.3 定制日志器

我们知道依赖注入的范围是模块级别的。首先，通过运行"nest g module logger --no-spec"命令来创建一个 LoggerModule 模块，操作十分简单，示例代码如下：

```
import { Module } from '@nestjs/common';
import { Logger } from './logger';

@Module({
  providers: [Logger],
  exports: [Logger]
})
export class LoggerModule {}
```

还需要创建一个 logger 服务提供者，用于扩展日志方法并获取配置服务的信息，如环境变量等应用信息。此时有两种方案：

（1）完全定制一套属于自己的日志器，需要实现更底层的 LoggerService 类；
（2）继承 ConsoleLogger 类，这样我们依然可以使用 Nest 内置的日志行为。示例代码如下：

```
import { ConsoleLogger, Injectable, Scope } from '@nestjs/common';

@Injectable()
export class Logger extends ConsoleLogger {
  error(message: any, stack?: string, context?: string) {
    // 在这里调用 configService 方法，例如加上环境标识
    ...
    message = message + ' 环境: dev'
    super.error(message, stack, context);
  }
}
```

此时，在 main.ts 中，通过 useLogger 方法将应用程序的默认日志器替换为定制的日志器实例。这样，当我们在应用程序的任何地方使用 this.logger.error()打印日志时，将调用自定义 Logger 实例中定义的方法。示例代码如下：

```
async function bootstrap() {
  const app = await NestFactory.create(AppModule, {
    bufferLogs: true,
  });
  app.useLogger(app.get(Logger));
  await app.listen(3000);
}
```

重新运行"curl -X GET http://localhost:3000/"命令请求接口，可以看到记录的自定义环境配置信息，如图 9-4 所示。

```
[Nest] 81470  - 2024/03/11 13:34:30     LOG [InstanceLoader] LoggerModule dependencies initialized
[Nest] 81470  - 2024/03/11 13:34:30     LOG [InstanceLoader] AppModule dependencies initialized
[Nest] 81470  - 2024/03/11 13:34:30     LOG [RoutesResolver] AppController {/}:
[Nest] 81470  - 2024/03/11 13:34:30     LOG [RouterExplorer] Mapped {/, GET} route
[Nest] 81470  - 2024/03/11 13:34:30     LOG [NestApplication] Nest application successfully started
[Nest] 81470  - 2024/03/11 13:34:34   ERROR [AppService] getHello failed 环境: dev
```

图 9-4 自定义日志信息

9.1.4 记录日志的正确姿势

在后端服务开发中，通常要求开发者及时对某些行为进行日志记录，以便应用上线后能及时追踪运行状态。为了确保日志记录的高效性和准确性，我们需要注意以下几点：

（1）日志行为不应该出现异常。当我们使用日志记录系统行为时，首先需要保证日志行为不能出现异常，如空指针异常等。示例代码如下：

```
this.logger.log('服务操作用户 id=${userService.getUser().id}')
```

其中，userService.getUser()的执行结果可能为 null，导致获取 id 操作抛出异常，应避免出现这种情况。

（2）日志行为不应该产生副作用。日志应该是无状态的，日志行为不应该对业务逻辑产生任何副作用，否则系统行为可能完全出乎意料。示例代码如下：

```
this.logger.log('保存用户成功${userRepository.save(user)}')
```

其中，执行 userRepository.save 后会在数据库中新增记录，应该杜绝这种情况。

（3）日志信息不应该包含敏感信息。日志用于记录程序的执行状态，例如调用了哪些函数、发生了什么错误等，需要确保不记录用户的账号和密码、银行卡等财务敏感信息。

（4）日志记录尽可能详细。日志是面向开发人员的，当描述错误时，应该提及具体操作及失败原因，并提醒接下来应该怎么做。示例代码如下：

```
this.logger.error('用户${id}调用 getUserInfo 方法失败，进行重试中', error)
getUserInfo()
```

9.1.5 第三方日志器 Winston

Nest 内置的日志器可以在开发过程中记录系统行为，而在生产环境中通常使用专用的日志统计模块，如 Winston。它能够满足特定的日志记录要求，包括高级过滤、格式化和集中日志记录。下一节将详细介绍如何在 Nest 中使用 Winston 进行日志管理。

9.2　Winston 日志管理实践

在实际的生产应用中，日志记录扮演着关键角色。它不仅用于服务问题排查，还广泛应用于业务数据分析和监控告警平台的建设。在这些场景中，我们通常不会使用传统的 console 模块来打印日志，因为它不支持持久化。相反，我们需要将日志写入数据库或文件中，以确保它们的长期保存和可追溯性。

Winston 是一个在 Node.js 领域广泛使用的日志框架，它提供了一套强大的功能来帮助我们实现这一目标。Winston 支持将日志传输到多种目的地，包括文件系统、控制台以及数据库等。此外，它还支持定义多种日志级别（如 DEBUG、INFO、WARN、ERROR 等）和自定义日志格式，以满足不同场景下的日志需求。

使用 Winston，我们可以轻松地配置和定制日志策略，确保日志记录的灵活性和有效性。无论是开发人员进行问题诊断还是运维团队进行系统监控，Winston 都能提供必要的支持。

9.2.1　Winston 的基础使用

为了更好地演示在 Nest 中集成 Winston，我们先来创建一个 Nest 项目。运行 "nest n nest-winston -p pnpm" 命令，效果如图 9-5 所示。

```
$ nest n nest-winston -p pnpm
⚡ We will scaffold your app in a few seconds..

CREATE nest-winston/.eslintrc.js (663 bytes)
CREATE nest-winston/.prettierrc (51 bytes)
CREATE nest-winston/README.md (3347 bytes)
CREATE nest-winston/nest-cli.json (171 bytes)
CREATE nest-winston/package.json (1953 bytes)
CREATE nest-winston/tsconfig.build.json (97 bytes)
CREATE nest-winston/tsconfig.json (546 bytes)
CREATE nest-winston/src/app.controller.spec.ts (617 bytes)
CREATE nest-winston/src/app.controller.ts (274 bytes)
CREATE nest-winston/src/app.module.ts (249 bytes)
CREATE nest-winston/src/app.service.ts (142 bytes)
CREATE nest-winston/src/main.ts (208 bytes)
CREATE nest-winston/test/app.e2e-spec.ts (630 bytes)
CREATE nest-winston/test/jest-e2e.json (183 bytes)

✓ Installation in progress... 🍺

🚀 Successfully created project nest-winston
```

图 9-5　创建项目

运行如下命令，生成 logger 模块和服务提供者，专门用于处理日志模块：

```
nest g module logger --no-spec
nest g provider logger --no-spec
```

其中，在 LoggerModule 中导出 Logger，以便在其他模块中共享它，代码如下：

```
import { Module } from '@nestjs/common';
import { Logger } from './logger';

@Module({
  providers: [Logger],
  // 导出 Logger 供其他模块使用
  exports: [Logger]
})
export class LoggerModule {}
```

接下来，重点是实现 logger.ts 的逻辑。在此之前，需要安装 winston、winston-daily-rotate-file、dayjs、chalk 这些依赖。其中，winston 是日志框架；winston-daily-rotate-file 可以根据日期和大小限制进行日志文件的轮转，旧日志可以根据计数或已用天数进行删除；dayjs 用于格式化日期；chalk 用于为控制台文本的着色。运行 "pnpm add winston winston-daily-rotate-file dayjs chalk -S" 命令即可安装这些依赖。

由于我们要定制一个全新的日志模块来替换系统内置的日志器，这里选择实现 Nest 底层的 LoggerService 类来编写 Logger 类，这意味着需要在 LoggerService 类下实现日志方法。首先，我们来实现在控制台打印日志，代码如下：

```
import { Injectable, LoggerService } from '@nestjs/common';
import 'winston-daily-rotate-file';
import {
  Logger as WinstonLogger,
  createLogger,
  transports,
} from 'winston';

@Injectable()
export class Logger implements LoggerService {
  private logger: WinstonLogger;
  constructor() {
    this.logger = createLogger({
      level: 'debug',
      transports: [
        // 打印到控制台,生产环境可关闭
        new transports.Console()
      ],
    });
  }
  log(message: string, context: string) {
    this.logger.log('info', message, { context });
  }
  info(message: string, context: string) {
    this.logger.info(message, { context });
  }
  error(message: string, context: string) {
    this.logger.error(message, { context });
  }
  warn(message: string, context: string) {
    this.logger.warn(message, { context });
  }
}
```

上述代码通过 createLogger 方法创建了一个日志管理器,并指定传输方式为控制台,同时实现了常用的 log、info、error、warn 方法。

Winston 支持以下几种日志级别统计,它们是:

```
const levels = {
  error: 0,
  warn: 1,
  info: 2,
  http: 3,
  verbose: 4,
  debug: 5,
  silly: 6
};
```

这里指定为 debug,意味着所有小于数字 5 的日志级别都会被统计并记录,只有 silly 级别不会被统计和记录。

然后,在 main.js 中指定使用 Winston 日志器,并关闭内置日志器,代码如下:

```
async function bootstrap() {
  const app = await NestFactory.create(AppModule, {
```

```
    logger: false,
  });
  // 使用自定义 Logger
  app.useLogger(app.get(Logger))
  await app.listen(3000);
}
```

在 AppService 中注入 Logger 类,并在 getHello 路由方法中调用日志方法打印结果,代码如下:

```
import { Injectable } from '@nestjs/common';
import { Logger } from './logger/logger';

@Injectable()
export class AppService {
  constructor(private logger: Logger) {}
  getHello(): string {
    this.logger.info('getHello info', AppService.name)
    this.logger.warn('getHello warn', AppService.name)
    this.logger.error('getHello error', AppService.name)
    return 'Hello World!';
  }
}
```

运行 "pnpm start:dev" 命令启动服务后,可以在控制台中看到最基础的日志记录,如图9-6所示。

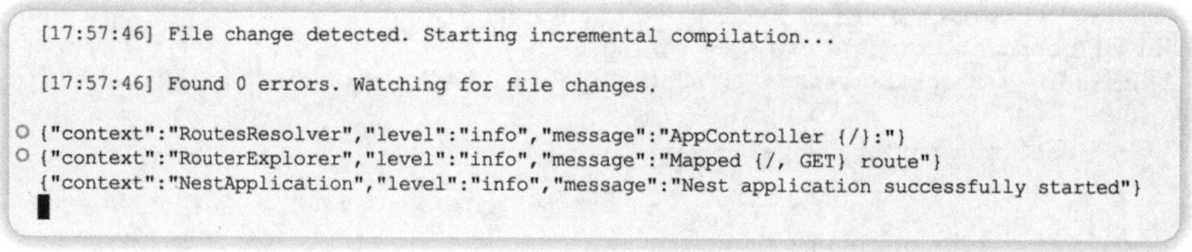

```
[17:57:46] File change detected. Starting incremental compilation...

[17:57:46] Found 0 errors. Watching for file changes.

{"context":"RoutesResolver","level":"info","message":"AppController {/}:"}
{"context":"RouterExplorer","level":"info","message":"Mapped {/, GET} route"}
{"context":"NestApplication","level":"info","message":"Nest application successfully started"}
```

图 9-6 在控制台中打印日志

运行 "curl -X GET http://localhost:3000/" 命令请求 getHello 接口后,可以在控制台看到成功打印出了日志,如图9-7所示。

```
{"context":"RoutesResolver","level":"info","message":"AppController {/}:"}
{"context":"RouterExplorer","level":"info","message":"Mapped {/, GET} route"}
{"context":"NestApplication","level":"info","message":"Nest application successfully started"}
{"context":"AppService","level":"info","message":"getHello info"}
{"context":"AppService","level":"warn","message":"getHello warn"}
{"context":"AppService","level":"error","message":"getHello error"}
```

图 9-7 自定义日志记录

9.2.2 本地持久化日志

讲到这里,有些读者可能会有疑惑:Nest 内置的日志系统缺少时间戳,并且日志在服务重启后会丢失,应该如何解决?

为了同时满足上述需求,我们需要改造 logger.ts 的逻辑,以实现以下几点要求:

- 日志记录加上时间戳,并格式化日志输出。

- 日志记录加上颜色进行美化。
- 根据日志级别进行分类统计,特别是区分错误日志,并按日实现滚动日志。
- 定期清理日志文件。

logger.ts 实现逻辑如下:

```ts
import { Injectable, LoggerService } from '@nestjs/common';
import 'winston-daily-rotate-file';
import {
  Logger as WinstonLogger,
  createLogger,
  format,
  transports,
} from 'winston';
import * as chalk from 'chalk';
import * as dayjs from 'dayjs';

@Injectable()
export class Logger implements LoggerService {
  private logger: WinstonLogger;
  constructor() {
    this.logger = createLogger({
      level: 'debug',
      transports: [
        // 打印到控制台,生产环境可关闭
        new transports.Console({
          format: format.combine(
            // 颜色
            format.colorize(),
            // 日志格式
            format.printf(({ context, level, message, timestamp }) => {
              const appStr = chalk.green(`[NEST]`);
              const contextStr = chalk.yellow(`[${context}]`);

              return `${appStr} ${timestamp} ${level} ${contextStr} ${message} `;
            }),
          ),
        }),
        // 保存到文件
        new transports.DailyRotateFile({
          // 日志文件文件夹
          dirname: process.cwd() + '/src/logs',
          // 日志文件名 %DATE% 会自动设置为当前日期
          filename: 'application-%DATE%.info.log',
          // 日期格式
          datePattern: 'YYYY-MM-DD',
          // 压缩文档,用于定义是否对存档的日志文件进行 gzip 压缩,默认值为 false
          zippedArchive: true,
          // 文件最大大小,可以是 Bytes、KB、MB、GB
          maxSize: '20M',
          // 最大文件数,可以是文件数,也可以是天数,天数加单位"d"
          maxFiles: '7d',
          // 格式定义,同 winston
          format: format.combine(
```

```
          format.timestamp({
            format: 'YYYY-MM-DD HH:mm:ss',
          }),
          format.json(),
        ),
        // 日志等级，如果不设置，所有日志将会记录在同一个文件中
        level: 'info',
      }),
      // 用上述方法区分 error 日志和 info 日志，保存在不同文件中，方便排查问题
      new transports.DailyRotateFile({
        dirname: process.cwd() + '/src/logs',
        filename: 'application-%DATE%.error.log',
        datePattern: 'YYYY-MM-DD',
        zippedArchive: true,
        maxSize: '20m',
        maxFiles: '14d',
        format: format.combine(
          format.timestamp({
            format: 'YYYY-MM-DD HH:mm:ss',
          }),
          format.json(),
        ),
        level: 'error',
      }),
    ],
  });
}
log(message: string, context: string) {
  const timestamp = dayjs(Date.now()).format('YYYY-MM-DD HH:mm:ss');
  this.logger.log('info', message, { context, timestamp });
}
info(message: string, context: string) {
  const timestamp = dayjs(Date.now()).format('YYYY-MM-DD HH:mm:ss');
  this.logger.info(message, { context, timestamp });
}
error(message: string, context: string) {
  const timestamp = dayjs(Date.now()).format('YYYY-MM-DD HH:mm:ss');
  this.logger.error(message, { context, timestamp });
}
warn(message: string, context: string) {
  const timestamp = dayjs(Date.now()).format('YYYY-MM-DD HH:mm:ss');
  this.logger.warn(message, { context, timestamp });
}
}
```

在上述代码中，笔者特别强调了日志系统的关键配置部分。为了将日志输出到文件，我们使用了 dailyRotateFile 传输方式，这允许日志按日进行轮转，从而便于管理和归档。此外，我们还使用了 Winston 的 format 方法来定义一组格式化日志输出的格式，包括：

[AppName] [Timestamp] [LogLevel] [Context] Message

输出的日志如图 9-8 所示。

```
[18:31:47] Starting compilation in watch mode...
[18:31:47] Found 0 errors. Watching for file changes.

[NEST] 2024-03-12 18:31:48 info [RoutesResolver] AppController {/}:
[NEST] 2024-03-12 18:31:48 info [RouterExplorer] Mapped {/, GET} route
[NEST] 2024-03-12 18:31:48 info [NestApplication] Nest application successfully started
```

图 9-8　格式化日志

同时，在文件目录中可以看到生成的 logs 文件夹，专门用于存放日志文件，如图 9-9 所示。

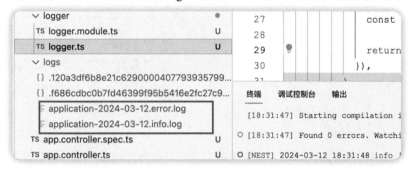

图 9-9　生成的.log 日志文件

在.info 日志文件中可以看到生成了日志记录，如图 9-10 所示。

```
src > logs > ≡ application-2024-03-12.info.log
  1  {"context":"RoutesResolver","level":"info","message":"AppController {/}:","timestamp":"2024-03-12
  2  {"context":"RouterExplorer","level":"info","message":"Mapped {/, GET} route","timestamp":"2024-03
  3  {"context":"NestApplication","level":"info","message":"Nest application successfully started","ti
  4
```

图 9-10　生成日志记录

再次运行"curl -X GET http://localhost:3000/"命令请求 getHello 接口后，可以在.error 日志文件中看到新增了一条错误日志，如图 9-11 所示。

```
src > logs > ≡ application-2024-03-12.error.log
  1  {"context":"AppService","level":"error","message":"getHello error","timestamp":
  2
```

图 9-11　生成错误日志

至此，我们已经完成了日志本地持久化的功能。在实际业务中，我们也可以将日志文件上传到远程服务器，或者通过指定 transports.Http 的方式将日志持久化到数据库中，以便进行后续的业务数据分析、数据过滤和数据清洗等操作。

9.3 面向切面日志统计实践

在实际业务环境中，日志记录是一种关键的监控和问题诊断工具。虽然在某些特殊业务场景中，开发人员可能会选择手动记录日志，但在处理请求、响应和捕获服务异常等常见场景时，手动记录日志可能效率较低。为了减少冗余的日志代码并统一日志格式，通常会采用全局日志记录策略，这基于面向切面编程（Aspect-Oriented Programming，AOP）的思想来实现。

本节将探讨如何在 Nest 应用程序中使用以下全局机制来自动化日志记录过程：全局中间件，用于自动记录每个请求的参数、方法和 IP 地址等信息；全局响应拦截器，用于统一应用程序的响应格式，确保响应的一致性；全局异常过滤器，用于捕获应用程序中的异常，并在合适的位置记录错误日志。

通过这些机制，我们可以减少日志记录的代码量，同时确保关键信息的记录，提高应用程序的可维护性和可监控性。

9.3.1 中间件日志统计

在 Nest 中，中间件是指在路由处理程序之前或之后执行的函数。它们可以操作请求和响应对象，或执行其他运行时确定的任务。通常，中间件用于收集请求参数、请求体、请求方法、IP 地址等信息，这些信息对于后续的问题排查至关重要。

为了在 nest-winston 项目中实现日志记录功能，我们需要新建一个文件 common/logger.middleware.ts。在这个文件中，我们将实现一个类中间件，用于记录每个请求的关键信息。示例代码如下：

```typescript
import { Inject, Injectable, NestMiddleware } from '@nestjs/common';
import { NextFunction, Request, Response } from 'express';
import { Logger } from 'src/logger/logger';

@Injectable()
export class LoggerMiddleware implements NestMiddleware {
  @Inject(Logger)
  private logger: Logger
  use(req: Request, res: Response, next: NextFunction) {
    const statusCode = res.statusCode;
    const logFormat = `
###############################################################################################
RequestOriginal: ${req.originalUrl}
Method: ${req.method}
IP: ${req.ip}
StatusCode: ${statusCode}
Params: ${JSON.stringify(req.params)}
Query: ${JSON.stringify(req.query)}
Body: ${JSON.stringify(req.body)}
###############################################################################################
`;
```

```
    next();

    if (statusCode >= 500) {
      this.logger.error(logFormat, 'Request LoggerMiddleware');
    } else if (statusCode >= 400) {
      this.logger.warn(logFormat, 'Request LoggerMiddleware');
    } else {
      this.logger.log(logFormat, 'Request LoggerMiddleware');
    }
  }
}
```

上述代码在中间件中注入了 Logger 对象，对不同的状态码进行判断，并用 Logger 收集指定格式的日志记录，然后在 AppModule 中将中间件应用于所有路由上，代码如下：

```
export class AppModule implements NestModule{
  configure(consumer: MiddlewareConsumer) {
    consumer.apply(LoggerMiddleware).forRoutes('*')
  }
}
```

运行"curl -X GET http://localhost:3000/"命令，可以看到在控制台中打印出了上面格式的日志记录，如图 9-12 所示。

```
[NEST] 2024-03-12 23:31:06 info [Request LoggerMiddleware]
################################################################################
####
RequestOriginal: /
Method: GET
IP: ::ffff:127.0.0.1
StatusCode: 200
Params: {"0":""}
Query: {}
Body: {}
################################################################################
####
```

图 9-12　中间件打印日志

我们成功地使用中间件收集了所有的 HTTP 请求信息，并将其持久化到指定目录中。由于每个请求都有对应的响应数据，因此通常我们使用拦截器来记录 HTTP 响应日志。

9.3.2　拦截器日志统计

在 common 目录下创建 response.interceptor.ts 文件，用于实现使用拦截器记录 HTTP 响应成功日志的功能。示例代码如下：

```
import {
  CallHandler,
  ExecutionContext,
  Injectable,
  NestInterceptor,
} from '@nestjs/common';
import { Observable } from 'rxjs';
import { map } from 'rxjs/operators';
import { Logger } from '../logger/logger';

@Injectable()
```

```typescript
export class ResponseInterceptor implements NestInterceptor {
  constructor(private readonly logger: Logger) {}

  intercept(context: ExecutionContext, next: CallHandler): Observable<any> {
    const req = context.getArgByIndex(1).req;
    return next.handle().pipe(
      map((data) => {
        const logFormat = `
################################################################################################
Request original url: ${req.originalUrl}
Method: ${req.method}
IP: ${req.ip}
Response data: ${JSON.stringify(data)}
################################################################################################
`;
        this.logger.info(logFormat, 'Response ResponseInterceptor');
        return data;
      }),
    );
  }
}
```

上述代码收集了请求 URL、请求方式、IP 地址和响应体数据，并在返回数据之前打印日志。然后在 AppModule 中注册该中间件：

```typescript
providers: [
  AppService,
  // 应用拦截器
  {
    provide: APP_INTERCEPTOR,
    useClass: ResponseInterceptor,
  }
],
```

运行"curl -X GET http://localhost:3000/"命令，可以看到在控制台中打印出了响应日志，如图 9-13 所示。

```
[NEST] 2024-03-12 23:56:30 info [Response ResponseInterceptor]
################################################################################################
##
Request original url: /
Method: GET
IP: ::ffff:127.0.0.1
Response data: "Hello World!"
################################################################################################
##
```

图 9-13　拦截器打印日志

此时可以在 logs 日志文件中看到最新记录，如图 9-14 所示。

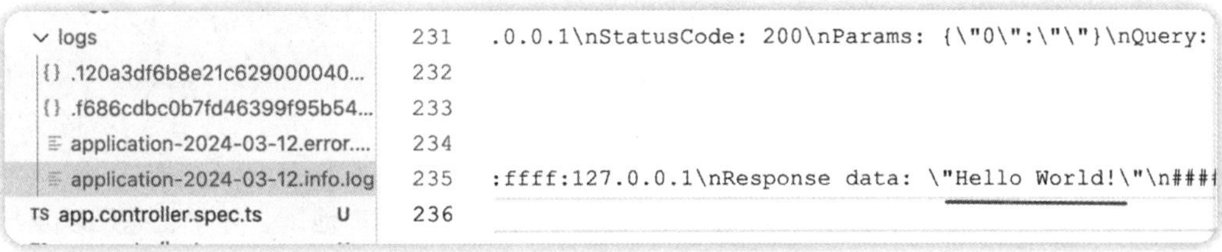

图 9-14 成功收集日志记录

除正常的 HTTP 请求响应外，我们还需要收集异常日志信息。接下来使用过滤器来实现这一功能。

9.3.3 过滤器日志统计

在 common 目录下新建 http-exceptions.filter.ts 文件，用于实现使用过滤器收集 HTTP 异常信息的功能。示例代码如下：

```
import {
  Catch,
  HttpException,
  ExceptionFilter,
  ArgumentsHost,
  Inject,
} from '@nestjs/common';
import { Logger } from 'src/logger/logger';

@Catch(HttpException)
export class HttpExceptionsFilter implements ExceptionFilter {
  @Inject(Logger)
  private loggger: Logger;
  catch(exception: any, host: ArgumentsHost) {
    const ctx = host.switchToHttp();
    const response = ctx.getResponse();
    const request = ctx.getRequest();
    const status = exception.getStatus();
    const exceptionResponse = exception.getResponse();
    const logFormat = `
################################################################################
###########################
Request original url: ${request.originalUrl}
Method: ${request.method}
IP: ${request.ip}
Status code: ${status}
Response: ${
        exception.toString() +
        `(${exceptionResponse?.message || exception.message})`
    }
################################################################################
###########################
`;
    this.loggger.error(logFormat, 'HttpException filter');
    response.status(status).json({
      code: status,
```

```
      error: exceptionResponse?.message || exception.message,
      msg: `${status >= 500 ? 'Service Error' : 'Client Error'}`,
    });
  }
}
```

上述代码记录了发生异常的 HTTP 请求的 URL、请求方法、IP 地址、异常状态码以及异常信息，并调用 Logger 的 error 方法统计日志。然后在 AppModule 中注册该过滤器：

```
providers: [
  AppService,
  // 应用拦截器
  {
    provide: APP_INTERCEPTOR,
    useClass: ResponseInterceptor,
  },
  // 应用过滤器
  {
    provide: APP_FILTER,
    useClass: HttpExceptionsFilter,
  },
],
```

接下来在 AppService 的 getHello 方法中抛出一个 HTTP 异常：

```
getHello(): string {
  throw new HttpException('getHello 请求异常', HttpStatus.EXPECTATION_FAILED)
  return 'Hello World!';
}
```

运行"curl -X GET http://localhost:3000/"命令，在控制台打印出异常日志，如图 9-15 所示。

```
[NEST] 2024-03-13 00:30:37 error [HttpException filter ]
################################################################################
##
Request original url: /
Method: GET
IP: ::ffff:127.0.0.1
Status code: 417
Response: HttpException: getHello 请求异常 (getHello 请求异常)
################################################################################
##
```

图 9-15　异常过滤器打印日志

在 .error 日志文件中收集到了异常信息，可以明确提示开发者异常发生的地方，帮助快速定位问题，如图 9-16 所示。

```
 8
     atus code: 417\nResponse: HttpException: getHello 请求异常 (getHello 请求异常) \n##
 9
10   atus code: 417\nResponse: HttpException: getHello 请求异常 (getHello 请求异常) \n##
11
12   atus code: 417\nResponse: HttpException: getHello 请求异常 (getHello 请求异常) \n##
13
```

图 9-16　成功收集异常日志

至此，我们完成了三种全局日志的统计和收集。读者可以利用 Nest 中面向切面的编程思想，实现无侵入式的日志统计效果。请在你的项目中尽情使用这些日志记录工具吧！

第 4 部分

Nest 项目实战篇

在本书的最后部分,我们将通过项目实战把前面学到的知识应用于实际开发。从产品需求分析到技术选型,再到编码实现,你将学习如何从零开始搭建一个完整的 Nest 应用。最后,我们将演示如何将项目成功部署到 Docker 环境中,实现从开发到部署的全流程覆盖。

第 10 章

数字门店管理平台开发

终于迎来了本书最后的实战部分。本章将深入探索数字化门店管理系统的开发与应用。如今，随着数字化工具的广泛普及，线下门店纷纷采用管理系统以优化业务流程，包括运营数据分析、用户行为可视化、数字化营销策略以及人工智能等。这些工具极大地提升了门店的运营效率和管理效能。

我们将在 Nest 框架中构建一个通用的门店管理平台，这不仅是对在前面章节中所学知识的综合运用，也是对项目实战技能的一次全面展示。通过本项目，读者将亲身体验到理论知识在实际开发中的应用。

需要特别指出的是，在本章的编码阶段，我们将重点讲解后端服务的实现细节。虽然前端交互也是系统开发的重要组成部分，但它并非本书的核心内容。因此，我们不会深入展开每个页面组件的具体实现，尤其是那些功能相似的列表组件、表单组件等。然而，对于系统中的核心功能，笔者仍将不吝篇幅，详细介绍其实现思路和过程，以确保读者能够深刻理解并掌握关键技术。

10.1 产品需求分析与设计

许多线下门店都拥有自己的运营管理系统。这些系统中通常包含多种角色，比如在餐饮行业中，有店长、收银员、服务员、厨师长和厨师等职位角色；而在健身行业中，有管理员、店长、前台、教练和销售等职位角色。以餐饮行业为例，管理员拥有最高权限，可以根据门店人员岗位的情况创建角色，并为每个角色分配相应的权限。不同的角色根据其所拥有的权限集合来访问和操作系统。

10.1.1 产品需求说明

普通员工可以注册账号。在注册过程中，他们通过邮件获取验证码，并在注册成功后使用账号登录系统。普通员工的账号将被分配最低权限的功能。管理员账号是系统内置的，管理员可以为用户指定角色并分配权限。

管理员与普通用户的功能权限如图 10-1 所示。

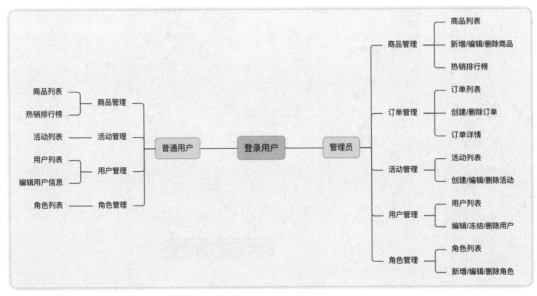

图 10-1　用户功能权限

系统整体功能可分为五大模块，分别为商品管理、订单管理、活动管理、用户管理和角色管理。普通用户默认只有访问数据列表和编辑个人信息的权限，而订单管理模块包含较为敏感的信息，普通用户无权查看该模块。

管理员拥有系统最高权限，能够对系统进行新增、修改和编辑等敏感操作。通常情况下，管理员会为指定的普通用户分配角色，如收银员，使得拥有这个角色的员工可以在客户点餐时生成待付款订单，并在付款完成后将订单改为已付款。

商品售出后，需要根据销售量进行统计，并展示出门店热销的商品，以便门店进行活动推广。

值得注意的是，管理员可以冻结指定用户的账号，被冻结的用户不能登录系统，需要管理员解冻后才能继续访问。

10.1.2　功能原型图

在分析需求后，接下来对功能模块进行原型设计，以展示系统最基础的交互效果。

1. 登录注册

首先在用户登录页面输入账号和密码进行登录，如图 10-2 所示。

如果忘记密码，可以通过邮箱验证找回密码，如图 10-3 所示。

图 10-2　登录功能

图 10-3　找回密码功能

在注册页面需要二次确认密码并填写邮箱，接收到邮箱验证码后，允许注册并登录，如图 10-4 所示。

图 10-4　注册功能

2. 商品管理

在商品管理模块，商品列表按照创建时间排序，列表中展示的字段包括商品编号、商品名称、商品图片、状态、价格、创建时间和操作，用户可以对商品进行上下架、编辑、删除操作，如图10-5所示。

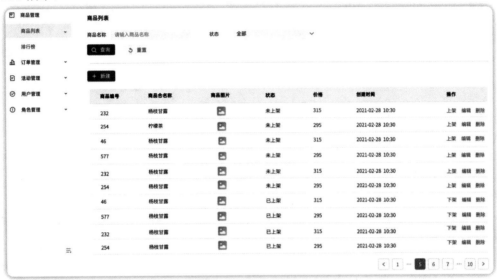

图 10-5　商品列表

单击"新建"按钮，可以新增门店的商品信息，如图10-6所示。

图 10-6　新增商品

在页面左侧单击"排行榜"选项，可以查看当前热销的商品情况，如图10-7所示。

图 10-7　商品排行榜

3. 订单管理

订单数据通常由收银员创建，根据状态可分为未付款、已付款和已取消状态，列表中展示的字段包括订单编号、商品名称、关联员工、状态、价格、创建时间和操作，如图 10-8 所示。

图 10-8　订单列表

单击"开单收银"按钮，创建订单，表单字段如图 10-9 所示。

单击首列"订单编号"，可以查看当前订单的详细信息，如图 10-10 所示。

第 10 章　数字门店管理平台开发　259

图 10-9　创建订单　　　　　　　　图 10-10　订单详情

4. 活动管理

店铺活动通常由店长创建，可以设置开始时间和结束时间，根据状态活动可分为未开始、进行中和已结束，店长可以设置活动提前结束。列表中展示的字段包括活动编号、活动名称、商品名称、状态、开始时间、结束时间和操作，如图 10-11 所示。

图 10-11　活动列表

单击"创建活动"按钮，可以新增店铺活动，并且指定一个商品参加活动，表单字段如图 10-12 所示。

图 10-12 创建活动

5. 用户管理

登录用户可以在"用户管理"中查看个人信息，列表字段如图 10-13 所示。

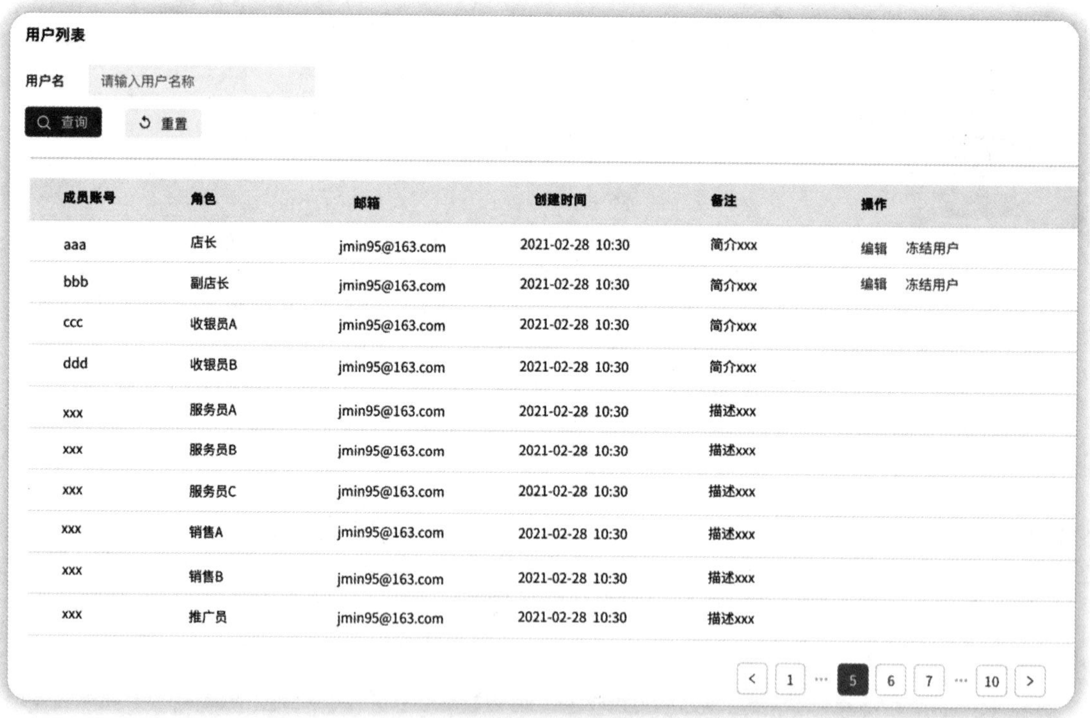

图 10-13 用户列表

单击"编辑"按钮，用户可以修改个人信息，而管理员可以修改任意用户的信息，如图 10-14 所示。

图 10-14 编辑用户信息

6. 角色管理

角色通常由管理员创建，每个角色对应一份权限集，列表字段如图 10-15 所示。

图 10-15 角色列表

管理员可以重新编辑角色来分配权限，如图 10-16 所示。

图 10-16 编辑角色

除模块的基本交互外，我们还需要定义一些全局交互共识，例如单击所有操作栏中的冻结、删除等敏感操作按钮，统一需要用户二次确认后才能进行操作，如图 10-17 所示。

图 10-17 二次确认

10.2　技术选型与项目准备

在 10.1 节中，我们从产品经理的角度分析了实际业务需求。本节将带领读者转换角色，从技术的角度出发，分别对前端和后端的技术方案进行选型，引导读者从 0 到 1 学会搭建全栈项目。

10.2.1　前端技术选型

为了与各大公司的前端技术栈契合，我们选择主流的 React 作为开发框架，并利用 Ant Design + ProComponents 快速编写 UI 页面。我们采用 TypeScript 作为类型检测工具，选择 Vite 作为构建工具，以获得极致的构建体验，并使用 Axios 作为 HTTP 请求库。在开发过程中，为了方便调试接口，我们采用 VS Code 插件中的 REST Client 作为接口请求工具。有关更多辅助依赖的详细信息，请查看项目中的 package.json 文件。

10.2.2　初始化前端项目

首先用 Vite 初始化 React 项目，运行"pnpm create vite store-web-frontend --template react"命令，结果如图 10-18 所示。

```
$ pnpm create vite store-web-frontend --template react
.../Library/pnpm/store/v3/tmp/dlx-36975  | Progress: resolved 1, reused 0, downl.../Library/pnpm
/store/v3/tmp/dlx-36975  |  +1 +
.../Library/pnpm/store/v3/tmp/dlx-36975  | Progress: resolved 1, reused 0, downl.../Library/pnpm
/store/v3/tmp/dlx-36975  | Progress: resolved 1, reused 0, downlPackages are hard linked from th
e content-addressable store to the virtual store.
  Content-addressable store is at: /Users/jmin/Library/pnpm/store/v3
  Virtual store is at:             ../../../Library/pnpm/store/v3/tmp/dlx-36975/node_modules/.pn
pm
.../Library/pnpm/store/v3/tmp/dlx-36975  | Progress: resolved 1, reused 0, downl.../Library/pnpm
/store/v3/tmp/dlx-36975  | Progress: resolved 1, reused 0, downloaded 1, added 1, done

Scaffolding project in /Users/jmin/development/nest-book/第十章/store-web-frontend...

Done. Now run:

  cd store-web-frontend
  pnpm install
  pnpm run dev
```

图 10-18　创建 React 项目

接下来，安装 "antd @ant-design/pro-components @ant-design/icons axios" 依赖。其中，antd 是基础 UI 框架版本，@ant-design/pro-components 是对 antd 的更高级别抽象，配置更高效；@ant-design/icons 是对应的字体图标库，axios 则是用于前端的请求库。运行 "pnpm add antd @ant-design/pro-components @ant-design/icons axios -S" 命令进行安装。安装成功后，在项目根目录下运行 "pnpm dev" 命令启动项目，效果如图 10-19 所示。

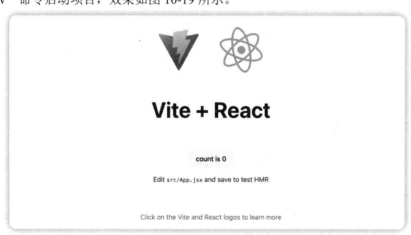

图 10-19　预览效果

10.2.3　前端架构设计

前端项目基于 React 框架和 Vite 构建，我们设计了如下的目录结构：

```
├── README.md
├── config
│   ├── defaultSettings.ts
│   ├── mock
│   ├── proxy.ts
│   └── routes
├── index.html
├── package.json
├── pnpm-lock.yaml
```

```
├── public
│   └── vite.svg
├── src
│   ├── 404.tsx
│   ├── App.tsx
│   ├── ErrorPage.tsx
│   ├── _defaultProps.tsx
│   ├── apis
│   ├── assets
│   ├── common
│   ├── components
│   ├── layout
│   ├── main.tsx
│   ├── pages
│   ├── server
│   ├── store
│   ├── types
│   ├── utils
│   └── vite-env.d.ts
├── tsconfig.json
└── vite.config.ts
```

其中，config 目录用于配置页面路由信息，管理不同开发环境下的代理配置，同时支持以 mock 方式模拟接口数据。在 src 目录中，根据模块职能设计了不同目录进行管理：apis 负责统一定义接口，common 用于管理全局通用属性，layout 用于搭建系统骨架，pages 用于管理每个页面组件。页面组件需要借助 components 中的积木组件完成搭建过程。为了更方便地管理请求工具，在 server 中封装了 Axios 逻辑，并使用 store 作为数据仓库进行公共数据管理。最后，结合 utils 和 types 提供编码所需的工具和类型支持。

10.2.4 后端技术选型

在本项目的后端开发中，我们精心挑选了以下技术栈来构建稳定、高效、易于维护的服务。

- Nest：选择 Nest 作为后端服务的开发框架，它提供了模块化和高性能的特点，非常适合构建企业级应用程序。
- MySQL + TypeORM：使用 MySQL 作为数据库存储解决方案，并结合 TypeORM 作为对象关系映射（ORM）工具，以简化数据库的增删改查（CRUD）操作。
- RBAC 模型：为了有效管理用户和权限，采用基于角色的访问控制（RBAC）模型，以实现灵活的权限分配和管理。
- Redis：考虑到性能优化和响应速度，引入了 Redis 作为数据缓存层，这不仅提升了性能，同时也减轻了数据库的压力。
- Winston：为了增强系统的稳定性并便于问题排查，采用 Winston 作为日志记录工具，它支持丰富的日志统计功能。
- Docker Compose：在开发完成后，使用 Docker Compose 来部署项目，它简化了容器化应用程序的部署和管理。
- PM2：采用 PM2 作为进程守护工具，确保后端服务的高可用性和稳定性。

- **Nginx**：使用 Nginx 作为网关层，它不仅托管系统的静态资源文件，还提供了负载均衡、请求转发等功能。

通过这些技术选型，可以确保后端服务的高性能、高稳定性以及良好的可维护性，为构建全栈项目打下坚实的基础。

10.2.5 初始化后端项目

首先用 CLI 初始化 Nest 项目，运行"nest n store-web-backend -p pnpm"命令，结果如图 10-20 所示。

```
$ nest n store-web-backend -p pnpm
⚡ We will scaffold your app in a few seconds..

CREATE store-web-backend/.eslintrc.js (663 bytes)
CREATE store-web-backend/.prettierrc (51 bytes)
CREATE store-web-backend/README.md (3347 bytes)
CREATE store-web-backend/nest-cli.json (171 bytes)
CREATE store-web-backend/package.json (1958 bytes)
CREATE store-web-backend/tsconfig.build.json (97 bytes)
CREATE store-web-backend/tsconfig.json (546 bytes)
CREATE store-web-backend/src/app.controller.spec.ts (617 bytes)
CREATE store-web-backend/src/app.controller.ts (274 bytes)
CREATE store-web-backend/src/app.module.ts (249 bytes)
CREATE store-web-backend/src/app.service.ts (142 bytes)
CREATE store-web-backend/src/main.ts (208 bytes)
CREATE store-web-backend/test/app.e2e-spec.ts (630 bytes)
CREATE store-web-backend/test/jest-e2e.json (183 bytes)

✔ Installation in progress... ☕

🚀 Successfully created project store-web-backend
👉 Get started with the following commands:

$ cd store-web-backend
$ pnpm run start
```

图 10-20　创建 Nest 项目

然后安装依赖，操作数据库需要用到 mysql2、typeorm、@nestjs/typeorm，实现接口缓存需要用到 Redis，实现请求数据校验需要用到 class-validator、class-transformer，实现本地策略验证和 JWT 认证需要用到@nestjs/passport、passport、passport-local、@types/passport-local、@nestjs/jwt passport-jwt、types/passport-jwt，实现文件上传需要用到@types/multer，实现密码哈希加密需要用到 bcryptjs @types/bcryptjs，实现路径比对需要用到 path-to-regexp。

执行下面两条命令：

```
pnpm add mysql2 typeorm @nestjs/typeorm redis class-validator class-transformer passport @nestjs/passport passport-local @nestjs/jwt passport-jwt bcryptjs path-to-regexp -S
pnpm add @types/passport-local @types/multer @types/bcryptjs -D
```

安装成功后，效果如图 10-21 所示。

```
dependencies:
+ @nestjs/jwt 10.2.0
+ @nestjs/passport 10.0.3
+ bcryptjs 2.4.3
+ class-transformer 0.5.1
+ class-validator 0.14.1
+ mysql2 3.9.2
+ passport 0.7.0
+ passport-jwt 4.0.1
+ passport-local 1.0.0
+ path-to-regexp 6.2.1
+ redis 4.6.13
+ typeorm 0.3.20

The integrity of 1386 files was checked. This might have caused installation to take longer.
Done in 3.2s
Packages: +5
+++++
Progress: resolved 738, reused 738, downloaded 0, added 5, done

devDependencies:
+ @types/bcryptjs 2.4.6
+ @types/multer 1.4.11
+ @types/passport-local 1.0.38

Done in 1.6s
```

图 10-21 安装后端依赖

在项目根目录下运行"pnpm start:dev"命令启动服务，并确保测试能够正常运行，如图 10-22 所示。

```
$ pnpm start:dev

> store-web-backend@0.0.1 start:dev /Users/jmin/development/nest-book/第十章/store-web-backend
> nest start --watch

[18:24:42] Starting compilation in watch mode...

[18:24:43] Found 0 errors. Watching for file changes.

[Nest] 37860  - 2024/03/14 18:24:43     LOG [NestFactory] Starting Nest application...
[Nest] 37860  - 2024/03/14 18:24:43     LOG [InstanceLoader] AppModule dependencies initialized +5ms
[Nest] 37860  - 2024/03/14 18:24:43     LOG [RoutesResolver] AppController {/}: +6ms
[Nest] 37860  - 2024/03/14 18:24:43     LOG [RouterExplorer] Mapped {/, GET} route +0ms
[Nest] 37860  - 2024/03/14 18:24:43     LOG [NestApplication] Nest application successfully started +1ms
```

图 10-22 运行项目

10.2.6 后端架构设计

服务端基于 Nest 及多种中间件开发，我们设计了如下的目录结构：

```
├── README.md
├── nest-cli.json
├── package.json
├── pnpm-lock.yaml
├── src
│   ├── activity
│   │   ├── activity.controller.ts
│   │   ├── activity.module.ts
│   │   ├── activity.service.ts
│   │   ├── dto
│   │   ├── entities
│   │   └── start-activity.service.ts
│   ├── app.controller.spec.ts
│   ├── app.controller.ts
│   ├── app.module.ts
│   ├── app.service.ts
```

```
│   ├── auth
│   │   ├── auth.module.ts
│   │   ├── auth.service.ts
│   │   ├── jwt-auth.guard.ts
│   │   ├── jwt.strategy.ts
│   │   └── role-auth.guard.ts
│   ├── common
│   │   ├── decorators
│   │   ├── enums
│   │   ├── http-exceptions.filter.ts
│   │   ├── logger
│   │   ├── logger.middleware.ts
│   │   ├── mail
│   │   ├── redis
│   │   ├── response.interceptor.ts
│   │   └── utils
│   ├── logs
│   │   ├── application-2024-04-05.error.log.gz
│   │   ├── application-2024-04-05.info.log.gz
│   │   ├── application-2024-04-06.error.log
│   │   ├── application-2024-04-06.info.log.gz
│   │   ├── application-2024-04-07.error.log
│   │   └── application-2024-04-07.info.log
│   ├── main.ts
│   ├── order
│   │   ├── dto
│   │   ├── entities
│   │   ├── order.controller.ts
│   │   ├── order.module.ts
│   │   └── order.service.ts
│   ├── permission
│   │   ├── dto
│   │   ├── entities
│   │   ├── permission.controller.ts
│   │   ├── permission.module.ts
│   │   └── permission.service.ts
│   ├── product
│   │   ├── dto
│   │   ├── entities
│   │   ├── hot-sales.service.ts
│   │   ├── product.controller.ts
│   │   ├── product.module.ts
│   │   └── product.service.ts
│   ├── role
│   │   ├── dto
│   │   ├── entities
│   │   ├── role.controller.ts
│   │   ├── role.module.ts
│   │   └── role.service.ts
│   ├── sys
│   │   ├── dto
│   │   ├── sys.controller.ts
│   │   ├── sys.module.ts
│   │   └── sys.service.ts
│   └── user
```

```
|           ├── dto
|           ├── entities
|           ├── user.controller.ts
|           ├── user.module.ts
|           └── user.service.ts
├── tsconfig.build.json
└── tsconfig.json
```

由于 Nest 具有优秀的架构设计理念，因此我们可以很轻松在此基础上搭建服务端项目。在 src 目录下，每个功能模块用单独的目录来管理自身的 DTO（验证规则）、entities（实体）、controller（控制器）、service（服务）和 module（模块）；common 目录用于管理全局公共拦截器、过滤器、守卫，它们会被注册到全局中，此外还有统一的装饰器、枚举类、工具方法、日志记录器、Redis 服务和邮件服务等，可以很方便地维护它们。

为了后续能够快速进行项目开发，基于前面学习的内容，我们完成了项目的基础模块建设，包含 RBAC 权限控制逻辑、自定义日志器、MySQL 配置和 Redis 配置等，详情可在 GitHub 中查看。

至此，我们完成了前后端技术方案的选型及项目初始化，下一节将设计 API 接口和数据库表结构。

10.3 API 接口及数据库表设计

技术方案确定之后，接下来设计数据表结构，梳理清楚表与表之间的关系，并按照功能划分 API 接口。

10.3.1 API 接口功能划分

按照功能划分，首先实现登录注册模块，包含用户登录、注册用户和找回密码功能，涉及的接口如表 10-1 所示。

表10-1 登录注册模块

请求路径	请求方式	描述
/sys/registry	POST	注册用户
/sys/login	POST	用户登录
/sys/forgot	POST	找回密码
/sys/sendMail	GET	发送邮箱验证码

接着实现用户管理模块，包含用户列表、删除用户和编辑用户信息等功能，涉及的接口如表 10-2 所示。

表10-2 用户管理模块

请求路径	请求方式	描述
/user/currentUser	GET	获取当前登录用户信息
/user/delete	GET	删除用户
/user/list	GET	用户列表
/user/edit	POST	编辑用户信息
/user/freeze	GET	冻结用户

餐饮行业会销售商品，对应商品管理模块，包含商品列表、新增/删除商品和编辑商品信息等功能，涉及的接口如表10-3所示。

表10-3 商品管理模块

请求路径	请求方式	描述
/product/create	POST	新增商品
/product/list	GET	商品列表
/product/edit	POST	编辑商品信息
/product/delete	GET	删除商品

商品销售出去会产生订单，所以有订单管理模块，包含创建订单、订单列表和订单详情等功能，涉及的接口如表10-4所示。

表10-4 订单管理模块

请求路径	请求方式	描述
/order/list	GET	订单列表
/order/detail	GET	订单详情
/order/create	POST	创建订单
/order/delete	DELETE	删除订单

店铺通常涉及营销，推出拓客活动来吸引顾客，对应活动管理模块，包含活动列表、创建/删除活动和编辑活动信息等功能，涉及的接口如表10-5所示。

表10-5 活动管理模块

请求路径	请求方式	描述
/activity/list	GET	活动列表
/activity/create	POST	创建活动
/activity/edit	POST	编辑活动
/activity/delete	GET	删除活动

基于RBAC模型，还要实现角色管理模块，包含角色列表、新增/删除角色和编辑角色信息等功能，涉及的接口如表10-6所示。

表10-6 角色管理模块

请求路径	请求方式	描述
/role/list	GET	活动列表
/role/create	POST	创建活动
/role/edit	POST	编辑活动
/role/delete	GET	活动详情

10.3.2 数据库设计

完成功能API接口的定义后，接下来设计数据库，理清表与表之间的关系。

基于RBAC模型，用户与角色是多对多关系，角色与权限也是多对多关系。在实际业务中，我们将不使用物理外键约束来处理表与表之间的关系，而是使用业务逻辑来处理。首先创建用户表，其字段如表 10-7 所示。

表10-7　用户表的字段

字　　段	类　　型	描　　述
id	int	用户 ID
username	varchar(30)	用户账号
password	varchar(50)	用户密码
salt	varchar(50)	哈希加密的盐
userType	int	用户类型
email	varchat(50)	用户邮箱
freezed	boolean	是否冻结
createTime	Date	创建时间
updateTime	Date	更新时间

角色表的字段如表 10-8 所示。

表10-8　角色表的字段

字　　段	类　　型	描　　述
id	int	角色 ID
name	varchar(50)	角色名称
desc	varchar(255)	角色描述
createTime	Date	创建时间
updateTime	Date	更新时间

用户与角色的关系需要一个中间表来维护，字段如表 10-9 所示。

表10-9　中间表的字段

字　　段	类　　型	描　　述
id	int	中间表 ID
userId	int	用户 ID
roleId	int	角色 ID

还要创建权限表，用于维护每个模块的权限信息，字段如表 10-10 所示。

表10-10　权限表的字段

字　　段	类　　型	描　　述
id	int	权限 ID
title	varchar	前端菜单栏的名称
code	varchar	菜单代码
type	int	菜单类型（菜单/页面/组件/按钮）

角色与权限集的关系用中间表来维护，字段如表10-11所示。

表10-11 中间表的字段

字　　段	类　　型	描　　述
id	int	中间表 ID
permissionId	int	权限集 ID
roleId	int	角色 ID

权限表中的每一条记录对应页面上的一个功能操作，等同于一个后端接口，我们用专门的表来维护这种关系，字段如表10-12所示。

表10-12 接口记录表的字段

字　　段	类　　型	描　　述
id	int	接口记录 ID
apiUrl	varchar	接口 URL
apiMethod	varchar	接口请求方式
permissionId	int	功能权限 ID

接下来创建商品表，用于存放店铺创建的商品信息，字段如表10-13所示。

表10-13 商品表的字段

字　　段	类　　型	描　　述
id	int	表自增 ID
name	varchar(255)	商品名称
price	bigint	商品价格
desc	varchar(255)	商品描述
images	ProductImage[]	商品图片
createTime	Date	创建时间
updateTime	Date	更新时间
status	int	商品状态（未上架、已上架、已下架）
avtivityId	int	商品对应的活动 ID

商品的图片会通过单独表来维护，属于一对多关系，字段如表10-14所示。

表10-14 商品的图片

字　　段	类　　型	描　　述
id	int	图片 ID
imageUrl	varchar	图片路径
productId	int	图片对应的商品 ID

前台收银后会生成订单，订单表信息的字段如表10-15所示。

表10-15　订单表信息的字段

字　　段	类　　型	描　　述
id	bigint	订单 ID
name	varchar(50)	商品名称
count	int	商品数量
status	int	订单状态
price	bigint	订单金额
discount	int	商品折扣
operator	varchar	操作员
createTime	Date	订单创建时间

商品与订单之间的关系通过中间表来维护，字段如表 10-16 所示。

表10-16　中间表的字段

字　　段	类　　型	描　　述
id	init	活动 ID
productId	int	商品 ID
orderId	bigint	订单 ID

最后，店铺经常会以做活动的形式对商品进行促销，同一时间段一个活动绑定一种商品，是一对一关系，通过在商品表中绑定 activity_id 可以找到活动信息，活动表的字段如表 10-17 所示。

表10-17　活动表的字段

字　　段	类　　型	描　　述
id	init	活动 ID
name	varchar（255）	活动名称
status	int	活动状态
type	int	活动类型
desc	text	活动描述
startTime	Date	活动开始时间
endTime	Date	活动结束时间
createTime	Date	创建时间
updateTime	Date	更新时间
productId	int	商品 ID

数据表结构设计完成了。可能有读者对表与表之间的关系并不清楚，没关系，我们通过图来展示各种表之间的关系，让读者更容易理解，如图 10-23 所示。

第 10 章　数字门店管理平台开发　273

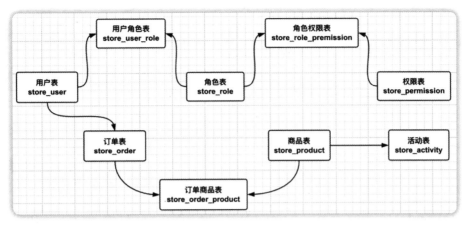

图 10-23　数据表关系

我们清楚了各表之间的关系和作用，从下一节开始，将进入搭积木阶段，逐一编写各模块的代码。

10.4　实现注册登录

在前文中，我们已经完成了项目需求分析和原型设计，同时设计了 API 接口和数据库表。在本节中，我们将进入实际编码阶段。前端将实现登录、注册和找回密码页面，而服务端则会实现相应页面的接口，并完成数据对接工作。

10.4.1　页面效果展示

首先来实现前端页面，用户通过用户名、密码、邮箱和邮箱验证码进行注册，邮箱验证码以 6 位数字的形式发送到用户邮箱中。该邮箱还将用于密码找回、发送通知等功能。用户注册页面如图 10-24 所示。

图 10-24　用户注册页面

注册页面路由组件的具体实现代码如下：

```
import { getCaptchaCode, login, registry } from '@/apis/login'
import { ComponTypeEnum } from '@/layout/BasicLayout'
import { storeGlobalUser } from '@/store/globalUser'
import { storage } from '@/utils/Storage'
import {
  LockOutlined,
  MailOutlined,
  UserOutlined
} from '@ant-design/icons'
import {
  LoginForm,
  ProFormCaptcha,
  ProFormInstance,
  ProFormText
} from '@ant-design/pro-components'
import { RouteType } from '@config/routes'
import { routers } from '@config/routes/routers'
import { message, Tabs } from 'antd'
import { useRef } from 'react'
import { useNavigate } from 'react-router-dom'
import logo from '@/assets/mouse.jpg'

const Registry = () => {
  const navigate = useNavigate()
  const formRef = useRef<ProFormInstance>()
  const backToLogin = () => {
    navigate('/login')
  }
  const handleLogin = async (val: Login.LoginEntity) => {
    const { data = {} } = await login(val)
    storage.set('token', data?.access_token)
    console.log('data', data)
    /** 跳转有权限的第一个菜单 */
    await storeGlobalUser.getUserDetail()
    const flattenRoutes: (routes: RouteType[]) => RouteType[] = (routes: RouteType[])
=> {
      const flattenedRoutes: RouteType[] = []
      function traverse(routes: RouteType[]) {
        routes.forEach(route => {
          flattenedRoutes.push(route)
          if (route.children) {
            traverse(route.children)
          }
        })
      }

      traverse(routes)

      return flattenedRoutes
    }
    const resRoutes = flattenRoutes(routers)
    const findPath =
      resRoutes?.[
```

```
        resRoutes?.findIndex(
          item =>
            item?.name ===
            storeGlobalUser?.userInfo?.menus?.filter(
              citem => citem?.type === ComponTypeEnum.MENU
            )?.[0]?.title
        )
      ]?.path
    navigate(findPath || '/')
  }
  const handleRegistry = async (val: Login.LoginEntity) => {
    const { success } = await registry(val)
    if (success) {
      handleLogin(val)
    }
  }
  return (
    <div style={{ backgroundColor: 'white', height: '100vh' }}>
      <LoginForm
        logo={logo}
        title="数字门店管理平台"
        subTitle="基于 NestJS + React 的全栈项目"
        formRef={formRef}
        onFinish={async (val: Login.LoginEntity) => {
          await handleRegistry(val)
        }}
        submitter={{
          searchConfig: {
            submitText: '注册'
          }
        }}
      >
        <>
          <Tabs
            centered
            activeKey={'registry'}
          >
            <Tabs.TabPane key={'registry'} tab={'用户注册'} />
          </Tabs>
          <>
            <ProFormText
              name="username"
              fieldProps={{
                size: 'large',
                prefix: <UserOutlined className={'prefixIcon'} />
              }}
              placeholder={'请输入账号/用户名'}
              rules={[
                {
                  required: true,
                  message: '请输入用户名!'
                }
              ]}
            />
            <ProFormText.Password
```

```jsx
  name="password"
  fieldProps={{
    size: 'large',
    prefix: <LockOutlined className={'prefixIcon'} />
  }}
  placeholder={'请输入注册密码'}
  rules={[
    {
      required: true,
      message: '请输入注册密码!'
    }
  ]}
/>
<ProFormText.Password
  name="confirmPassword"
  fieldProps={{
    size: 'large',
    prefix: <LockOutlined className={'prefixIcon'} />
  }}
  placeholder={'请输入确认密码'}
  rules={[
    {
      required: true,
      message: '请输入确认密码!'
    }
  ]}
/>
<ProFormText
  name="email"
  fieldProps={{
    size: 'large',
    prefix: <MailOutlined />,
  }}
  placeholder={'请输入邮箱'}
  rules={[
    {
      required: true,
      message: '请输入用户名!'
    }
  ]}
/>
<ProFormCaptcha
  name="code"
  rules={[
    {
      required: true,
      message: '请输入邮箱验证码',
    },
  ]}
  placeholder="请输入邮箱验证码"
  onGetCaptcha={async () => {
    // 获取验证码
    const { success } = await getCaptchaCode({ email: formRef.current?.getFieldValue('email') })
    if (success) {
```

```
              message.success('验证码已发送至邮箱')
            }
          }}
        />
      </>
    </>
    <div
      style={{
        marginBlockEnd: 24,
      }}
    >
      <a onClick={backToLogin}>
        返回登录
      </a>
    </div>
    </LoginForm>
  </div>
  )
}

export default Registry
```

如果已经注册过了，则单击"返回登录"按钮进入登录页面，输入账号和密码进行登录，如图10-25所示。

图 10-25　用户登录页面

为了方便读者预览效果，可以使用默认的管理员账号和密码（admin/admin）来登录以体验效果。在实际业务中，应严格保护好用户的账号和密码（隐私保护）。具体登录组件代码实现如下：

```
import { login } from '@/apis/login'
import { ComponTypeEnum } from '@/layout/BasicLayout'
import { storeGlobalUser } from '@/store/globalUser'
import { storage } from '@/utils/Storage'
import { LockOutlined, UserOutlined } from '@ant-design/icons'
import { LoginForm, ProFormText } from '@ant-design/pro-components'
import { RouteType } from '@config/routes'
import { routers } from '@config/routes/routers'
```

```typescript
import { Tabs } from 'antd'
import { useState } from 'react'
import { useNavigate } from 'react-router-dom'
import logo from '@/assets/mouse.jpg'

type LoginType = 'username' | 'email'

const Login = () => {
  const [loginType, setLoginType] = useState<LoginType>('username')
  const navigate = useNavigate()

  const handleForgot = () => {
    navigate('/forgetPassword')
  }
  const handleLogin = async (val: Login.LoginEntity) => {
    const { data = {} } = await login(val)
    storage.set('token', data?.access_token)
    console.log('data', data)
    /** 跳转有权限的第一个菜单 */
    await storeGlobalUser.getUserDetail()
    const flattenRoutes: (routes: RouteType[]) => RouteType[] = (routes: RouteType[]) => {
        const flattenedRoutes: RouteType[] = []
        function traverse(routes: RouteType[]) {
          routes.forEach(route => {
            flattenedRoutes.push(route)
            if (route.children) {
              traverse(route.children)
            }
          })
        }

        traverse(routes)

        return flattenedRoutes
      }
    const resRoutes = flattenRoutes(routers)
    const findPath =
      resRoutes?.[
        resRoutes?.findIndex(
          item =>
            item?.name ===
            storeGlobalUser?.userInfo?.menus?.filter(
              citem => citem?.type === ComponTypeEnum.MENU
            )?.[0]?.title
        )
      ]?.path
    navigate(findPath || '/')
  }
  const handleRegistry = () => {
    navigate('/registry')
  }
  return (
    <div style={{ backgroundColor: 'white', height: '100vh' }}>
      <LoginForm
```

```jsx
      logo={logo}
      title="数字门店管理平台"
      subTitle="基于 NestJS + React 的全栈项目"
      onFinish={async (val: Login.LoginEntity) => {
        await handleLogin(val)
      }}
    >
      <>
        <Tabs
          centered
          activeKey={loginType}
          onChange={(activeKey: LoginType) => setLoginType(activeKey as LoginType)}
        >
          <Tabs.TabPane key={'username'} tab={'账号密码登录'} />
        </Tabs>
        {loginType === 'username' && (
          <>
            <ProFormText
              name="username"
              fieldProps={{
                size: 'large',
                prefix: <UserOutlined className={'prefixIcon'} />
              }}
              placeholder={'用户名: admin'}
              rules={[
                {
                  required: true,
                  message: '请输入用户名!'
                }
              ]}
            />
            <ProFormText.Password
              name="password"
              fieldProps={{
                size: 'large',
                prefix: <LockOutlined className={'prefixIcon'} />
              }}
              placeholder={'密码: admin'}
              rules={[
                {
                  required: true,
                  message: '请输入密码!'
                }
              ]}
            />
          </>
        )}
      </>
      <div
        style={{
          marginBlockEnd: 24
        }}
      >
        <a onClick={handleRegistry}>立即注册</a>
        <a style={{ float: 'right' }} onClick={handleForgot}>
```

```
            忘记密码
          </a>
        </div>
      </LoginForm>
    </div>
  )
}

export default Login
```

在登录过程中,如果用户忘记密码,可以通过单击"忘记密码"按钮来重新设置密码。用户需要使用注册时填写的邮箱来接收验证码,如果账号与邮箱不匹配,系统将拒绝该提交请求。找回密码的页面如图 10-26 所示。

图 10-26 用户找回密码的页面

最后,在路由管理文件 routers.tsx 中注册这 3 个页面组件,代码如下:

```
{
  path: '/login',
  name: '登录',
  element: <Login />
},
{
  path: '/registry',
  name: '注册',
  element: <Registry />
},
{
  path: '/forgetPassword',
  name: '找回密码',
  element: <ForgetPassword />
}
```

前端页面组件编写完毕后,接下来实现服务端的接口逻辑。

10.4.2 接口实现

在实现接口之前，需要定义用户数据实体 userEntity，代码如下：

```typescript
import { Column, CreateDateColumn, Entity, PrimaryGeneratedColumn } from "typeorm";

@Entity('store_user')
export class UserEntity {
  @PrimaryGeneratedColumn({type: 'int'})
  id: number;

  @Column({ type: 'varchar', length: 32, comment: '用户登录账号' })
  username: string;

  @Column({ type: 'varchar', length: 200, nullable: false, comment: '用户登录密码' })
  password: string;

  @Column({ type: 'varchar', length: 50, nullable: false, comment: '哈希加密的盐' })
  salt: string;

  @Column({ type: 'int', comment: '用户类型 0 管理员 1 普通用户', default: 1 })
  userType: number;

  @Column({ type: 'varchar', comment: '用户邮箱', default: '' })
  email: string;

  @Column({ type: 'int', comment: '是否冻结用户 0 不冻结 1 冻结', default: 0 })
  freezed: number;

  @Column({ type: 'varchar', comment: '用户头像', default: '' })
  avatar: string;

  @Column({ type: 'varchar', comment: '用户备注', default: '' })
  desc: string;

  @CreateDateColumn({ type: 'timestamp', comment: '创建时间' })
  createTime: Date
}
```

创建 sys 模块，用于管理用户注册、登录、找回密码和发送验证码接口等公共路由方法，代码如下：

```typescript
import { Controller, Get, Post, Body, Inject, Query } from '@nestjs/common';
import { SysService } from './sys.service';
import { LoginUserDto } from './dto/login-user.dto';
import { CreateUserDto } from 'src/user/dto/create-user.dto';
import { UserService } from 'src/user/user.service';
import { ForgotUserDto } from './dto/forgot-user.dto';
import { AllowNoToken } from 'src/common/decorators/token.decorator';
@Controller('sys')
export class SysController {
  @Inject(UserService)
  private userService: UserService;

  constructor(private readonly sysService: SysService) {}
```

```typescript
// 用户注册
@Post('registry')
@AllowNoToken()
registry(@Body() createUserDto: CreateUserDto) {
  return this.userService.registry(createUserDto);
}

// 用户登录
@Post('login')
@AllowNoToken()
login(@Body() loginUserDto: LoginUserDto) {
  return this.userService.login(loginUserDto);
}

// 找回密码
@Post('forgot')
@AllowNoToken()
forgot(@Body() forgotUserDto: ForgotUserDto) {
  return this.userService.updatePassword(forgotUserDto);
}
// 发送找回密码邮箱验证码
@Get('sendEmailForGorgot')
@AllowNoToken()
sendEmailForGorgot(@Query() dto: { email: string }) {
  return this.sysService.sendEmailForGorgot(dto.email);
}

// 发送注册邮箱验证码
@Get('sendEmailForRegistry')
@AllowNoToken()
sendEmailForRegistry(@Query() dto: { email: string }) {
  return this.sysService.sendMail(dto.email,'注册验证码');
}
}
```

接下来，在生成的 user 模块中，实现具体的 registry 方法，代码如下：

```typescript
/**
 * 注册新用户
 *
 * @param createUserDto 创建用户 DTO
 * @returns 返回用户信息
 * @throws 如果用户名或注册邮箱已存在，则返回冲突错误
 * @throws 如果两次输入的密码不一致，则返回预期失败错误
 * @throws 如果验证码有误或已过期，则返回预期失败错误
 */
async registry(createUserDto: CreateUserDto) {
  const { username, email } = createUserDto;
  // 1. 判断用户是否存在，参数为邮箱或者用户名，使用 createQueryBuilder 一次性查询两个字段
  const user = await this.userRepository
    .createQueryBuilder('su')
    .where('su.username = :username OR su.email = :email', { username, email })
    .getOne()
  // 2. 若存在则返回错误信息
```

```
    if (user) {
      throw new HttpException('用户名或注册邮箱已存在，请重新输入', HttpStatus.CONFLICT);
    }
    if (createUserDto.confirmPassword !== createUserDto.password) {
      throw new HttpException('两次输入的密码不一致，请重新输入', HttpStatus.EXPECTATION_FAILED);
    }
    // 3. 校验注册验证码
    const codeRedisKey = getRedisKey(RedisKeyPrefix.REGISTRY_CODE, createUserDto.email)
    const code = await this.redisService.get(codeRedisKey)
    if (!code) {
      throw new HttpException('验证码有误或已过期', HttpStatus.EXPECTATION_FAILED);
    }
    // 4. 哈希加密
    const salt = await genSalt()
    createUserDto.password = await hash(createUserDto.password, salt);

    const newUser = plainToClass(UserEntity, { salt, ...createUserDto }, {ignoreDecorators: true});
    // 5. 若不存在则创建用户
    const { password, salt: salter, ...rest } = await this.userRepository.save(newUser);
    // 6. 缓存用户信息
    const redisKey = getRedisKey(RedisKeyPrefix.USER_INFO, rest.id)
    await this.redisService.hSet(
      redisKey,
      rest
    )
    return rest;
  }
```

在注册接口中，我们对用户密码进行加密操作，同时把加密的盐持久化到数据库中。当需要修改用户密码时，可以使用这个盐对新密码进行加密。保存到数据库后，将用户信息缓存到 Redis 中，以便下一次访问时直接从 Redis 中获取，最后，返回除密码和盐外的非敏感数据给客户端。

在登录接口逻辑中，首先根据用户名查找用户是否存在。若用户存在，则使用参数中的密码和加密后的哈希密码进行比较，判断是否相同。如果用户已经被冻结，则禁止登录。所有判断通过后，将使用用户信息生成 access_token，JWT 策略会把用户信息存储在 Request 对象中，以便后续使用。最后，将 access_token 返回到客户端。具体代码如下：

```
  /**
   * 登录方法
   *
   * @param loginUserDto 登录用户信息
   * @returns 返回生成的 Token
   * @throws 当账号或密码错误时，抛出 HttpException 异常
   * @throws 当账号被冻结时，抛出 HttpException 异常
   */
  async login(loginUserDto: LoginUserDto) {
    const user = await this.userRepository.findOne({
      where: {
        username: loginUserDto.username,
      },
    });
    // 1. 判断用户是否存在
    if (!user) {
```

```
      throw new HttpException('账号或密码错误', HttpStatus.EXPECTATION_FAILED);
    }
    // 2. 判断密码是否正确
    const checkPassword = await compare(loginUserDto.password, user.password);
    if (!checkPassword) {
      throw new HttpException('账号或密码错误', HttpStatus.EXPECTATION_FAILED);
    }
    // 3. 判断用户是否被冻结
    if (user.freezed) {
      throw new HttpException('账号已被冻结，请联系管理员', HttpStatus.EXPECTATION_FAILED);
    }
    // 4. 生成 Token
    const { password, salt, ...rest } = user
    const access_token = this.generateAccessToken(rest)
    return {
      access_token
    };
  }
```

用户忘记密码并希望重新设置密码以便登录时，同样地，需要判断用户是否存在、是否已经冻结。同时需要验证客户端提供的验证码 code 与缓存在 Redis 中缓存的邮箱验证码是否匹配，并检查验证码是否在 5 分钟的有效期内。更新密码成功后，最后删除 Redis 中旧的用户信息及使用过的验证码。具体代码如下：

```
/**
 * 更新用户密码
 *
 * @param dtoForgotUserDto 用户信息对象
 * @returns 修改成功信息
 * @throws 验证码错误或已过期，用户不存在或已删除，用户已被冻结，两次输入的密码不一致，修改失败
 */
async updatePassword(dto: ForgotUserDto) {
  const { password, confirmPassword, code, username } = dto
  if (password !== confirmPassword) {
    throw new HttpException('两次输入的密码不一致', HttpStatus.EXPECTATION_FAILED);
  }
  // 1. 判断用户是否存在
  const exists = await this.userRepository.findOneBy({ username })
  if (!exists) {
    throw new HttpException('用户不存在或已删除', HttpStatus.EXPECTATION_FAILED);
  }
  const { id, freezed } = exists
  if (freezed) {
    throw new HttpException('用户已被冻结，请解冻后再修改', HttpStatus.EXPECTATION_FAILED);
  }
  // 2. 验证 Redis 中的验证码与用户输入的验证码是否一致
  const cacheCode = await this.redisService.get(
    getRedisKey(RedisKeyPrefix.PASSWORD_RESET, id)
  );
  if (cacheCode !== code) {
    throw new HttpException('验证码错误或已过期', HttpStatus.EXPECTATION_FAILED);
  }
```

```
    // 3. 更新密码
    const newPassword = await hash(password, exists.salt)
    const { affected } = await this.userRepository.update({ id }, { password: newPassword });
    if (!affected) {
      throw new HttpException('修改失败，请稍后重试', HttpStatus.EXPECTATION_FAILED);
    }
    // 4. 删除 Redis 缓存的验证码和用户信息
    const codeRedisKey = getRedisKey(RedisKeyPrefix.PASSWORD_RESET, id)
    this.redisService.del(codeRedisKey)

    const userRedisKey = getRedisKey(RedisKeyPrefix.USER_INFO, id)
    this.redisService.del(userRedisKey)

    return '修改成功，请重新登录'
}
```

我们在多个场景中需要使用邮箱验证码功能，可以通过 nodemailer 包来实现邮件收发，并将 MailService 抽象为 Provider 注入需要的模块中。MailService 的具体实现代码如下：

```
import { HttpException, HttpStatus, Injectable } from "@nestjs/common";
import { ConfigService } from "@nestjs/config";
import * as nodemail from "nodemailer";

@Injectable()
export class MailService {
  private transporter: nodemail.Transporter;
  constructor(private config: ConfigService) {
    this.transporter = nodemail.createTransport({
      host: config.get<string>('EMAIL_HOST'),
      port: config.get<Number>('EMAIL_PORT'),
      secure: config.get<string>('EMAIL_SECURE'),
      auth: {
        user: config.get<string>('EMAIL_USER'),
        pass: config.get<string>('EMAIL_PASS'),
      },
    });
  }

  /**
   * 发送邮件
   *
   * @param email 收件人邮箱
   * @param subject 邮件主题
   * @param html 邮件正文，可选
   * @returns 返回一个 Promise 对象，解析后得到一个包含验证码和邮件信封信息的对象
   * @throws HttpException 当邮件发送失败时，会抛出一个 HttpException 异常
   */
  sendMail(email: string, subject: string, html?: string): Promise<Record<string, string>> {
    const code = Math.random().toString().slice(-6);
```

```typescript
    const mailOptions = {
      from: this.config.get<string>('EMAIL_USER'),
      to: email,
      text: '用户验证码为：${code}，有效期为 5 分钟，请及时使用！',
      subject,
      html,
    };
    return new Promise((resolve, reject) => {
      this.transporter.sendMail(mailOptions, (error: Error, info: { envelope: Record<string, string[]> }) => {
        if (error) {
          reject(new HttpException('发送邮件失败:${error}', HttpStatus.INTERNAL_SERVER_ERROR));
        } else {
          resolve({
            code,
            ...info.envelope
          })
        }
      });
    })
  }
}
```

邮件发送相关配置使用 .env 配置文件进行管理，配置及解释如下：

```
# 邮箱配置
EMAIL_PASS=EBBOFYAQDWBLOTZY       # 邮箱授权码，需要从邮箱平台中申请
EMAIL_HOST=smtp.163.com           # 邮件服务器地址
EMAIL_PORT=465                    # 邮件服务器端口，默认为 465
EMAIL_SECURE=true                 # 是否使用默认的邮箱端口
EMAIL_USER=jmin95@163.com         # 发送给用户邮件的邮箱号
EMAIL_ALIAS=store_web_project     # 邮箱别名
```

在注册和找回密码的过程中，我们需要实现发送邮箱验证码的逻辑。具体过程包括生成验证码，将其存储在 Redis 中，并使用不同的键（key）来区分不同的用户或场景。为了确保安全性，这些验证码设置了一个 5 分钟的有效时间。以下是实现这一功能的示例代码：

```typescript
import { HttpException, HttpStatus, Inject, Injectable } from '@nestjs/common';
import { InjectRepository } from '@nestjs/typeorm';
import { RedisKeyPrefix } from 'src/common/enums/redis-key.enum';
import { MailService } from 'src/common/mail/mail.service';
import { RedisService } from 'src/common/redis/redis.service';
import { getRedisKey } from 'src/common/utils';
import { UserEntity } from 'src/user/entities/user.entity';
import { Repository } from 'typeorm';

@Injectable()
export class SysService {
  @Inject(MailService)
  private mailService: MailService;
```

```typescript
@Inject(RedisService)
private redisService: RedisService;

@InjectRepository(UserEntity)
private userRepository: Repository<UserEntity>;

/**
 * 发送找回密码邮件
 *
 * @param email 用户邮箱
 * @returns 返回验证码已发送至邮箱的提示信息
 * @throws 当邮箱为空时,抛出 HttpException 异常
 * @throws 当前用户未绑定该邮箱时,抛出 HttpException 异常
 */
async sendEmailForGorgot(email: string) {
  if (!email) {
    throw new HttpException('邮箱不能为空', HttpStatus.EXPECTATION_FAILED);
  }
  const exists = await this.userRepository.findOneBy({email})
  // 1. 判断当前的邮箱与用户注册邮箱是否一致
  if (!exists) {
    throw new HttpException('当前用户未绑定该邮箱,请检查后重试', HttpStatus.EXPECTATION_FAILED);
  }
  // 2. 发送邮箱验证码
  const { code } = await this.mailService.sendMail(email, '找回密码验证码')

  // 缓存 Redis
  const redisKey = getRedisKey(RedisKeyPrefix.PASSWORD_RESET, exists.id);
  await this.redisService.set(redisKey, code, 60*5); // 5 分钟有效
  return '验证码已发送至邮箱,请注意查收'
}


/**
 * 发送注册邮件验证码
 *
 * @param email 邮箱地址
 * @param text 邮件内容
 * @returns 返回发送成功信息
 */
async sendMailForRegistry(email: string, text: string) {
  const { code } = await this.mailService.sendMail(email, text)
  // 缓存 Redis
  const redisKey = getRedisKey(RedisKeyPrefix.REGISTRY_CODE, email);
  await this.redisService.set(redisKey, code, 60*5);
  return '发送成功';
}
}
```

完成编码后，接下来测试一遍流程，可以使用 REST Client 模拟用户请求来获取注册验证码，如图 10-27 所示。

图 10-27　获取注册验证码

使用这个验证码发送注册请求，成功注册之后，返回用户信息，如图 10-28 所示。

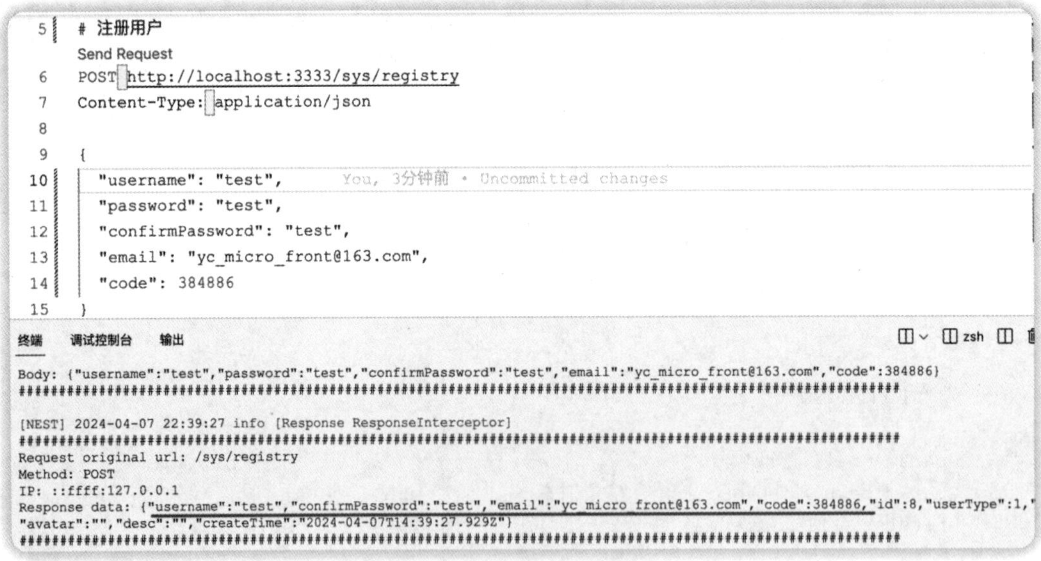

图 10-28　用户注册账号

注册完毕后，接下来使用账号和密码进行登录操作，成功换取了 access_token，如图 10-29 所示。

图 10-29　用户登录

同理，找回密码之前，需要获取邮箱验证码，如图 10-30 所示。

图 10-30　获取验证码

用获取的验证码重置密码，将密码从 test 改为 test2，接口返回"修改成功，请重新登录"，如图 10-31 所示。

图 10-31　找回密码

至此，我们完成了登录注册模块中前端页面组件的编写和服务端接口的实现。关于本节更详细的接口代码实现，请访问提供的 GitHub 源码。下一节将继续实现用户模块的功能。

10.5 实现用户与角色模块

用户与角色是 RABC 权限设计中必不可少的部分，本节将实现用户与角色管理模块。用户管理功能包含获取当前登录用户信息、获取用户列表、编辑用户信息、冻结用户和删除用户。角色管理包含获取角色列表、编辑角色信息和删除角色功能。在操作权限上，普通用户不能越权操作，例如不能修改管理员信息，冻结管理员账号或删除管理员账号。此外，普通用户也不能冻结或删除自己的账号。系统内置的角色不允许删除，并且所有列表支持模糊查询。

10.5.1 页面效果展示

在实现用户与角色管理模块之前，我们需要先实现左侧的多级菜单栏功能，如图 10-32 所示。

图 10-32　系统菜单栏

用户登录成功后，系统将获取当前用户的详细信息，包括用户的基本信息、角色以及权限菜单数据。前端应用将根据用户路由表中预设的配置信息与服务端返回的权限数据进行对比。通过计算这两个数据集的交集，前端将筛选出用户具有相应权限的菜单项，并据此渲染菜单选项。核心实现代码如下：

```
    /** 处理菜单权限隐藏菜单 */
    const reduceRouter = (routers: RouteType[]): RouteType[] => {
      // 过滤出属于菜单和页面类型的数据
      const authMenus = storeGlobalUser?.userInfo?.menus
        ?.filter(item => item?.type === ComponTypeEnum.MENU || item?.type === ComponTypeEnum.PAGE)
        ?.map(item => item?.title)

      return routers?.map(item => {
```

```
    if (item?.children) {
      const { children, ...extra } = item
      return {
        ...extra,
        routes: reduceRouter(item?.children),
        hideInMenu: item?.hideInMenu || !item?.children?.find(citem => authMenus?.includes(citem?.name))
      }
    }
    return {
      ...item,
      hideInMenu: item?.hideInMenu || !authMenus?.includes(item?.name)
    }
  }) as any
}
```

接下来，在菜单组件中绑定路由数据，示例代码如下：

```
<ProLayout
  ...
  route={reduceRouter(router?.routes)?.[1]}
  ...
>
  ...
</ProLayout>
```

ProLayout 组件接收 route 等属性并自动渲染出页面框架。除了图 10-32 中展示的用户列表外，单击"编辑"按钮可以对用户信息进行修改，如图 10-33 所示。

图 10-33　编辑用户信息

在列表上方，用户可以通过输入用户名来执行模糊搜索。此外，列表还提供了操作按钮，如单击"删除"按钮可以删除指定用户，单击"冻结"按钮可以冻结用户。被冻结的用户将无法登录系统，只有在执行"解冻"操作后才能恢复登录权限。

角色模块中的列表字段布局和功能与此类似，如图 10-34 所示。

图 10-34　角色管理

在项目中，列表页面的交互基于 ProComponents 的 ProTable 来实现。我们在此基础上封装了更加简单、高效且可复用的业务组件 ExcelTable，使得可以用更少的代码实现表格交互效果。以下是封装的核心示例代码：

```
<ExcelTable
  columns={[
    {
      title: '角色名称',
      dataIndex: 'name',
      hideInTable: true
    },
    /** search */
    {
      title: '角色名称',
      dataIndex: 'name',
      hideInSearch: true
    },
    {
      title: '描述',
      dataIndex: 'desc',
      hideInSearch: true
    },
    {
      title: '是否系统内置',
      dataIndex: 'isSystem',
      hideInSearch: true,
      render(_, entity) {
        return <Tag color={entity.isSystem ? 'green' : 'red'}>{entity.isSystem ? '是' : '否'}</Tag>
      }
    },
    {
      title: '创建时间',
      dataIndex: 'createTime',
      hideInSearch: true,
      valueType: 'dateTime'
```

```jsx
      },
      {
        title: '操作',
        key: 'option',
        valueType: 'option',
        render: (_, record) => (!record.isSystem && [
          <Button key="edit" type="link" onClick={() => showModal(record)}>
            编辑
          </Button>,
          <Popconfirm
            key="delete"
            placement="topRight"
            title="确定要删除吗?"
            onConfirm={async () => {
              const res = await delRole({ id: record?.id })
              if (res?.code === 200) {
                message.success('删除成功')
                actionRef?.current?.reloadAndRest?.()
                return Promise.resolve()
              }
              return Promise.reject()
            }}
            okText="确定"
            okType="danger"
            cancelText="取消"
          >
            <Button type="link" danger key="delete">
              删除
            </Button>
          </Popconfirm>
        ])
      }
    ]}
    requestFn={async (params) => {
      const data = await getRoleList(params)
      return data
    }}
    actionRef={actionRef}
    rowSelection={false}
    toolBarRenderFn={() => [
      <Button key="add" type='primary' onClick={() => showModal()}>
        新增角色
      </Button>
    ]}
  />
```

关于自定义组件更详细的实现过程，请查看 GitHub 项目中的源码部分。

单击"新增角色"按钮，用户可以自定义角色并为其分配初始权限，如图 10-35 所示。

图 10-35　新增角色

单击"编辑"按钮，用户可以对角色信息进行二次修改和分配权限，如图 10-36 所示。

图 10-36　编辑角色

角色权限支持树形交互选择，勾选后可以为角色分配不同的权限，如图 10-37 所示，用户拥有相关权限才能进行相应的操作。

图 10-37　分配权限的交互过程示例

我们可以使用 Modal 组件轻松实现以上弹窗的交互效果，示例代码如下：

```
Modal.confirm({
  title: record ? '编辑角色' : '新增角色',
```

```jsx
        onOk: async () => onSubmit(record),
        okText: '确定',
        cancelText: '取消',
        maskClosable: true,
        width: 800,
        content: (
          <ProForm
            labelCol={{ span: 6 }}
            wrapperCol={{ span: 12 }}
            submitter={false}
            layout="horizontal"
            initialValues={{
              status: 1,
              ...record,
              permissions: record?.permissions?.map(item => item?.id)
            }}
            formRef={modalFormRef}
          >
            <ProFormText label="角色名称" name="name" rules={[{ required: true }]} />
            <ProFormTreeSelect
              label="角色权限"
              name="permissions"
              fieldProps={{
                multiple: true,
                showCheckedStrategy: TreeSelect.SHOW_ALL,
                fieldNames: {
                  label: 'title',
                  value: 'id'
                },
                treeCheckable: true
              }}
              allowClear
              rules={[{ required: true, message: '请选择' }]}
              request={async () => {
                const { data } = await getPermissionList()
                return data.list
              }}
            />
            <ProFormTextArea label="描述" name="desc" />
          </ProForm>
        )
      })
```

编写完前端交互页面后，接下来实现服务端的接口逻辑。

10.5.2 表关系设计

获取用户信息通常涉及多表联合查询。在 RBAC 权限系统中，除了用户表，还包括角色表和权限表，这些表通过中间表来维护它们之间的多对多关系。角色表实体如下：

```typescript
import { Column, CreateDateColumn, Entity, PrimaryGeneratedColumn, UpdateDateColumn } from "typeorm";

@Entity('store_role')
```

```typescript
export class RoleEntity {
  @PrimaryGeneratedColumn()
  id: number

  @Column({ type: 'varchar', length: 50, comment: '角色名称' })
  name: string

  @Column({ type: 'varchar', length: 255, comment: '角色描述'})
  desc: string

  @CreateDateColumn({ type: 'timestamp', comment: '创建时间' })
  createTime: Date

  @UpdateDateColumn({ type: 'timestamp', comment: '更新时间' })
  updateTime: Date

  @Column({ type: 'int', comment: '是否为系统内置 0 否 1 是', default: 0 })
  isSystem: number
}
```

系统通常会预设一些默认角色，例如管理员、店长、收银员等，但这些角色的设置并非固定不变，而是根据具体的应用场景来确定。在 RBAC 权限系统中，用户与角色之间的关系通过一个中间表 store_user_role 来维护，实体字段定义如下：

```typescript
import { Column, Entity, PrimaryGeneratedColumn } from "typeorm";

@Entity('store_user_role')
export class UserRoleEntity {
  @PrimaryGeneratedColumn()
  id: number

  @Column({ type: 'int', comment: '用户 id' })
  userId: number

  @Column({ type: 'int', comment: '角色 id' })
  roleId: number
}
```

权限表用于存储系统中包含哪些菜单、页面、按钮或组件的权限信息，通常也被称为"菜单表"，其实体字段如下：

```typescript
import { PermissionType } from "src/common/enums/common.enum";
import { Column, Entity, PrimaryGeneratedColumn } from "typeorm";

@Entity('store_permission')
export class PermissionEntity {
  @PrimaryGeneratedColumn()
  id: number

  @Column({ type: 'varchar', length: 10, comment: '权限名称（菜单名称）' })
  title: string

  @Column({ type: 'varchar', length: 50, comment: '权限码' })
  code: string
```

```
  @Column({ type: 'int', comment: '权限类型 0 菜单 1 页面 2 组件 3 按钮' })
  type: PermissionType

  @Column({ type: 'int', comment: '父级 id', default: 0 })
  parentId: number
}
```

在设计 RBAC 权限系统时,需要注意权限数据具有多种类型,这样的设计有助于实现更细粒度的权限控制。组件或按钮级别的细粒度权限通常不会直接展示在菜单栏中,而是通过权限码在页面中进行单独控制。例如,权限码可以传递给高阶组件,该组件将自动判断当前用户是否有权限渲染对应的组件或按钮。

对于 Vue 开发者来说,一种常见的做法是使用自定义指令 v-permission 来接收权限码,并据此实现权限控制。

权限表定义好后,角色与权限之间还需要一个中间表 store_role_permission 来关联,实体字段如下:

```
import { Column, Entity, PrimaryGeneratedColumn } from "typeorm";

@Entity('store_role_permission')
export class RolePermissionEntity {
  @PrimaryGeneratedColumn()
  id: number

  @Column({ type: 'int', comment: '角色 id' })
  roleId: number

  @Column({ type: 'int', comment: '权限集 id' })
  permissionId: number
}
```

至此,我们大致完成了 RBAC 基本的表关系设计。菜单权限将会分配给不同的角色,而不同的角色又将分配给不同的用户,从而实现了不同用户只能操作特定的权限功能。这一设计看起来没问题,但实际上在服务端还存在一个隐患。

尽管在用户页面上看不到具体的功能页面或按钮,无法直接进行操作,但某些功能对应的接口却没有进行权限控制,使得一些用户可以通过手动发送 HTTP 请求的方式访问需要权限的接口因此,我们还需要一个表(store_permission_api),用来维护服务端所有的路由接口。该表的主要字段如下:

```
import { Column, Entity, PrimaryGeneratedColumn } from "typeorm";
@Entity('store_permission_api')
export class PermissionApiEntity {
  @PrimaryGeneratedColumn()
  id: number

  @Column({ type: 'varchar', length: 50, comment: '权限名称' })
  apiUrl: string

  @Column({ type: 'varchar', length: 100, comment: '权限码' })
  apiMethod: string
```

```
@Column({ type: 'int', comment: '功能id' })
PermissionId: number
}
```

这样设计的目的在于，当用户每一次请求接口时，我们需要判断当前用户发送的请求 URL 是否在他的角色控制范围内，从而确定是否放行该客户端的请求。我们定义了一个全局守卫（RoleAuthGuard）来处理这个逻辑，代码如下：

```typescript
import { CanActivate, Inject, ExecutionContext, Injectable, ForbiddenException } from '@nestjs/common'
import { Reflector } from '@nestjs/core'
import { pathToRegexp } from 'path-to-regexp'
import { ALLOW_NO_PERMISSION } from 'src/common/decorators/permission.decorator'
import { PermissionService } from 'src/permission/permission.service'
import { ALLOW_NO_TOKEN } from 'src/common/decorators/token.decorator'
import { UserType } from 'src/common/enums/common.enum'

@Injectable()
export class RoleAuthGuard implements CanActivate {
  constructor(
    private readonly reflector: Reflector,
    @Inject(PermissionService)
    private readonly permissionService: PermissionService,
  ) {}

  async canActivate(ctx: ExecutionContext): Promise<boolean> {
    // 若函数请求头配置了@AllowNoToken()装饰器，则无须验证 Token 权限
    const allowNoToken = this.reflector.getAllAndOverride<boolean>(ALLOW_NO_TOKEN, [ctx.getHandler(), ctx.getClass()])
    if (allowNoToken) return true

    // 若函数请求头配置了@AllowNoPermission()装饰器，则无须验证权限
    const allowNoPerm = this.reflector.getAllAndOverride<boolean>(ALLOW_NO_PERMISSION, [ctx.getHandler(), ctx.getClass()])
    if (allowNoPerm) return true

    const req = ctx.switchToHttp().getRequest()
    const user = req.user
    // 没有携带 Token 直接返回 false
    if (!user) return false
    // 管理员拥有所有接口权限，不需要判断
    if (user.userType === UserType.ADMIN_USER) return true

    // 获取该用户所拥有的接口权限
    const userApis = await this.permissionService.getPermApiList(user)
    console.log('当前用户拥有的接口权限集：', userApis);

    const index = userApis.findIndex((route) => {
      // 请求方法类型相同
      if (req.method.toUpperCase() === route.method.toUpperCase()) {
        // 比较当前请求 URL 是否在用户接口权限集中
        const reqUrl = req.url.split('?')[0]
        console.log('当前请求 URL：', reqUrl);

        return !!pathToRegexp(route.url).exec(reqUrl)
```

```
      }
      return false
    })
    if (index === -1) throw new ForbiddenException('您无权限访问该接口')
    return true
  }
}
```

在代码实现中，通过注释明确了几种特殊情况：管理员账户、无需进行 token 验证的账户以及无需进行权限验证的账户，这些账户将被直接允许访问。对于其他用户的请求，则必须通过 API 权限校验。如果校验失败，守卫将阻止访问。在梳理清楚权限关系后，接下来实现具体的接口逻辑。

10.5.3　接口实现

用户管理与角色管理的接口实现极其相似，本节以用户模块为例，当用户登录成功后，首先获取用户信息，包含用户基本信息以及对应的角色和权限数据。具体实现代码如下：

```
/**
 * 获取当前用户信息
 *
 * @param currentUser 当前用户实体
 * @returns 返回当前用户信息、用户角色以及权限
 */
async getCurrentUser(currentUser: UserEntity) {
  // 同时查询用户角色和权限
  const queryBuilder = this.dataSource
    .createQueryBuilder()
    .select([
      'user.id AS userId',
      'user.username AS userName',
      'role.id AS roleId',
      'role.name AS roleName',
      'p.id AS permissionId',
      'p.title AS permissionTitle',
      'p.type AS permissionType',
      'p.code AS permissionCode'
    ])
    .from('store_user', 'user')
    .leftJoin('store_user_role', 'userRole', 'user.id = userRole.userId')
    .leftJoin('store_role', 'role', 'userRole.roleId = role.id')
    .leftJoin('store_role_permission', 'rolePerm', 'role.id = rolePerm.roleId')
    .leftJoin('store_permission', 'p', 'rolePerm.permissionId = p.id')
    .where('user.id = :userId', { userId: currentUser.id })
    .orderBy('p.id', 'ASC');

  const enrichedData = await queryBuilder.getRawMany();
  // 这里，我们需要对 enrichedData 进行一些后处理，以将其转换为所需的格式
  const roles = enrichedData.reduce((acc, row) => {
    if (!acc.includes(row.roleId)) {
      acc.push(row.roleId);
    }
    return acc;
  }, [] as number[]);
```

```
  const permissions = enrichedData.map(row => ({
    id: row.permissionId,
    title: row.permissionTitle,
    type: row.permissionType,
    code: row.permissionCode
  }));

  return {
    ...currentUser,
    menus: permissions,
    roles: roles.map((roleId: number) => ({ id: roleId }))
  };
}
```

在上述代码中,我们首先根据当前登录用户的 ID,使用 createQueryBuilder 方法查询并获取用户的权限和角色信息,然后将这些信息返回给客户端。客户端在请求成功后,会根据返回的用户权限渲染相应的菜单栏,并将用户信息存储到全局状态管理中,以便于后续的重复使用。

为了增强用户体验和系统性能,在查询用户列表时,我们实现了模糊查询和分页功能。此外,还需要查询并返回每个用户对应的角色集合。接下来详细查看具体的代码实现。

```
/**
 * 获取用户列表
 *
 * @returns 返回用户列表、用户总数、当前页码
 */
async getUserList(dto: UserListDto) {
  const { username, page, pageSize } = dto
  const skipCount = pageSize * (page - 1)
  let queryBuilder = this.dataSource
    .createQueryBuilder('store_user', 'u')
    .leftJoin('store_user_role', "ur", "ur.userId = u.id")
    .leftJoin('store_role', "r", "r.id = ur.roleId")
    .select(['u.*', "JSON_ARRAYAGG(JSON_OBJECT('id', r.id, 'name', r.name)) as roles"])
    .groupBy('u.id')

  // 如果有模糊检索条件
  if (username) {
    queryBuilder = queryBuilder.where('u.username Like :username').setParameter('username', `%${username}%`)
  }
  const list = await queryBuilder.skip(skipCount).take(pageSize).getRawMany()

  return {
    list: list.map(user => {
      Reflect.deleteProperty(user, 'password')
      Reflect.deleteProperty(user, 'salt')
      return user
    }),
    total: list.length,
    page
  }
}
```

在查询用户及其关联角色的数据时，代码使用了 JSON_ARRAYAGG 和 JSON_OBJECT 语句来生成一个 JSON 格式的 roles 数组。此外，根据提供的查询条件，通过使用 LIKE 语句实现模糊搜索。为了优化性能和用户体验，代码中还结合了 skip 和 take 语句来实现分页功能。在返回查询结果之前，确保对敏感数据进行了适当的清理。

对于用户信息的二次编辑，除了更新数据库中用户的基本信息之外，还需要考虑同步更新用户角色的中间表，并及时刷新 Redis 中缓存的用户信息，以确保数据的一致性。示例代码如下：

```
/**
 * 更新用户信息
 *
 * @param updateUserDto 更新用户 DTO
 * @param currentUser 当前用户
 * @returns 更新成功提示
 * @throws 用户不存在异常
 * @throws 没有权限修改管理员信息异常
 */
async update(updateUserDto: UpdateUserDto, currentUser: UserEntity) {
  // 1. 判断用户是否存在
  const user = await this.userRepository.findOne({where: { id: updateUserDto.id }});
  if (!user) {
    throw new HttpException('用户不存在', HttpStatus.EXPECTATION_FAILED);
  }
  // 2. 普通用户不能修改管理员信息
  if (user.userType === UserType.ADMIN_USER && currentUser.userType === UserType.NORMAL_USER) {
    throw new HttpException('你没有权限修改管理员信息', HttpStatus.FORBIDDEN);
  }
  // 3. 更新数据
  const { password, ...rest } = await this.userRepository.save(
    plainToClass(UserEntity, {
      ...user,
      ...updateUserDto,
    }, { ignoreDecorators: true })
  );
  // 4. 更新用户角色表
  if (updateUserDto.roleIds) {
    // 先删除用户角色表
    await this.userRoleRepository.delete({ userId: user.id })
    const roles = updateUserDto.roleIds.map(item => ({ userId: user.id, roleId: item }))
    const result = await this.userRoleRepository.save(roles)
    if (!result) {
      throw new HttpException('更新失败，请稍后重试', HttpStatus.EXPECTATION_FAILED);
    }
  }
  // 5. 更新 Redis 缓存
  const redisKey = getRedisKey(RedisKeyPrefix.USER_INFO, updateUserDto.id)
  await this.redisService.hSet(
    redisKey,
```

```
      classToPlain(rest)
    );
    return '更新成功';
  }
```

有权限的用户可以对其他用户的账号执行冻结操作。一旦用户账号被冻结，该用户将无法登录系统，直到有权限的用户执行解冻操作。重要的是，在执行冻结操作后，必须及时更新 Redis 缓存，以确保系统的响应与数据状态保持一致。示例代码如下：

```
/**
 * 更新用户冻结状态
 *
 * @param id 用户 ID
 * @param freezed 是否冻结
 * @param currUserId 当前用户 ID
 * @returns 操作成功
 * @throws 用户不能冻结自己
 * @throws 用户不存在
 * @throws 你没有权限修改管理员信息
 * @throws 冻结或解冻失败，请稍后重试
 */
async updateFreezedStatus(id: number, freezed: number, currUserId: number) {
  // 1. 用户不能冻结自己
  if (id === currUserId) {
    throw new HttpException('你不能冻结自己', HttpStatus.EXPECTATION_FAILED);
  }
  // 2. 判断用户是否存在
  const user = await this.userRepository.findOne({where: {id}});
  if (!user) {
    throw new HttpException('用户不存在', HttpStatus.EXPECTATION_FAILED);
  }
  // 3. 管理员是有最高权限的，不能被冻结
  if (user.userType === UserType.ADMIN_USER) {
    throw new HttpException('你没有权限修改管理员信息', HttpStatus.FORBIDDEN);
  }
  // 4. 更新数据
  const { affected } = await this.userRepository.update({ id }, { freezed });
  if (!affected) {
    throw new HttpException(`${freezed ? '冻结' : '解冻'}失败，请稍后重试`, HttpStatus.EXPECTATION_FAILED);
  }
  // 5. 更新 Redis 缓存
  const redisKey = getRedisKey(RedisKeyPrefix.USER_INFO, id)
  const { password, ...rest } = user
  await this.redisService.hSet(
    redisKey,
    classToPlain({ ...rest, freezed })
```

    );

    return '操作成功';
}
```

管理员可以删除指定用户,用户被删除之后,需要及时更新用户角色表及 Redis 缓存,代码如下:

```
/**
 * 删除指定 ID 的用户
 *
 * @param id 用户 ID
 * @returns 返回删除结果,成功返回'删除成功',失败抛出异常
 * @throws 当用户不存在时,抛出'用户不存在'的 HTTP 异常
 * @throws 当删除失败时,抛出'删除失败,请稍后重试'的 HTTP 异常
 */
async delete(id: number) {
  // 1. 判断用户是否存在
  const user = await this.userRepository.findOne({where: {id}});
  if (!user) {
    throw new HttpException('用户不存在', HttpStatus.EXPECTATION_FAILED);
  }
  if (user.userType === UserType.ADMIN_USER) {
    throw new HttpException('你没有权限删除管理员', HttpStatus.FORBIDDEN);
  }
  const { affected } = await this.userRepository.delete({id})
  if (!affected) {
    throw new HttpException('删除失败,请稍后重试', HttpStatus.EXPECTATION_FAILED);
  }
  // 2. 删除角色关联表
  const result = await this.userRoleRepository.delete({userId: id})
  if (!result) {
    throw new HttpException('删除失败,请稍后重试', HttpStatus.EXPECTATION_FAILED);
  }
  // 3. 删除 Redis 缓存
  const redisKey = getRedisKey(RedisKeyPrefix.USER_INFO, id)
  await this.redisService.del(redisKey)

  return '删除成功'
}
```

至此,已完成了用户模块接口逻辑的实现。相信读者已经学会了如何编辑角色模块的接口。下一节将继续开发门店系统中核心的商品与订单模块。

10.6 实现商品与订单模块

商品模块是门店经营系统中的核心模块。商家在经营过程中会销售各种各样的商品,收银员
```

为指定商品开单后会产生一条销售订单。门店系统可以根据销售订单数据进行统计分析，以便进行销售提成计算、门店财务统计以及时调整经营策略。

10.6.1 页面效果展示

商品管理菜单衍生出商品列表和热销排行两个子菜单。在商品列表页面实现对商品表的操作，如查询、新增、编辑、删除商品，以及对商品进行上下架操作，如图 10-38 所示。

图 10-38 商品列表

单击"新增商品"按钮，用户可以新增门店对外销售的商品信息，如图 10-39 所示。

图 10-39 新增商品

需要注意的是，如果在创建商品时启用了"是否上架"按钮，这意味着可以对商品进行开单操作，未上架的商品不会出现在开单商品列表中。

单击"编辑"按钮，用户可以对商品信息进行二次修改后保存，如图 10-40 所示。

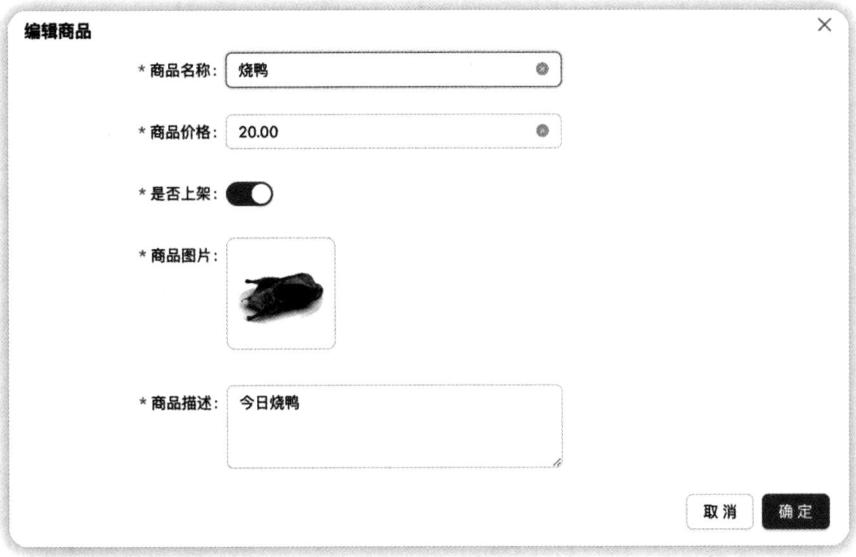

图 10-40　编辑商品

另外，用户可以直接在列表中更新商品的上下架状态和删除指定商品信息。

在完成商品列表模块的开发之后，接下来实现订单管理模块，效果如图 10-41 所示。

图 10-41　订单列表

用户可以根据订单编号查询指定的订单信息，同时支持通过订单状态筛选数据。收银员根据客户付款情况选择操作对应的订单状态，包括未付款、已付款和已取消。

单击"开单收银"按钮可以为客户购买某个商品进行开单，如图 10-42 所示。

图 10-42　创建订单

在开单时，系统要确保所选商品不包括任何已下架的商品。用户选定商品、数量和折扣后，系统将自动计算出商品的总价。此外，选择一个指定的关联员工是必要的，这将使当前订单的销售额与该员工的业绩挂钩，为可能需要的员工业绩统计或提成计算提供依据。

订单创建完成后，用户可以通过单击订单列表中的"订单编号"按钮来查看订单的详细信息，如图 10-43 所示。

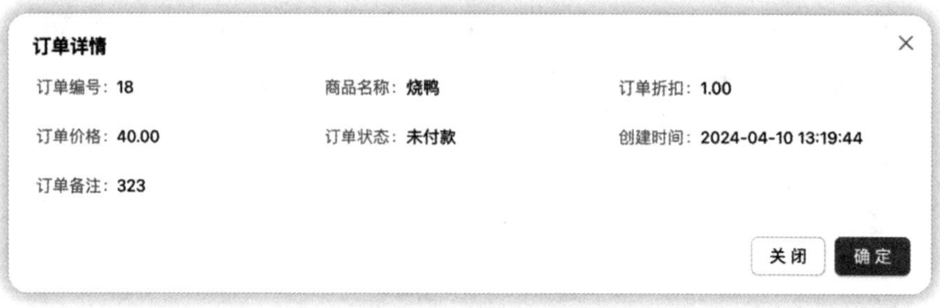

图 10-43　订单详情

完成前端页面的编写后，接下来实现服务端的接口逻辑。

10.6.2　表关系设计

在本项目中，商品与订单之间存在一对多关系，这种关联关系在不同的应用场景中存在差异，比如在众多的电商系统中，商品与订单之间存在多对多关系。首先定义商品实体表字段，代码如下：

```
import { Column, CreateDateColumn, Entity, PrimaryGeneratedColumn, UpdateDateColumn } from "typeorm";

@Entity('store_product')
export class ProductEntity {
```

```typescript
  @PrimaryGeneratedColumn()
  id: number;

  @Column({ type: 'varchar', length: 50, comment: '商品名称'})
  name: string;

  @Column({ type: "decimal", precision: 10, scale: 2, comment: '商品价格'})
  price: number

  @Column({ type: 'simple-array', comment: '商品图片', nullable: true })
  images: string[]

  @Column({ type: 'text',  comment: '商品描述' })
  desc: string

  @Column({ type: 'int', comment: '商品状态 0 未上架  1 已上架 2 已下架'})
  status: number

  @CreateDateColumn({ type: 'datetime', comment: '创建时间'})
  createTime: Date

  @UpdateDateColumn({ type: 'datetime', comment: '更新时间'})
  updateTime: Date
}
```

10.6.3　接口实现

首先完成商品模块的接口开发，在 ProductController 中定义了增删改查相关的路由方法，代码如下：

```typescript
import {
  Controller,
  Get,
  Post,
  Body,
  Patch,
  Param,
  Query,
} from '@nestjs/common';
import { ProductService } from './product.service';
import { CreateProductDto } from './dto/create-product.dto';
import { UpdateProductDto } from './dto/update-product.dto';
import { ProductListDto } from './dto/product-list.dto';
import { HotSalesService } from './hot-sales.service';

@Controller('product')
export class ProductController {
  constructor(
    private readonly productService: ProductService,
    private readonly hotSalesService: HotSalesService,
  ) {}

  @Post('create')
```

```typescript
  create(@Body() createProductDto: CreateProductDto) {
    return this.productService.create(createProductDto);
  }

  @Get('list')
  getProductList(@Query() productListDto: ProductListDto) {
    return this.productService.getProductList(productListDto);
  }

  @Patch('edit')
  update(@Body() updateProductDto: UpdateProductDto) {
    return this.productService.update(updateProductDto);
  }

  @Get('delete/:id')
  delete(@Param('id') id: number) {
    return this.productService.delete(id);
  }

  @Patch('updateStatus')
  updateStatus(@Body() updateProductDto: UpdateProductDto) {
    return this.productService.updateStatus(
      updateProductDto.id,
      updateProductDto.status,
    );
  }
}
```

新增商品比较简单，只需将客户端发送的信息保存到数据库中即可。实现 create 方法的代码如下：

```typescript
/**
 * 创建商品
 *
 * @param createProductDto 创建商品的 DTO
 * @returns 返回创建成功的消息
 * @throws 当创建失败时，抛出 HttpException 异常
 */
async create(createProductDto: CreateProductDto) {
  const product = await this.productRepository.save(createProductDto)
  if (!product) {
    throw new HttpException('创建失败，请稍后重试', HttpStatus.EXPECTATION_FAILED)
  }
  return '创建成功'
}
```

需要注意的是，在并发请求大的场景下，可以考虑将商品数据保存到 Redis 中。在需要获取商品详情时，首先从缓存中获取，以减少对数据库的访问次数。

接下来，编写获取商品列表数据的接口。该接口需要支持用户输入商品名称进行模糊查询，并根据指定的商品状态筛选数据，最后返回经过分页和排序处理的数据，实现代码如下：

```typescript
/**
 * 获取产品列表
 *
```

```
 * @param productListDto 产品列表 DTO
 * @returns 返回产品列表和总数
 */
async getProductList(productListDto: ProductListDto) {
  const { name, status, page, pageSize } = productListDto
  const where = {
    ...(name ? { name: Like(`%${name}%`) } : null),
    ...(status ? { status } : null),
  }
  const [list, total] = await this.productRepository.findAndCount({
    where,
    order: { id: 'DESC' },
    skip: pageSize * (page - 1),
    take: pageSize,
  });
  return {
    list,
    total
  }
}
```

在上述代码中,根据商品名称和商品状态组合查询条件(where),调用实体方法 findAndCount 查询出列表(list)并统计数据条数(total)。

接下来,对商品进行二次修改。根据客户端传递的商品 id 判断是否存在被编辑的记录,如果存在,则将数据库信息更新为新传递的商品信息。示例代码如下:

```
/**
 * 更新商品信息
 *
 * @param updateProductDto 更新商品信息所需的数据传输对象
 * @returns 返回更新成功的信息
 * @throws 如果商品不存在或已删除,则抛出 HttpException 异常
 * @throws 如果更新失败,则抛出 HttpException 异常
 */
async update(updateProductDto: UpdateProductDto) {
  const { id } = updateProductDto
  const exists = await this.productRepository.findOneBy({ id });
  if (!exists) {
    throw new HttpException('商品不存在或已删除', HttpStatus.EXPECTATION_FAILED)
  }
  const { affected } = await this.productRepository.update({ id }, updateProductDto)
  if (!affected) {
    throw new HttpException('更新失败,请稍后重试', HttpStatus.EXPECTATION_FAILED)
  }
  return '更新成功';
}
```

除了编辑商品信息外,系统还允许对商品执行快速上下架操作。为了更精细地控制这一操作的权限,我们将其设计为一个独立的接口进行管理,代码如下:

```
/**
 * 更新商品状态
 *
 * @param id 商品 ID
```

```
 * @param status 商品状态，1 表示上架，2 表示下架
 * @returns 返回一个字符串，表示上架或下架是否成功
 * @throws 如果商品不存在或已删除，则抛出 HttpException 异常
 * @throws 如果更新状态失败，则抛出 HttpException 异常
 */
async updateStatus(id: number, status: 1 | 2) {
  const exists = await this.productRepository.findOneBy({ id });
  if (!exists) {
    throw new HttpException('商品不存在或已删除', HttpStatus.EXPECTATION_FAILED)
  }
  const { affected } = await this.productRepository.update({ id }, { status })
  const text = status === 1 ? '上架' : '下架'
  if (!affected) {
    throw new HttpException(`${text}失败，请稍后重试`, HttpStatus.EXPECTATION_FAILED)
  }
  return '${text}成功';
}
```

然后在 store_web.sql 中执行这行 SQL 语句，向 store_permission 表中插入一条数据：

```
INSERT INTO 'store_permission' VALUES (17, '上下架商品', 'updateStatus:product', 3, 2);
```

此时权限表中多了一条数据，可以在编辑角色的弹窗中看到"上下架商品"选项，为指定的角色分配权限，如图 10-44 所示。

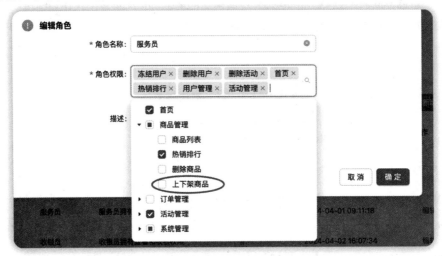

图 10-44 分配"上下架商品"权限

除此之外，我们还需要在 store_permission_api 表中增加 updateStatus 接口数据：

```
INSERT INTO `store_permission_api` VALUES (24, '/product/updateStatus', 'PATCH', 17);
```

这条数据关联 store_permission 表中 id=17 的记录，此时守卫会在每次请求时，使用当前用户角色对应的权限集校验是否有该接口的请求权限，以此决定是否放行，实现接口层的权限控制。

同理，商品模块中的删除接口的权限控制也是如此，这里不再赘述，接口的具体实现代码如下：

```
/**
 * 删除商品
```

```
 * @param id 商品ID
 * @returns 返回删除结果，成功返回'删除成功'，失败抛出异常
 * @throws 当商品不存在或已删除时，抛出 HttpException 异常，状态码为 HttpStatus.EXPECTATION_FAILED
 * @throws 当删除失败时，抛出 HttpException 异常，状态码为 HttpStatus.EXPECTATION_FAILED
 */
async delete(id: number) {
  const exists = await this.productRepository.findOneBy({ id });
  if (!exists) {
    throw new HttpException('商品不存在或已删除', HttpStatus.EXPECTATION_FAILED)
  }
  const { affected } = await this.productRepository.delete({ id })
  if (!affected) {
    throw new HttpException('删除失败，请稍后重试', HttpStatus.EXPECTATION_FAILED)
  }
  return '删除成功';
}
```

删除商品与编辑商品类似，首先根据 id 判断商品是否存在，如果存在，则按照正常删除逻辑删除商品即可。

前面介绍了商品模块的接口实现，接下来看订单模块的接口实现过程。前面提到了商品与订单是一对多关系，在设计上，订单表中会记录当前订单对应商品的 id，通过 id 就能查询商品信息，具体实体表字段如下：

```
import { Column, CreateDateColumn, Entity, PrimaryGeneratedColumn } from "typeorm";

@Entity('store_order')
export class OrderEntity {
  @PrimaryGeneratedColumn()
  id: number;

  @Column({ type: 'varchar', length: 50, comment: '商品名称' })
  name: string;

  @Column({ type: 'int', default: 1, comment: '商品数量' })
  count: number;

  @Column({ type: "decimal", precision: 5, scale: 2, default: 1, comment: '订单折扣' })
  discount: number;

  @Column({ type: "decimal", precision: 10, scale: 2, comment: '订单价格' })
  price: number

  @Column({ type: "decimal", precision: 10, scale: 2, comment: '订单折扣价' })
  discountPrice: number

  @Column({ type: 'int', comment: '订单状态 0 未付款 1 已付款 2 已取消' })
  status: number;

  @Column({ type: 'varchar', comment: '操作员' })
  operator: string;

  @CreateDateColumn({ type: 'timestamp', comment: '创建时间' })
```

```typescript
  createTime: Date;

  @Column({ type: 'text', comment: '订单备注', nullable: true })
  desc: string;

  @Column({ type: 'int', comment: '商品id' })
  productId: number;
}
```

其中，订单价格表示在未参与折扣时的订单总价，而订单折扣价表示在参与折扣后客户应该实付的价格。定义完实体表后，在 OrderController 中定义需要实现的路由接口，代码如下：

```typescript
import { Controller, Get, Post, Body, Patch, Param, Req, Query } from '@nestjs/common';
import { OrderService } from './order.service';
import { CreateOrderDto } from './dto/create-order.dto';
import { UpdateOrderDto } from './dto/update-order.dto';
import { OrderListDto } from './dto/order-list.dto';

@Controller('order')
export class OrderController {
  constructor(private readonly orderService: OrderService) {}

  @Post('create')
  create(@Body() createOrderDto: CreateOrderDto) {
    return this.orderService.create(createOrderDto);
  }

  @Get('list')
  getOrderList(@Query() orderListDto: OrderListDto) {
    return this.orderService.getOrderList(orderListDto);
  }

  @Get('detail/:id')
  getOrderDetail(@Param('id') id: string) {
    return this.orderService.getOrderDetail(+id);
  }

  @Patch('updateOrder')
  updateOrder(@Body() updateOrderDto: UpdateOrderDto, @Req() req) {
    return this.orderService.updateOrder(updateOrderDto, req.user);
  }

  @Get('delete/:id')
  delete(@Param('id') id: string, @Req() req ) {
    return this.orderService.delete(+id, req.user);
  }
}
```

接下来，根据上面的方法分别实现服务接口逻辑。在实际经营场景中，一般收银员、店长或管理员才拥有开单权限，在本项目中，我们将此权限分配给收银员这个角色。开单接口的具体实现代码如下：

```typescript
/**
```

```
 * 创建订单
 *
 * @param createOrderDto 创建订单所需的参数
 * @returns 返回成功信息
 * @throws HttpException 抛出 HTTP 异常
 */
async create(createOrderDto: CreateOrderDto) {
  const { productId, discount = 1, status = 0, count } = createOrderDto
  // 获取商品信息
  const product = await this.productRepository.findOneBy({ id: productId })
  if (!product) {
    throw new HttpException('商品不存在', HttpStatus.NOT_FOUND)
  }
  // 获取开单商品计算价格
  let totalPrice = product.price * count
  let discountPrice = totalPrice * discount
  let orderItem = {
    ...createOrderDto,
    name: product.name,
    price: totalPrice,
    status,
    count,
    productId,
    discountPrice,
    discount
  }
  const order = await this.orderRepository.save(plainToClass(OrderEntity, orderItem))
  if (!order) {
    throw new HttpException('开单失败', HttpStatus.INTERNAL_SERVER_ERROR)
  }
  return '开单成功';
}
```

在上述代码中，首先验证所选商品是否存在。如果商品存在，系统将基于商品的单价、数量以及订单折扣重新计算订单的总价。这一步骤至关重要，因为它确保了数据的准确性，而不是简单地保存客户端发送的价格。

完成订单总价的计算后，我们使用 plainToClass 方法将包含订单详情的普通对象转换成 OrderEntity 类的实例。这一转换确保了订单对象拥有 OrderEntity 类定义的所有属性，从而可以被正确地保存到数据库中。

订单创建成功后，获取订单列表的操作就变得相对简单。这一过程与之前获取商品列表的操作类似。订单列表获取接口的代码如下：

```
/**
 * 获取订单列表
 *
 * @param orderListDto 订单列表 DTO
 * @returns 返回一个包含订单列表和总条数的对象
 */
async getOrderList(orderListDto: OrderListDto) {
  const { page, pageSize, id, status } = orderListDto
  const where = {
    ...(id ? { id } : null),
```

```
      ...(status ? { status } : null)
    }
    const [list, total] = await this.orderRepository.findAndCount({
      where,
      order: { id: 'DESC' },
      skip: pageSize * (page - 1),
      take: pageSize
    });
    return {
      list,
      total
    }
}
```

在通常情况下，列表中不能够完全展示订单的所有字段，为此可以通过查看订单详情看到更详细的订单信息，接口代码如下：

```
/**
 * 获取订单详情
 *
 * @param id 订单ID
 * @returns 返回订单详情对象，包含订单信息和商品信息
 * @throws 当订单不存在时，抛出HttpException异常，状态码为 NOT_FOUND
 * @throws 当查询订单商品失败时，抛出HttpException异常，状态码为 NOT_FOUND
 */
async getOrderDetail(id: number) {
  const order = await this.orderRepository.findOneBy({ id })
  if (!order) {
    throw new HttpException('订单不存在', HttpStatus.NOT_FOUND)
  }
  const product = await this.productRepository.findOneBy({ id: order.productId })
  if (!product) {
    throw new HttpException('查询订单商品失败', HttpStatus.NOT_FOUND)
  }
  return {
    ...order,
    product
  };
}
```

在上述代码中，先根据订单号获取对应的订单信息，由于订单中绑定着商品 id，再根据商品 id 查询出关联的商品信息，重新组装后再返回给客户端。

收银员可以为客户生成未付款订单，随后对用户订单进行收款或取消订单操作，接口的实现代码如下：

```
/**
 * 更新订单状态及描述
 *
 * @param updateOrderDto 更新订单信息
 * @param currentUser 当前用户信息
 * @returns 返回更新结果
 * @throws 当订单不存在或已删除时，抛出HttpException异常，状态码为 NOT_FOUND
 * @throws 当修改失败时，抛出HttpException异常，状态码为 INTERNAL_SERVER_ERROR
 */
```

```
async updateOrder(updateOrderDto: UpdateOrderDto, currentUser: UserEntity) {
  const { id, status, desc = '' } = updateOrderDto
  const exists = this.orderRepository.findOneBy({ id })
  if (!exists) {
    throw new HttpException('订单不存在或已删除', HttpStatus.NOT_FOUND);
  }
  const { affected } = await this.orderRepository.update({ id }, { status, desc })
  if (!affected) {
    throw new HttpException('修改失败', HttpStatus.INTERNAL_SERVER_ERROR);
  }
  return '修改成功';
}
```

在上述代码中，允许对订单的状态和备注进行修改。然而，根据业务规则，一旦订单状态被更改，该变更通常是不可逆的。例如，一旦订单被取消，就无法将其状态重新修改为"待付款"。这种设计旨在确保交易过程的完整性，并防止可能的不当商家行为，如滥用订单状态更改进行欺诈或其他不当操作。

此外，收银员有权根据订单 id 删除已创建的订单记录。这通常在订单因错误或其他原因需要撤销时进行。订单状态修改和订单记录删除的示例代码如下：

```
/**
 * 删除订单
 *
 * @param id 订单 ID
 * @param currentUser 当前用户
 * @returns 删除成功或失败的信息
 * @throws 如果订单不存在或已删除，则抛出 404 错误；如果删除失败，则抛出 500 错误
 */
async delete(id: number) {
  const exists = await this.orderRepository.findOneBy({ id })
  if (!exists) {
    throw new HttpException('订单不存在或已删除', HttpStatus.NOT_FOUND);
  }
  const { affected } = await this.orderRepository.delete({ id })
  if (!affected) {
    throw new HttpException('删除失败', HttpStatus.INTERNAL_SERVER_ERROR);
  }
  return '删除成功';
}
```

至此，我们完成了商品与订单模块的开发。在此基础上，我们可以新增几行代码逻辑，实现用 Redis 来统计热销的商品排行榜，详细的实现过程将在下一节进行介绍。

10.7　基于 Redis 实现商品热销榜

排行榜是日常生活中普遍存在的一种列表形式，例如在新闻平台如今日头条、游戏排行榜、热门文章榜单以及商品销量排行榜中都能找到它们的身影。这类排行榜在很大程度上展现了用户对不同内容或产品的关注热度。通过对这些数据进行细致的统计分析，企业可以及时调整其营销和运

营策略，从而有效提高营销的转化效率。

学习本节内容后，读者将能够理解并掌握在多种场景下实现排行榜的原理和方法。

10.7.1 页面效果展示

商品热销排行榜的前端页面比较简单，只需要一个列表即可展示，如图10-45所示。

图10-45　商品热销排行榜

列表页面的实现代码如下：

```
import { getHotProductList } from '@/apis/product'
import ExcelTable from '@/components/exportExcel'
import { Tag, Image } from 'antd'
import { observer } from 'mobx-react'
import { ProductStatus } from '@/common/enums'

const HotProductList: React.FC = () => {
  return (
    <>
      <ExcelTable
        hideSearch
        columns={[
          {
            title: '排名',
            key: 'index',
            render(_, __, index) {
              return index + 1
            }
          },
          {
            title: '商品名称',
            dataIndex: 'name',
            hideInSearch: true
          },
          {
            title: '商品图片',
            dataIndex: 'images',
            hideInSearch: true,
```

```
            render(_, record) {
              return record.images?.length ? record.images?.map((item: string) => (
                <Image width={60} src={item} alt={item} key={item} />
              )) : '-'
            },
          },
          {
            title: '状态',
            dataIndex: 'status',
            hideInSearch: true,
            render(_, record) {
              const green = record.status === ProductStatus.ON_SALE
              const text = record.status === ProductStatus.NOT_ON_SALE ? '未上架' :
                record.status === ProductStatus.ON_SALE ? '已上架' : '已下架'
              return <Tag color={green ? 'green' : 'red'}>{ text }</Tag>
            },
          },
          {
            title: '价格',
            dataIndex: 'price',
            hideInSearch: true
          },
          {
            title: '销量',
            dataIndex: 'score',
            hideInSearch: true
          },
          {
            title: '创建时间',
            dataIndex: 'createTime',
            hideInSearch: true,
            valueType: 'dateTime'
          }
        ]}
        requestFn={async () => {
          const data = await getHotProductList({ topN: 10 })
          return data
        }}
        rowSelection={false}
      />
    </>
  )
}

export default observer(HotProductList)
```

上述代码定义了排行榜相关的列字段，并在 requestFn 方法中请求了 top10 的商品销量数据，展示销量从高到低的排行数据。

10.7.2 接口实现

服务端想要实现这种排名效果，通常会使用 Redis 中的 Sorted Set，也就是有序集合。有序集合的每个成员都与一个分数相关联，并根据这个分数进行排序，因此非常适合这种场景。

首先创建一个单独的热销服务类 HotSalesService，并实现一个添加统计商品销量的接口 addProductSales，当每次对商品进行开单时，将商品的数量添加到 Redis 中，代码如下：

```
/**
 * 添加产品销量
 *
 * @param productId 产品 ID
 * @param saleCount 销量
 * @returns Promise<void>
 */
async addProductSales(productId: string, saleCount: number) {
  const client = this.redisService.getClient()
  const redisKey = getRedisKey(RedisKeyPrefix.HOT_SALES)
  await client.zIncrBy(redisKey, saleCount, productId)
}
```

在上述代码中，接口接收商品 id（productId）与商品销量（saleCount）两个参数，并定义了名为 HOT_SALES 的 Redis 键，调用 zIncrBy 方法新增或更新指定成员（商品）的分数（销量），最终 Redis 中缓存的销量数据如图 10-46 所示。

图 10-46　销量数据

在图 10-46 中，左侧的 Member 代表商品 id，右侧的 Score 代表商品对应的销量数。

接下来，在开单接口中新增一行代码，调用 addProductSales 方法来统计数据，代码如下：

```
/**
 * 创建订单
 *
 * @param createOrderDto 创建订单所需的参数
 * @returns 返回成功信息
 * @throws HttpException 抛出 HTTP 异常
 */
async create(createOrderDto: CreateOrderDto) {
  const { productId, discount = 1, status = 0, count } = createOrderDto
  // 获取商品信息
```

```
    const product = await this.productRepository.findOneBy({ id: productId })
    if (!product) {
      throw new HttpException('商品不存在', HttpStatus.NOT_FOUND)
    }
    // 获取开单商品计算价格
    let totalPrice = product.price * count
    let discountPrice = totalPrice * discount
    let orderItem = {
      ...createOrderDto,
      name: product.name,
      price: totalPrice,
      status,
      count,
      productId,
      discountPrice,
      discount
    }
    const order = await this.orderRepository.save(plainToClass(OrderEntity, orderItem))
    if (!order) {
      throw new HttpException('开单失败', HttpStatus.INTERNAL_SERVER_ERROR)
    }
+   // 缓存销量到 Redis 做排行榜
+   await this.hotSalesService.addProductSales(String(productId), count)
    return '开单成功';
  }
```

有了 Redis 销量排行数据后,接下来实现接口获取 topN 的热销商品数据,代码如下:

```
/**
 * 获取热销产品列表前 n 个
 *
 * @param n 获取数量
 * @returns 返回包含热销产品列表的对象,每个产品包含 id、name、price、score 等属性,按照 score 从高到低排序
 */
async getTopNProducts(n: number) {
  const client = this.redisService.getClient()
  const redisKey = getRedisKey(RedisKeyPrefix.HOT_SALES)
  const scoreList = await client.zRangeWithScores(redisKey, 0, -1)
  const hotIdList = scoreList.reverse().slice(0, n).map(item => item.value)
  const hotList = await this.productRepository.find({where: { id: In(hotIdList) }})
  return {
    list: hotList.map((item, index) => ({ ...item, score: scoreList[index].score })).sort((a, b) => b.score - a.score)
```

        }
    }

由于我们使用 node-redis 包来操作 Redis 客户端，为了获取 topN 排行数据，代码中用 zRangeWithScores 方法来获取商品 id 及销量数据（其他 Redis 包可能有现成的 API 直接调用）。在默认情况下，这些数据按销量升序排列。

查询出 topN 的商品 id 列表后，最后从数据库中查询指定 id 的商品数据，根据分数由高到低排序并返回客户端。

至此，我们使用 Redis 中一个高效的集合完成了商品热销排行的功能。除此之外，按日排行或按月排行无非是将 redisKey 加入一个时间的维度来实现存储不同的排行数据，如 daily_ranking:${year}-${month}-${day} 或 monthly_ranking:${year}-${month} 等。

10.8　实现活动模块与定时任务

在本项目的最终模块中，我们将聚焦于活动管理模块的开发。商家为了提升销售额，经常推出多种营销策略，比如团购、限时优惠或买一赠一等促销活动。在创建这些活动时，可以指定参与的商品并设置活动的有效期，包括具体的开始和结束时间。

为了自动化管理活动的时间限制，我们计划引入定时任务功能，以便在预设时间自动启动或终止活动。此外，我们还将开发一个展示当前活动列表的功能，并允许用户编辑活动、更改活动状态或删除活动。

10.8.1　页面效果展示

活动列表页面的交互方式与商品列表类似，区别在于对列表数据执行的操作不同，如图 10-47 所示。

图 10-47　活动列表

用户可以通过输入活动名称的部分或全部内容来对活动列表进行模糊搜索。系统也允许用户提前结束正在进行中的活动，但请注意，活动一旦结束，其状态将无法更改。另外，用户还可以通过单击"创建活动"按钮来添加新的门店活动，如图10-48所示。

图 10-48　创建活动

同样地，用户可以对未结束的活动详情进行二次编辑操作，如图10-49所示。

图 10-49　编辑活动

在图10-48的创建表单与图10-49的编辑表单中都可以选择参与活动的商品，未上架的商品不会在此显示，同时活动时间需要设置在合理的范围内，例如活动结束时间不能够早于当前时间，否则服务端会拒绝请求，详细的实现过程将在后文实现接口时进行介绍。

最后，为了预防用户误操作导致数据丢失，系统将在用户单击"删除"按钮时出现一个二次确认的交互提示。这样可以确保用户在删除前再次确认，如图10-50所示。

图 10-50　删除操作的交互过程

10.8.2　表关系设计

在营销活动中需要绑定一个参与活动优惠的商品，在活动表中记录一个商品 id，实体字段定义如下：

```typescript
import { ActivityStatus } from "src/common/enums/common.enum";
import { Column, CreateDateColumn, Entity, PrimaryGeneratedColumn, UpdateDateColumn } from "typeorm";

@Entity('store_activity')
export class ActivityEntity {
  @PrimaryGeneratedColumn()
  id: number;

  @Column({ type: 'varchar', length: 30, comment: '活动名称' })
  name: string;

  @Column({ type: 'int', comment: '活动状态 0 未开始 1 进行中 2 已结束', default: 0 })
  status: ActivityStatus

  @Column({ type: 'int', comment: '活动类型 0 普通活动 1 拼团活动' })
  type: number

  @Column({ type: 'text', comment: '活动描述' })
  desc: string

  @Column({ type: 'timestamp', comment: '活动开始时间' })
  startTime: Date

  @Column({ type: 'timestamp', comment: '活动结束时间' })
  endTime: Date

  @CreateDateColumn({ type: 'timestamp', comment: '创建时间' })
  createTime: Date

  @UpdateDateColumn({ type: 'timestamp', comment: '更新时间' })
  updateTime: Date

  @Column({ type: 'int', comment: '参与活动的商品 id' })
  productId: number

}
```

10.8.3 接口实现

门店活动的创建通常由店长或管理员负责。在创建活动的过程中，需要判断活动时间的有效性。结束时间小于当前时间或者开始时间大于结束时间，活动将无法成功创建。此外，如果开始时间已经大于当前时间，则应该把活动状态设置为已开始。具体的实现代码如下：

```typescript
/**
 * 创建活动
 *
 * @param createActivityDto 创建活动的 DTO
 * @param currentUser 当前用户
 * @returns 返回创建成功提示
 * @throws 当用户类型不是管理员时，抛出权限异常
 * @throws 当结束时间小于当前时间时，抛出期望失败异常
 * @throws 当开始时间大于结束时间时，抛出期望失败异常
 * @throws 当创建活动失败时，抛出内部服务器错误异常
 */
async create(createActivityDto: CreateActivityDto, currentUser: UserEntity) {
  if (currentUser.userType !== UserType.ADMIN_USER) {
    throw new HttpException('您没有权限创建活动,请联系管理员', HttpStatus.FORBIDDEN)
  }
  const { startTime, endTime } = createActivityDto
  let status = ActivityStatus.NOT_START
  if (new Date(endTime).getTime() < Date.now()) {
    throw new HttpException('结束时间不能小于当前时间', HttpStatus.EXPECTATION_FAILED)
  }
  if (startTime > endTime) {
    throw new HttpException('开始时间不能大于结束时间', HttpStatus.EXPECTATION_FAILED)
  }
  if (new Date(startTime).getTime() < Date.now()) {
    status = ActivityStatus.IN_PROGRESS
  }
  const activity = await this.activityRepository.save({...createActivityDto, status})
  if (!activity) {
    throw new HttpException('创建失败', HttpStatus.INTERNAL_SERVER_ERROR)
  }
  return '创建成功';
}
```

前文在设计活动表时记录了商品 id。在获取活动列表时，需要同时查询出 id 对应的商品信息。由于我们没有使用@OneToMany 或@ManyToOne 声明活动表与商品表的关联关系，因此在查询时需要手动使用 QueryBuilder 进行连接。具体的实现代码如下：

```typescript
/**
 * 获取活动列表
 *
 * @param dto 活动列表查询参数
 * @returns 返回活动列表及总数
 */
async getActList(dto: ActivityListDto) {
  const { page, pageSize, name, type, status } = dto
  const where = {
    ...(name ? { name: Like(`%${name}%`) } : null),
```

```
      ...(type ? { type } : null),
      ...(status ? { status } : null),
    }
    const queryBuilder = this.dataSource
      .createQueryBuilder('store_activity', 'act')
      .leftJoinAndSelect('store_product', 'p', 'p.id = act.productId')
      .select(['act.*', "JSON_OBJECT('name', p.name, 'price', p.price) as product"])

    const total = await queryBuilder.getCount()

    const list = await queryBuilder
      .where(where)
      .skip(pageSize * (page - 1))
      .take(pageSize)
      .orderBy('act.id', 'DESC')
      .getRawMany()

    return {
      list,
      total
    }
  }
```

在代码中，首先根据查询条件定义了 where 语句。接着创建了一个查询活动表的 queryBuilder，并使用 leftJoinAndSelect 方法查询出对应的 product 信息，将其命名为对象 product，其中包含 name 和 price 两个字段。最后，通过分页和排序后将结果返回给客户端。

当用户对活动进行"提前结束"或"编辑"操作时，需要重新修改活动信息。此时需要判断活动时间是否在合理范围内，并及时更新活动状态。具体的实现代码如下：

```
/**
 * 更新活动信息
 *
 * @param updateActivityDto 更新活动信息的 DTO
 * @returns 返回更新成功的提示信息
 * @throws 当活动不存在或已删除时，抛出 HTTP 异常，状态码为 404
 * @throws 当活动已结束时，抛出 HTTP 异常，状态码为 417
 * @throws 当结束时间小于当前时间时，抛出 HTTP 异常，状态码为 417
 * @throws 当更新失败时，抛出 HTTP 异常，状态码为 417
 */
async update(updateActivityDto: UpdateActivityDto) {
  const { id, startTime, endTime } = updateActivityDto
  const exists = await this.activityRepository.findOneBy({id})
  if (!exists) {
    throw new HttpException('活动不存在或已删除', HttpStatus.NOT_FOUND)
  }
  if (exists.status === ActivityStatus.END) {
    throw new HttpException('活动已结束，不能修改', HttpStatus.EXPECTATION_FAILED)
  }
  if (new Date(endTime).getTime() < Date.now()) {
    throw new HttpException('结束时间不能小于当前时间', HttpStatus.EXPECTATION_FAILED)
  }
  let newStatus = updateActivityDto.status || exists.status
  if (new Date(startTime).getTime() < Date.now()) {
    newStatus = ActivityStatus.IN_PROGRESS
```

```
    const activity = await this.activityRepository.save(plainToClass(ActivityEntity, {
...exists, ...updateActivityDto, newStatus }))
    if (!activity) {
      throw new HttpException('更新失败', HttpStatus.EXPECTATION_FAILED)
    }

    return '更新成功'
}
```

最后，具备删除权限的成员根据活动 id 对指定活动进行删除。实现过程比较简单，代码如下：

```
/**
 * 删除活动
 *
 * @param id 活动 ID
 * @returns 返回删除成功的字符串
 * @throws HttpException 当用户类型不是管理员时，抛出权限不足异常
 * @throws HttpException 当活动不存在或已删除时，抛出未找到异常
 * @throws HttpException 当删除失败时，抛出期望失败异常
 */
async delete(id: number) {
  const exists = await this.activityRepository.findOneBy({id})
  if (!exists) {
    throw new HttpException('活动不存在或已删除', HttpStatus.NOT_FOUND)
  }
  const { affected } = await this.activityRepository.delete({ id })
  if (!affected) {
    throw new HttpException('删除失败，请稍后重试', HttpStatus.EXPECTATION_FAILED)
  }
  return '删除成功';
 }
}
```

至此，我们完成了所有模块的前端页面开发和接口代码逻辑的实现，相信读者完全有能力编写属于自己的模块。下一节将进入项目部署阶段，将开发完毕的项目部署到服务器上运行。

10.9 使用 Docker Compose 部署项目

在之前的章节中，我们探讨了如何利用 Docker 将 Nest 服务部署到服务器上。然而，在实际应用项目中，还涉及 MySQL 和 Redis 服务。如果通过运行单独的"docker run"命令来启动这些服务，不仅效率低下，还需要考虑服务之间启动的先后顺序。例如，如果 Nest 服务尝试启动时，MySQL 或 Redis 服务尚未启动，程序就会直接报错。Docker Compose 能有效地解决这类问题。本节将使用 Docker Compose 将数字门店项目部署到本地环境。

10.9.1 编写后端 Docker Compose 文件

通常情况下，后端各种中间件会部署到测试、预发布和线上环境中，这里以 Docker 作为测试环境进行演示。在 Nest 中构建时，需要区分不同环境的端口号、MySQL 和 Redis 配置，使用.env

和 .env.docker 配置文件来管理这些配置。.env 配置如下：

```
# JWT 配置
JWT_SECRET=store-web-secret
JWT_EXPIRE_TIME=7d

# MySQL 配置
MYSQL_HOST=localhost
MYSQL_PORT=3306
MYSQL_USER=root
MYSQL_PASSWORD=jminjmin
MYSQL_DATABASE=store_web_project

# Redis 配置
REDIS_HOST=localhost
REDIS_PORT=6379

# 邮箱配置
EMAIL_PASS=EBBOFYAQDWBLOTZY
EMAIL_HOST=smtp.163.com
EMAIL_PORT=465
EMAIL_SECURE=true
EMAIL_USER=jmin95@163.com
EMAIL_ALIAS=store_web_project

API_PREFIX=/api
APP_PORT=3332
```

在上述代码中，在开发环境下，MySQL 与 Redis 的 HOST 指向的是 localhost，同时 Nest 服务启动在 3332 端口下。当项目部署在 Docker 容器中时，同样会用 localhost 来访问。为了避免开发与测试环境端口的冲突，需要进行适当区分。接下来，请看 .env.docker 的配置：

```
# JWT 配置
JWT_SECRET=store-web-secret
JWT_EXPIRE_TIME=7d

# MySQL 配置
MYSQL_HOST=mysql
MYSQL_PORT=3306
MYSQL_USER=root
MYSQL_PASSWORD=jminjmin
MYSQL_DATABASE=store_web_project

# Redis 配置
REDIS_HOST=redis
REDIS_PORT=6379

# 邮箱配置
EMAIL_PASS=EBBOFYAQDWBLOTZY
EMAIL_HOST=smtp.163.com
EMAIL_PORT=465
```

```
EMAIL_SECURE=true
EMAIL_USER=jmin95@163.com
EMAIL_ALIAS=store_web_project

API_PREFIX=/api
APP_PORT=3333
```

与 .env 不同的是，Docker 中的 Nest 服务启动在 3333 端口，以确保 Docker 中的 Nest 服务与宿主机上的 Nest 服务同时运行时不会发生冲突。此外，MySQL 的 HOST 指向名为 mysql 的容器，而 Redis 的 HOST 指向名为 redis 的容器。这两个容器将在 docker-compose.yml 中定义并运行。docker-compose.yml 的配置如下：

```yaml
version: "3"

services:

  servers:
    build:
      context: ./
      dockerfile: ./Dockerfile
    restart: always
    ports:
      - 3333:3333
    networks:
      - nest-net
    volumes:
      - upload:/upload
    depends_on:
      - mysql
      - redis

  mysql:
    container_name: mysql
    image: mysql:8
    command: mysqld --character-set-server=utf8mb4 --collation-server=utf8mb4_unicode_ci
    restart: always
    environment:
      - MYSQL_ROOT_PASSWORD=jminjmin
      - MYSQL_DATABASE=store_web_project
    networks:
      - nest-net
    volumes:
      - mysql:/var/lib/mysql

  redis:
    container_name: redis
```

```
      image: redis:latest
      restart: always
      networks:
        - nest-net
      volumes:
        - redis:/data

networks:
  nest-net:
    driver: bridge

volumes:
  mysql:
  redis:
  upload:
```

在上述配置中可以看到,在 services 下定义了三个镜像配置,分别为 Nest 服务镜像、MySQL 镜像和 Redis 镜像。它们都被指定同处于 nest-net 网络下,以保证 Nest 服务可以正常访问 MySQL 和 Redis 服务。连接方式采用桥接网络。最后,还声明了各镜像对应的数据卷。

首先来看 Nest 镜像配置,指定了使用 Dockerfile 为 Nest 的镜像文件,该镜像向宿主机暴露 3333 端口用于连接。在 networks 中定义了名为 nest-net 的网络,MySQL 和 Redis 通过这个网络名进行连接。最后,指定了数据卷的路径,并声明 Nest 镜像依赖于 mysql 和 redis 镜像。只有在这两者启动后,Nest 服务才会启动。Nest 的 Dockerfile 配置如下:

```
FROM node:18-alpine3.18
WORKDIR /servers
COPY . .
ENV TZ=Asia/Guangzhou
RUN npm set registry=https://registry.npmmirror.com
RUN npm i -g pnpm && pnpm i && pnpm run build
EXPOSE 3333
CMD pnpm start:docker
```

需要特别说明的是,容器启动时会执行"pnpm start:docker"命令来启动 Nest 服务。本质上,这个命令执行的是以下的 Node 命令,它在 package.json 的 scripts 中定义:

```
"start:docker": "cross-env NODE_ENV=docker node dist/main"
```

设置 Node 环境变量为 docker 后,启动服务,此时 Nest 就会获取 .env.docker 中的配置信息。

MySQL 和 Redis 的镜像配置过程大致相同,唯一的区别是这两个服务不需要将端口号暴露给外部宿主机或其他服务进行连接。我们只需确保 Nest 服务能够与它们连接即可。另外,请不要忘记配置 .dockerignore 文件:

```
.vscode
.prettierrc
.eslintrc.js
node_modules
dist
package-lock.json
```

```
npm-debug.log
```

一切就绪后，运行"docker-compose up"命令启动容器，效果如图 10-51 所示。

```
$ docker-compose up
[+] Running 15/20
 ⠿ mysql 10 layers [..........]  11.96MB/112.1MB Pulling                 27.6s
   ⠿ c6a0976a2dbe Downloading [=====>                  ]  ...             21.5s
   ✔ 8dd4f8e415ca Download complete                                       10.6s
   ✔ 6e01a6ece3af Download complete                                       13.0s
   ✔ 6cfdeffd9140 Download complete                                       14.0s
   ✔ 73fed55ee93c Download complete                                       15.0s
   ✔ 83404f4e4847 Download complete                                       15.5s
   ⠿ aad53405df78 Downloading [====>                   ]  ...             21.5s
   ✔ d9c5f6f4cc6e Download complete                                       17.0s
   ⠿ e04d803ff9c7 Waiting                                                 21.5s
   ⠿ f06a309d43da Waiting                                                 21.5s
 ✔ redis 8 layers [##########]  0B/0B    Pulled                           19.7s
   ✔ 22d97f6a5d13 Already exists                                           0.0s
   ✔ c7b117eba408 Pull complete                                            1.5s
   ✔ 3549e2a23473 Pull complete                                            1.5s
   ✔ ccadc22b76d5 Pull complete                                            2.4s
   ✔ d4ed6d335745 Pull complete                                           10.9s
   ✔ 00c30f88f8b5 Pull complete                                            3.2s
```

图 10-51　compose 启动容器

在此过程中，会优先创建 MySQL 和 Redis 的镜像并启动相应的容器，而 Nest 服务将在最后启动。一旦完成，我们将看到三个服务都已顺利启动，如图 10-52 所示。

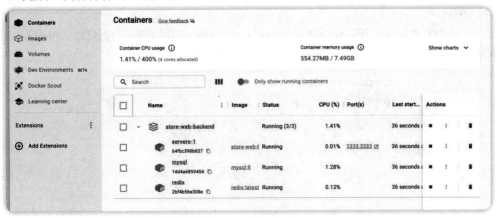

图 10-52　容器启动成功

在控制台下显示 Nest 服务已正常启动，如图 10-53 所示。

```
: '/var/run/mysqld/mysqld.sock'  port: 3306  MySQL Community Server - GPL.
store-web-backend-servers-1  | APP_PORT 3333
store-web-backend-servers-1  | [NEST] 2024-04-29 07:15:31 info [RoutesResolver] AppController {/api}:
store-web-backend-servers-1  | [NEST] 2024-04-29 07:15:31 info [RouterExplorer] Mapped {/api, GET} route
store-web-backend-servers-1  | [NEST] 2024-04-29 07:15:31 info [RoutesResolver] UserController {/api/user}:
store-web-backend-servers-1  | [NEST] 2024-04-29 07:15:31 info [RouterExplorer] Mapped {/api/user/list, GET} route
store-web-backend-servers-1  | [NEST] 2024-04-29 07:15:31 info [RouterExplorer] Mapped {/api/user/currentUser, GET} route
store-web-backend-servers-1  | [NEST] 2024-04-29 07:15:31 info [RouterExplorer] Mapped {/api/user/info/:id, GET} route
store-web-backend-servers-1  | [NEST] 2024-04-29 07:15:31 info [RouterExplorer] Mapped {/api/user/edit, PATCH} route
store-web-backend-servers-1  | [NEST] 2024-04-29 07:15:31 info [RouterExplorer] Mapped {/api/user/delete/:id, GET} route
store-web-backend-servers-1  | [NEST] 2024-04-29 07:15:31 info [RouterExplorer] Mapped {/api/user/update/freezed, PATCH} route
store-web-backend-servers-1  | [NEST] 2024-04-29 07:15:31 info [RouterExplorer] Mapped {/api/user/sendMail, GET} route
store-web-backend-servers-1  | [NEST] 2024-04-29 07:15:31 info [RoutesResolver] SysController {/api/sys}:
store-web-backend-servers-1  | [NEST] 2024-04-29 07:15:31 info [RouterExplorer] Mapped {/api/sys/registry, POST} route
store-web-backend-servers-1  | [NEST] 2024-04-29 07:15:31 info [RouterExplorer] Mapped {/api/sys/login, POST} route
store-web-backend-servers-1  | [NEST] 2024-04-29 07:15:31 info [RouterExplorer] Mapped {/api/sys/forgot, POST} route
```

图 10-53　服务正常运行

我们编写的 Dockerfile 和 docker-compose.yml 等配置文件是可以复用的，预发布和线上环境的部署类似。在部署完后端服务后，接下来部署前端静态资源。

10.9.2 编写 Dockerfile 文件

在前后端分离的开发模式中，前端采用单独部署的方式，将打包后的静态资源托管到 Nginx 服务下。首先来编写 Dockerfile 文件：

```
FROM nginx:stable

COPY ./nginx.conf /etc/nginx/nginx.conf

COPY ./dist /usr/share/nginx/html

EXPOSE 4444

ENV TZ=Asia/Guangzhou
```

构建镜像的过程相对简单，首先指定一个稳定的 nginx 基础镜像，然后将 nginx.conf 文件和打包后的 dist 文件夹复制到 nginx 的指定目录下。在这里，nginx.conf 文件用于配置端口和代理等重要信息，而 /usr/share/nginx/html 文件夹则是 nginx 默认的静态文件存储位置。最后，需要将 4444 端口暴露给宿主机以便访问。

此外，还需要在项目的根目录下创建 nginx.conf 文件，并填写如下配置信息。特别需要关注的是 http 块下的 server 部分：

```
error_log /var/log/nginx/error.log notice;
pid /var/run/nginx.pid;

events {
  worker_connections 1024;
}


http {
  include /etc/nginx/mime.types;
  default_type application/octet-stream;

  map $time_iso8601 $logdate {
    '~^(?<ymd>\d{4}-\d{2}-\d{2})' $ymd;
    default       'date-not-found';
  }

  log_format main '$remote_addr [$time_local] "$request" '
  '$status $body_bytes_sent "$http_referer"';

  root /var/log/nginx;
  access_log /var/log/nginx/access-$logdate.log main;

  sendfile on;
```

```
    #tcp_nopush     on;

    keepalive_timeout 65;

    #gzip  on;

    server {
        listen 4444;
        server_name localhost;

        location / {
            root /usr/share/nginx/html;
            index index.html index.htm;
            try_files $uri $uri/ /index.html;
        }

        location /static {
            root /usr/share/nginx/html;
        }

        location /api/ {
            proxy_pass http://localhost:3333;
        }
    }
}
```

在上述代码中，server 中指定了需要监听 4444 端口，localhost 是本机的服务名，location 指定请求的处理逻辑，会先请求指定的路由资源，而 index.html 是作为兜底访问的文件。/api/ 是反向代理，表示以 /api/ 开头的请求会被转发到 3333 端口服务，用于跨域请求 Nest 服务。

为了保证在生产环境中能够请求正确的接口路径，在 .env.product 配置文件中还需要定义接口请求路径为 VITE_APP_URL，以便 Axios 能够访问，配置代码如下：

```
VITE_MODE='production'
VITE_APP_URL='http://localhost:3333'
VITE_APP_NAME='store-web-frontend'
```

然后在 Axios 的配置中引用这个变量，如图 10-54 所示。

```
41    const Axios = new VAxios({
42      baseURL: import.meta.env.VITE_APP_URL,
43      timeout: 100 * 1000,
44      // 接口前缀
45      prefixUrl: urlPrefix,
```

图 10-54　设置请求路径

在构建镜像之前，运行"pnpm build"命令进行打包，生成 dist 文件后，再运行"docker build

-t store-web-frontend"命令。完成之后，打开 Docker Desktop，可以看到新增了前端项目镜像，如图 10-55 所示。

图 10-55　构建前端镜像

单击 Run 按钮填写参数，把镜像运行起来，然后设置端口和数据卷信息，如图 10-56 所示。

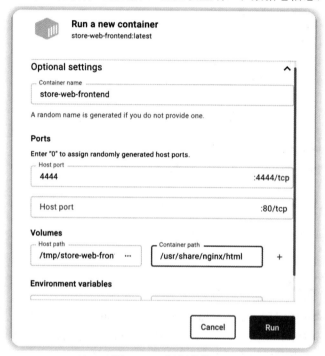

图 10-56　运行镜像

运行成功后，在浏览器输入 http://localhost:4444/activity 链接访问任意页面，如图 10-57 所示。

图 10-57　成功访问项目

至此，我们已经成功完成了前端和后端项目的运维部署。通过学习本节内容，读者们应能将项目部署至个人云服务器，并通过互联网实现外网访问。

本节作为本书项目实战章节的终章，展示了如何从基础开始，利用 Nest 和 React 技术栈构建一个数字门店管理平台。所有相关代码已经推送至笔者的 GitHub 仓库，并将持续更新，包括 Nest 技术的最新进展。鼓励并欢迎各位读者 Fork 仓库并给予 Star 支持。

完结语：是终点，更是新的起点

首先，恭喜你坚持到了最后。请记住，坚持的过程可能会孤独、枯燥，甚至痛苦，但坚持本身就是一种酷。我们都在做着很酷的事情。

一个小小的决定

当清华大学出版社的编辑询问我是否愿意尝试撰写技术图书时，我内心既感到新奇又充满疑问：我能否讲好一门技术？我能否提供超越书籍价值的知识？这需要不断的学习和实践。经过深思熟虑，我决定立即行动。于是我与编辑一拍即合，决定撰写这本书。

我在孩子四个月大时开始创作本书，现在他已经十一个月了。我用了七个月的时间完成这本书的创作，比预期多用了 1.5 个月。原因是我在中途决定重新打磨每个章节，丰富内容，优化表达，并在实战项目中改进了代码结构。每一次小小的决定，每一块时间的碎片，最终汇聚成了书中的每一页。

时间的杠杆

时间是一种每个人都能利用的杠杆。在时间的作用下，每一个微小的投入或改变都将汇聚成可观的收益。

与许多作者不同，这本书是时间杠杆的产物。超过一半的内容是在地铁上构思和编写的，因为我每天需要花费 3 小时以上通勤。这并没有给我带来困扰，反而让我享受这个过程，因为它为我提供了足够的时间来构思和编写本书的内容。加上其余时间的碎片投入，最终产生了累积效应，并呈现给每位读者，这或许就是复利的力量。

结　语

技术学习不是短跑，而是一场马拉松。本书的完成标志着一个阶段的圆满，同时也预示着新的开始。愿每位读者带着从书中汲取的知识与启示，寻找属于自己的那片璀璨。